WINERY UTILITIES

The Chapman & Hall Enology Library

Principles and Practices of Winemaking by Roger B. Boulton,
Vernon L. Singleton, Linda F. Bisson, and Ralph E. Kunkee

Wine Microbiology by Kenneth C. Fugelsang

Winery Utilities
Planning, Design and Operation by David R. Storm

Wine Analysis and Production by Bruce W. Zoecklein,
Kenneth C. Fugelsang, Barry H. Gump, and Fred S. Nury

Forthcoming Titles
Winemaking
From Grape Growing to Marketplace by Richard P. Vine,
Bruce Bordelon, Ellen M. Harkness, Theresa Browning and
Cheri Wagner

WINERY UTILITIES
Planning, Design and Operation

David R. Storm
Storm Engineering, Inc.

CHAPMAN & HALL

I(T)P® International Thomson Publishing

New York • Albany • Bonn • Boston • Cincinnati • Detroit • London • Madrid • Melbourne
Mexico City • Pacific Grove • Paris • San Francisco • Singapore • Tokyo • Toronto • Washington

JOIN US ON THE INTERNET
WWW: http://www.thomson.com
EMAIL: findit@kiosk.thomson.com

thomson.com is the on-line portal for the products, services and resources available from International Thomson Publishing (ITP). This Internet kiosk gives users immediate access to more than 34 ITP publishers and over 20,000 products. Through *thomson.com* Internet users can search catalogs, examine subject-specific resource centers and subscribe to electronic discussion lists. You can purchase ITP products from your local bookseller, or directly through *thomson.com*.

Visit Chapman & Hall's Internet Resource Center for information on our new publications, links to useful sites on the World Wide Web and an opportunity to join our e-mail mailing list. Point your browser to: **http://www.chaphall.com** or **http://www.chaphall.com/chaphall/foodsci.html** for Food Science

A service of **I (T) P**®

Cover design: Saïd Sayrafiezadeh, emDASH inc.
Art Direction: Andrea Meyer

Copyright © 1997 by Chapman & Hall

Printed in the United States of America

Chapman & Hall
115 Fifth Avenue
New York, NY 10003

Chapman & Hall
2-6 Boundary Row
London SE1 8HN
England

Thomas Nelson Australia
102 Dodds Street
South Melbourne, 3205
Victoria, Australia

Chapman & Hall GmbH
Postfach 100 263
D-69442 Weinheim
Germany

International Thomson Editores
Campos Eliseos 385, Piso 7
Col. Polanco
11560 Mexico D.F
Mexico

International Thomson Publishing–Japan
Hirakawacho-cho Kyowa Building, 3F
1-2-1 Hirakawacho-cho
Chiyoda-ku, 102 Tokyo
Japan

International Thomson Publishing Asia
221 Henderson Road #05-10
Henderson Building
Singapore 0315

1 2 3 4 5 6 7 8 9 10 XXX 01 00 99 98 97

Library of Congress Cataloging-in-Publication Data

Storm, David R.,
    Winery utilities : planning, design & operation / David R. Storm.
      p.   cm. -- (The Chapman & Hall enology library)
    Includes bibliographical references and index.
    ISBN 0-412-06601-7 (alk paper)
    1. Wineries–Environmental engineering. I. Title. II. Series.
  TH6057.W55S76   1996
    658 .2'6–dc20   96-27520
                                       CIP

British Library Cataloguing in Publication Data available

To order this or any other Chapman & Hall book, please contact **International Thomson Publishing, 7625 Empire Drive, Florence, KY 41042.** Phone: (606) 525-6600 or 1-800-842-3636. Fax: (606) 525-7778. e-mail: order@chaphall.com.

For a complete listing of Chapman & Hall titles, send your request to **Chapman & Hall, Dept. BC, 115 Fifth Avenue, New York, NY 10003.**

To wine lovers everywhere, whose
enthusiasm for and continued loyalty to
the remarkable beverage, fuels the
wine industry engine.

TABLE OF CONTENTS

PREFACE

This book has been written for an eclectic audience of winery developers (owners), winemakers with utility responsibilities (real or implied), winery design professionals (architects and engineers), and university-level enology professors, all of whom at sometime in their careers must address the subject of winery site utilities as a distinct and important element of their jobs.

Wine and other fermented beverages in one form or another are produced commercially in almost all temperate zones of the world. Utility requirements for wineries, which use grapes as the fermentable sugar source, are the focus of this reference book, although similarities in fundamental production processes for other subdivisions of the fermented beverage industry may find useful reference information in the chapters which follow.

Wine production methods may differ somewhat from country to country, but the sizing, need for reliability, ease of operation, and cost-effectiveness of water, wastewater, electrical, fire protection, and other support systems remain nearly universally constant. Of necessity, the author's past planning and design experience with nearly 60 winery utility systems, will emphasize contemporary design fundamentals related to the U.S. wine industry. However, where possible, opportunities will be taken to relate American practice to, for example, European, Australian, and South American wine industries where discrete differences in utility systems have been observed by the author or discovered in the literature research that was part of the production effort for this volume.

A glossary of terms has been included with each chapter, although the structure of the text, illustrations, and references presumes a limited technical knowledge or experience with utilities nomenclature on the part of the reader.

Finally, this book is not meant to serve as a substitute for the services of competent design professionals. Architects and engineers must still play an important role in the planning, design, and construction of new wineries and their utility systems, and in the expansion of existing facilities to meet new and enlarged wine production goals.

The last decade has, more than any other time in recent history, coalesced worldwide thought to develop principles for the universal protection of freshwater, marine, terrestrial, and atmospheric resources. The wine industry has an opportunity to lead in the development of utility systems which both conserve and protect land, air, and water resources, and at the same time remain cost competitive in domestic and world markets.

ACKNOWLEDGMENTS

Special thanks to my family for their patience and forbearance during the preparation of the manuscript, and to Sally, my Springer Spaniel, who made absolutely no contribution, but was a silent companion during all the long hours of composition. My gratitude to Lorraine Lancaster for her exceptional word processing skills and to David and Kelly Gaskill for bringing the illustrations to life. Finally, thanks to Professor Roger Boulton, U.C. Davis, for his thoughtful review of the manuscript.

WINERY UTILITIES

CHAPTER 1

INTRODUCTION

WINERY CLASSIFICATION

Major differences exist in winery utility requirements because of location (i.e., urban or rural) and other more subtle differences that occur by virtue of differences in product and production methods.

Urban wineries are really the exception in the United States, but there are enough to deserve an abbreviated discussion of their design and function from a utility perspective. The utter simplicity of receiving water and sewerage service for a winery from a municipality can be a distinct development advantage and is an option that should not be overlooked from an economic and winery growth standpoint.

The textbook example of an urban winery, which received not only municipal water supply and sewerage services but also electric service, was Clos du Bois winery in Healdsburg, California. Until the demise of a large apple processing facility, Clos du Bois shared the honor of being the largest waste discharger to the city's waste treatment and disposal system. The loss in city revenue from the closure of the apple processing plant required the city to generate a new and higher wastewater rate schedule to balance its operating cost budget. The immediate economic impact on Clos du Bois was felt, but, more importantly, a planned expansion of the facility at the urban site appeared less attractive under the new and higher rate structure. Currently, Clos du Bois is located outside of Healdsburg city limits, operating on its own water and wastewater systems.

Almaden Winery (the original estate facility) near San Jose, California was, until about 1986, also a wastewater discharger to the San Jose regional wastewater treatment plant located on the southern margins of San Francisco Bay. The cost for sewerage service was levied on both hydraulic and solids loading bases and became onerous as new waste treatment plant capacity and improvements had to be financed by the City of San Jose. The winery was closed by the owners, Hublein, Inc., and plant capacity was transferred to other corporate facilities. J. Lohr, a very large premium table winery, is also located in the metropolitan but industrial sector of San Jose. J. Lohr has reduced the economic impact of urban wastewater treatment and disposal costs by transferring some crushing and fermentation capacity to rural satellite facilities in San Luis Obispo and Monterey counties, some 100 miles south of San Jose. Other urban-class prototype wineries exist in Southern California with combinations of (a) municipal water supplies and on-site wastewater treatment and disposal and (b) local groundwater supplies and municipal waste treatment and disposal. The latter situation is often created by the limited space available for wastewater system construction and the need to expand wine production in a cost-effective manner.

Rural wineries make up, by far, the largest percentage of wineries in the United States and elsewhere, driven in part by the marketing leverage given to estate appellations printed on wine labels and, in part, because the vineyard land costs (historically, at least) were considerably less than urban acreage for winery construction. Developed vineyard real estate in California's Napa Valley has hovered around the $50,000 per acre unit cost since about 1986, when the famous Winery Lake Vineyard sold for that record-setting price. Also not to be overlooked are the retail sales of wines at the winery location, which for many Napa Valley, California wineries is sometimes a make or break positive cash-flow proposition. Underlying all of the economic and wine marketing considerations in selecting rural sites is the perceived philosophy of most winery owners of the desire for utility flexibility and control and maximum independence in the management and operation of their business affairs.

For rural winery locations, all site utilities with the exception of telephone and electrical service are customarily developed on site. Fuel sources for boiler fuel can be compressed natural gas (CNG), liquified petroleum gas (LPG) or other economical fuels, dictated by the local availability and cost. The type of winery end product may also influence the selection and design of the appropriate site utility system combinations. Table wineries which crush their own or purchased grapes are, for equal size, the producers of the largest volume and highest strength liquid waste (suspended and dissolved solids). Table wineries which either field crush or purchase settled juice from another processor produce a smaller volume of wastewater and with a reduced solids content, just because the on-site crushing/destemming step has been eliminated. Machine harvesting also offers this advantage. Obviously, there are variations to the two subcategories of table wineries. For example, if the preponderance of wine production was white or blush wine from settled juice, the process wastewater solids concentration would be less than if the highest percentage of production was red varieties from field-crushed must. The pressing step would produce another component to add to the winery's waste stream that would occur in the white or blush wine example. The logic follows that wineries with less wastewater production may have reduced water supply demand requirements. Other utility requirements such as water-cooled condensers for refrigeration can distort the water supply demand values for a given winery. Although the blow-down (replacement water for the evaporative loss) does not contain carbonaceous waste, the concentration of mineral salts in the water caused by evaporation, requires special attention, particularly if the water is earmarked for vineyard irrigation reuse. These differences will be discussed in detail in Chapter 6, "Potable Water Supply Systems."

A bonded cellar is another winery subcategory that by federal permit is prohibited from crushing or fermenting grapes but is allowed to receive bulk wines in bond for blending, aging, and bottling for eventual sale. Bonded cellars often fall into the urban location category because they have no marketing or other pressure to be "estate" or to even be near vineyards. A tax-paid tasting room can be part of the bonded cellar operation. Thus, the tourist draw of vineyards as a venue for wine purchases may encourage bonded cellars to at least locate near vineyards and become part of the rural scene. Water use and wastewater production for bonded cellars are probably the lowest of all winery categories. Table 6.7, Chapter 6, details unit water use as a function of winery production (12–750 ml bottles per case) and clearly shows that bonded cellars have the lowest estimated unit consumption for any of the winery subtypes.

Sparkling wine facilities have a unique position among the several categories of producers. The Charmat process (bulk sparkling wine production) is very similar to table wine production in terms of process steps. The long (2-year) bottle fermentations are avoided, as is the disgorging step, which in *methodé champenoise* sparkling facilities can be a significant source of alcohol, yeast lees and metabolites in the waste stream. Sparkling wine production by *methodé champenoise* does require that the fermentation and clarification of the base wine occur rapidly so that the bottling and tirage step may proceed as quickly as possible. Separation of the base wine from solids occurs in a very short time and is not as long and drawn-out a process as are the many racking steps required in traditional red wine wineries, which barrel age for a period of time and separate the supernatant or clear wine from the naturally sedimented lees. Thus, most of the liquid waste loading for sparkling wineries occurs in the months shortly after crush, again when the new wine is blended and placed in glass for secondary fermentation, and at the completion of bottle-aging step when the disgorgement of secondary fermentation lees occurs.

Reference will be made periodically throughout subsequent chapters to "unit water consumption" [gallons/liters of water required to produce a unit of product (i.e., a case of wine)] and the relationship of unit water use to winery size or annual production capacity in terms of "wine gallons" or cases of 12–750 ml bottles. Unit water use is a significant factor in establishing reasonable and rational design parameters for a winery's water supply and wastewater systems. This parameter also reflects the degree of commitment by the winery's operating staff to water conservation. The information becomes critically important when a winery is contemplating expansion and historical water use records are compared to historical winery production. A very sound corollary can be given for monitoring unit water use and for taking water conservation means and methods seriously—water saved in wine production use is at the same time saving dollars in waste treatment. A typical table winery utility system schematic showing inputs and outputs is shown in Figure 1.1.

REQUIRED UTILITY SERVICES

Electrical Power

Electrical service that is reliable and affordable is the cornerstone of the winery utility program. Almost every system and subsystem associated with wine production requires electric power. Chapter 2 will discuss in considerable detail the key elements of the electrical system and design concepts, as related to other site utilities (i.e., electric motors, emergency lighting, and standby emergency power for critical system components).

Energy conservation is indeed as important as water conservation not only because of the economic implications but, more importantly, because the principal fuels for electrical energy sources (oil, gas, and coal) are not renewable commodities. The focus of electrical systems design must include, as a controlling factor in equipment selections, the optimum energy efficiency for lighting, heating, and air conditioning (1). Not to be over-

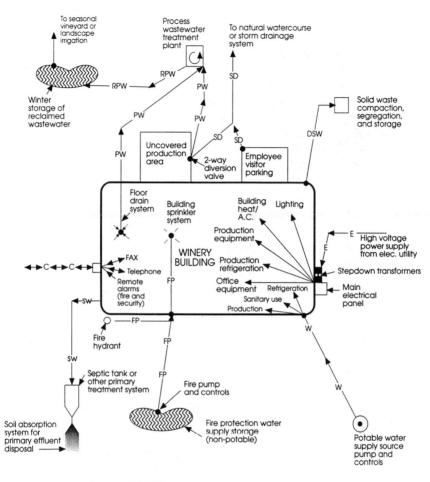

Fig. 1.1   Winery utility system schematic.

looked are the related architectural decisions to be made concerning winery building site planning and solar orientation to maximize winter passive solar heating and to minimize summer heat loads (2). Current energy conservation codes are often tailored to residential construction needs and do not highlight such low-energy benefits such as night-air cooling, which can offset a winery's daily heat load.

In the wet environments that generally prevail in wineries are the risks to production personnel from poorly or improperly grounded electrical equipment, motor control centers, and convenience outlets. The *National Electrical Code 1996 Edition*, the *National Electrical Safety Code 1996 Edition*, and the federal Occupational Safety and Health Act of 1970, as amended, into which the National Electrical Code has been incorporated clearly identifies minimum standards for industrial safety for electrical systems.

Refrigeration Ventilation and Air Conditioning

The temperate climate zones which are conducive to successful viticulture and which produce the high summer temperatures for proper grape ripening and maturity require that winery structures be artificially cooled in the summer and for office and laboratory spaces, at least, be heated in the winter. Separate from the winery space air-conditioning system should be a refrigeration unit that is used solely for wine production purposes, such as for the circulation of glycol or other working substances for (a) temperature control during fermentation, (b) cold stabilization of white wines for potassium bitartrate removal, and (c) bottle neck freezing to permit disgorging of sediment from bottle-fermented sparkling wine and generally holding wines in tanks at controlled temperatures.

In most Region I and II grape-growing zones (3), nighttime and daytime temperatures may differ by 40–50°F (5–10°C). Wineries in these locations frequently use that temperature difference to help balance the accumulated daytime heat load within the building to some reasonable value for a very small expenditure of electrical energy. The mechanical equipment combinations that are possible using two-way carbon dioxide exhaust fans that reverse to become night air supply will be discussed in Chapter 5.

Bermed-earth or cave structures, made popular in California's Napa Valley in the last decade, have distinct environmental advantages in a thermodynamic sense. Using the earth's relatively constant soil mantle temperature of about 56°F (13°C) to maintain lower ambient summer temperatures in all production and storage spaces reduces the size, expense, and complexity of the artificial air conditioning and tank-chilling systems required for conventional aboveground winery structures. A carefully prepared engineering cost analysis could demonstrate the economic advantages or disadvantages of a subterranean over a conventional winery structure. Improved tunnel-boring machines that in geologically and structurally sound rock can excavate 12-ft (3.7-m)-diameter holes for approximately $700–$800 per lineal foot ($2258 to $2625 per meter), November 1991 price levels, bring the capital cost of underground space into the realm of economic reality. To give the higher capital cost for underground structures an unbiased opportunity to be cost-effective, the engineering analysis must reflect the future expected higher cost of fossil fuels for electrical power generation.

Environmentally safe refrigerants will be discussed in detail in Chapter 5. The controversy continues to rage among atmospheric physicists over the precise role of dichlorodifluoromethane-based refrigerants (CFCs) in the depletion of the earth's protective ozone layer (5). Freon-12 was the refrigerant of choice for nearly 50 years, replacing early 20th-century ammonia vapor expansion systems, whose toxic fumes were dangerous to workers and maintenance personnel alike. The U.S. Navy is credited with promoting the development of the Freon-12 substitute while they were researching for an alternative refrigerant to use on warships during World War II to replace ammonia and prevent onboard casualties from ammonia leaks during sea battles (6).

Telecommunications

The ability of the winery to communicate with the ever-changing marketplace for its products is probably only second in importance to the emergency remote fire alarm and security signals that give early warning of a fire or unauthorized entry inside the building envelope. Conventional telephone service systems are used for day-to-day routine business contacts as well as an intercom system for communications between employees and management. Facsimile (FAX) machines, which use telephone circuitry, are a must in the contemporary world of digital imaging and transmission. Other signals and alarms of importance to winery operations can be transmitted during nonworking hours by an automatic telephone dialer, which can receive four to six separate alarm signals, which, when activated, dials a preset sequence of four telephone numbers to preassigned winery or emergency repair personnel. The signals might include (a) high temperatures recorded in the fermentors, (b) low water level in the fire protection storage pond, (c) high water level in the reclaimed wastewater storage pond, and (d) fire protection water supply pump failure.

Digital computer systems are routinely installed in new wineries and retrofitted to older wineries for document preparation, wine inventory control, bookkeeping, payroll, and general accounting. Computer system specialists, who market both hardware and software tailored to a winery's peculiar needs, are the best source for this important element of the winery's data processing system. Computer systems will be dealt with only peripherally in this book.

Sanitation, Steam, and Hot Water

All food and beverage processors rely on a near-sterile environment in which to produce their products. Outside of the crush period in the annually repeated winemaking cycle, the other greatest need for water for cleaning and sanitation is during the bottling step (which includes filling, corking, labeling, closure placement, and casing for storage and shipment). Some wineries have their bottling lines operating throughout the year, which means a twice daily washing with very hot water for parts of the bottling line in direct contact with the wine, and conveyor lubricant and hot water for the balance of the bottling system. For a near-sterile environment, steam or hot water at 212°F (100°C) is used for all equipment, hoses, and fittings that will come in contact with a white wine that has received sterile filtration and where postfiltration microbiological contamination is an unacceptable quality control occurrence. High-temperature steam or hot water is often produced by small steam generators using kerosene or electric-resistance heater-fired boilers. The economy and flexibility of using several portable steam units instead of a fixed natural gas-fired or liquified petroleum gas-fired, standard commercial boiler seems obvious. The heavily insulated steam distribution plumbing to points of use adds a significant extra to both capital and future maintenance cost. The kerosene-fired, portable steam units are not usable in closed, interior spaces even though well ventilated, although liquified petroleum gas (LPG)-fueled, portable steam generators are available commercially. LPG systems are discussed in detail in Chapter 9. The portable electric resistance-fired boiler units are slower to reach operating temperatures, but if the cleaning work is carefully planned, the steam generator can be warming up while other cleanup tasks are being performed.

Commercial hot-water heaters with relatively high thermal efficiency are available in a number of sizes and hydraulic capacities.

Potable Water Supply

Even though water does not enter the product stream at any time during the winemaking process as it does with the brewing of beer, to be without a highly reliable water supply of good mineral quality at the proper working pressure and volume would reduce modern winemaking to a nearly impossible task. Water resource engineers have historically referred to a "firm" water supply as being available irrespective of the perturbations of climate and hydrologic cycles. Nowhere in the last 100 years has this precept been as severely tested as in California and some adjacent non-wine-producing western states where an extended 7-year (1986–1993) drought brought available groundwater and surface water resources to their most critical record lows.

Before a winery development group makes a final selection on a site, a thorough water resource analysis should be made by a competent engineer and/or geohydrologist. A description of a typical prepurchase water resource assessment program is contained in Chapter 6.

Estimating water demand (maximum daily flow, peak hourly flow, and annual water use) can be reduced to an empirical formula to represent an average condition for all winery types and annual capacities or it can be related to historical water use for wineries of similar production capacity and product line or processes. The author is inclined to favor the use of record data adjusted to the particular nuances of process differences that prevail from winery to winery. Aggressive water conservation programs are often an operating necessity brought about by a combination of drought and/or limited water supplies. The use of water-cooled condenser systems for production refrigeration and chillers can add a significant volume of water to the potable water supply system. Water use in the latter case has two components, that is, evaporation during cooling and the need to "blow-down" or send water to the waste stream when the total salinity reaches a specified upper limit. In California, at least, water-cooled condensers appear to be the refrigeration cycle subunit of choice, based on capital and operating cost advantages over air-cooled condensers (4).

Meters are a relatively inexpensive method of providing records of water use. From the simplest totalizing meter to the more sophisticated remote recording, hard-copy or PC datalink systems that give a 24-h permanent record of water consumption, the information becomes valuable as an operating tool (verifying the effectiveness of water conservation programs) and for evaluating historical use versus production capacity in the event that the winery plans include an expansion. It is much easier to convince a regulatory agency of good intentions with an expansion if the winery can offer hard data that water use (and hence wastewater production) has not been excessive.

Irrigation Water Supply and Reclaimed Wastewater

Recycling of reclaimed wastewater is almost as old as civilization itself. The practice strongly recommended for wineries in this book is to separate the waste streams of process or production wastewater from that of sanitary waste. For all but the largest wineries with hundreds of employees, the daily and annual production of sanitary wastewater is quite small in comparison to process wastewater. The extra precautions and the restrictive recycling/reuse guidelines stipulated in all state laws are demanded universally by

Environmental Health agencies to protect public health. This fact alone makes the reclamation of sanitary wastewater for vineyard or landscape irrigation less than desirable from a cost standpoint and from a liability risk perspective. The timing and occurrence of the largest volumes of process wastewater (mid-August through October in the Northern Hemisphere) require that the reclaimed wastewater be stored and regulated for reuse as vineyard irrigation during the active growing season of April through October. Nothing is added to production wastewater that would render it unfit for vineyard irrigation. Neutralization of pH might be required, but then that must occur anyway in order to achieve waste stream treatability (see Chapter 8). Caustic tank cleaners with a chlorine base are generally quite dilute because of the large volumes of rinse water which follow. Open-storage reservoirs also receive the equivalent rainfall for the region on their open surfaces, but the quantity is reduced in volume by the surface evaporation rate for the same climatic zone. Monitoring of the mineral quality of the reclaimed wastewater and possible blending schemes can give the winery the largest flexibility in landscape or vineyard reuse programs. Cautions regarding the salvage and reuse of refrigeration evaporative condenser cooling blow-down have been previously cited.

Sanitary Wastewater

Some care should be exercised by the winery owner during its planning stages to provide the design engineer and the planning body, which issues its land use permit, a fairly precise estimate of the maximum number of employees plus visitors that might be expected on any given day or for a special winery event, if that is the controlling head count. Soil absorption systems for sanitary wastewater are highly influenced in their performance by the hydraulic loading at any given time. Regulatory agencies try to find a balance in approving systems in soils which absorb effluent too quickly (sand and gravels) and those which have difficulty in transmitting water at all, such as silts and clays and combinations thereof. Backup leachfields, dosing siphons, and proper septic tank maintenance can, with overall prudent system design, assist in making a workable sanitary wastewater system in spite of a world blessed with imperfect and nonhomogeneous soils. The details of sanitary wastewater system alternatives and their respective performances are described in Chapter 8.

Process Wastewater

Following closely behind fire protection systems in terms of ranking in utility importance and cost is the process wastewater treatment and disposal system.

Always working against a well-designed system is the inescapable fact that the winery process wastewater is highly variable in rate of flow and solids content. Also lacking are the abundant microbiological populations that assist in treating municipal wastewaters. Further, the metabolic processes of yeast and malolactic bacteria remove most of the bionutrients from grapes during fermentation, often creating problems of treatability. The principal acid components in grape juice (citric, tartaric, and malic) give the wastewater a decidedly low-pH character. Neutralizing this acidic condition (pH 6–7) to encourage and sustain the benign microbiological flora for metabolism of the carbonaceous waste components is a necessary step to achieve wastewater treatability.

Soil absorption systems for process wastewater have been used by both large and small wineries with a wide-ranging degree of success and operational and regulatory agency headaches. The high percentage of suspended solids in winery wastewater (70–80% of total settleable and suspended solids content) renders a septic tank ineffective in removing solids unless natural sedimentation is augmented with flocculents or very long detention times are called for. The carryover of these solids to the leachfield system is a predictable precursor for eventual failure of the disposal field from soil clogging. For winery sites which lack space for aboveground biological treatment systems, leachlines have been placed between vine rows. Root incursion into the perforated pipe is the rule with eventual clogging of the leachlines. The response of the vineyard is generally unbridled growth of the overlying and overwatered grape vines.

Liquified Petroleum Gas Systems

Fuel for both forklifts and possibly steam boilers can be supplied through liquified petroleum gas (LPG). Storage tanks can be replenished at intervals by a commercial supplier of the product. LPG is often the fuel of choice for standby emergency electrical generator systems, which can power critical utility system components in electrical utility service areas where power outages and/or summer brownouts are to be expected. LPG, like other fuel products, presents an explosion hazard if it is improperly stored and used. The storage, transportation, and installation of LPG in the United States must be in accordance with the National Fire Protection Association (NFPA Standard No 58).

Recreational-boat owners have learned by tragic experience that careless use of LPG aboard vessels (space heaters and stoves) created explosion hazards of unimaginable severity. The heavier gas from leaks often migrated to bottom or bilge spaces because of its greater density than atmospheric air. Compressed natural gas (CNG) is now the fuel of choice for many boaters because it is lighter than air and any incidental leakage would dissipate by convection through the vessel's vent system or open super-structure. Unfortunately, CNG is currently not as readily available to the commercial/industrial market as LPG. Fuel scarcity or the ratio of price to availability is very likely to change drastically before the end of the century, and CNG may eventually become an important factor in the fuel equation for rural industrial facilities, such as wineries. Chapter 9 explores design concepts and winery safety in the use of LPG and CNG.

Fire Protection Systems

The water supply for the winery's exterior hydrant and interior sprinkler system, as specified by National Fire Protection Association Code (NFPA), must be single purpose. The only allowable dual use of a fire protection water storage facility would be as a possible visual amenity in the winery's landscape plan. Water quality is not high on the specifications list for fire-fighting suitability. Excessive algae (plankton) growth, or "blooms" as they are called, can become a problem in fire storage ponds, particularly if there is an overabundance of attached algae species (periphyton) that could impair the hydraulics of the fire pump intake structure.

The volume of "reserve storage" and the rate of "fire flow" (duration of fire flow as a function of reserve storage) is determined by a formula, which historically included in its list of variables the volume of the winery structure (both storage and production space) and the degree of fire haz-ard, as determined by the type of industry. Newly issued editions of the Fire Code use floor area and building construction materials as the principal criteria. Wineries are classified as "moderate fire hazard" structures as defined by the NFPA and Uniform Building Codes.

The selection and design approval of the prime mover (drive motor) for the fire pump is governed by the state or appointed local fire official's judgment on (a) the reliability of the local electrical power supply and distribution network during a major emergency or disaster and (b) the proximity of the winery structure to a fully manned 24-h fire station (and how quickly fire-fighting equipment could be brought to the site). Diesel-powered pumps seem to be favored by regulatory fire agencies for remote locations, where major wild fires could disrupt electrical power supplies. The use of an electric-power-driven pump as the prime mover, augmented with a backup emergency diesel-powered system, would offer the maxi-mum reliability, but also the maximum capital cost. Motor-generator backup systems are often the system of choice for occupant safety in met-ropolitan high-rise buildings, where special alternative emergency electric power is required within the building envelope for elevator, emergency lighting, and booster fire pumps as a minimum. It is often difficult, on the other hand, to justify the additional safety factor for fire protection in wineries because they are generally single-story structures, which are easily evacuated in the event that a serious fire does occur.

Insurance underwriters are able to quickly justify lower insurance premiums for fire suppression and alarm systems that have a high degree of reliability. It is also easy to demonstrate the amortization of the initial capital cost of a fire protection system in terms of annual insurance premiums foregone over the economic life of the winery complex.

Selection of fire protection equipment by designers and consultants is carefully regulated by the provisions of the NFPA Codes and Standards. Only equipment that has undergone testing by Underwriters Laboratories or Factory Mutual Research Corporation test programs and that have received their certification will be approved for installation.

Solid Waste Systems

All wineries, large and small, generate solid waste that must be stored, so as not to impair winery traffic patterns or create a visual blight near the winery. Architectural or landscape screens can be judiciously placed to hide debris boxes or recycling bins. A balance has to be struck between long haul from winery to storage site and enough space being provided for large solid waste compactor trucks to service the containers without interfering with other winery operations.

The luxury of not having to think about streamlining the winery's solid waste program is no longer an affordable or responsible management posture. Early 1990s federal legislation targeted for giving greater longevity to sanitary landfills is requiring that all states adopt solid waste diversion programs that will reduce the total volume reaching landfills by 25% for 1995 and 50% by the year 2000. One of the effective ways to achieve these reductions will be to divert larger and larger quantities of disposable materials to the recycling stream. Wineries have already vigorously embraced recycling, partly from pressure exerted by environmental groups and partly because it makes good economic sense to be in the recycle loop by not only providing raw materials but also to attempt wherever possible to establish a strong winery purchase program that seeks out good quality products that are in part or wholly made from postconsumer materials. The solid waste compliance squeeze might emerge for some wineries if they "peak" too soon with their waste diversion programs. When the local solid waste jurisdiction asks for a 25% or 50% reduction, there must still be an increment of waste diversion remaining to meet the spirit and letter of the law. It is prudent for a winery to begin to maintain records of solid waste generation production and disposition to provide a second framework for management decisions, as solid waste issues reach the critical stage in the last years of this century.

Hazardous solid material disposal is currently tightly regulated. The winery's so-called "Business Plan," statutorily required in California at least (and elsewhere in a very short time), documents the quantity of hazardous materials on hand at any given time, including where and how they are stored. If there are residuals, the details must be given for their accountability until final approved disposal is certified. Spent lubricants, dry cell batteries, pesticide and herbicide containers, laboratory reagent bottles, and paint cans all require special handling and proper disposal.

Winery Utilities and Health and Safety Programs

Winery safety and illness-prevention programs are no longer just good business practices, they are mandated by federal and state statutes, with severe penalties for failure to document a plan, train supervisors and cellar staff, and perform the necessary administrative and record-keeping tasks. Chapter 12, which documents pertinent laws, selected safety equipment, and winery/vineyard hazards, presents a model health and safety plan which can be modified to conform to a winery's location, operational methods, and building envelope assets and liabilities.

Also included are subsections which relate industrial noise, indoor air quality, and confined space safety, as seen from a winery perspective. Each chapter, which details a specific utility system and for which inherent and unavoidable health or accident risks have been identified, is related to the model health and safety plan framework and its employee training and education provisions to reduce injury or illness to the lowest possible probability.

Typical accident statistics are presented for California agriculture and farming activity. Emphasis is given for those elements of the safety plan which focus on the reduction in worker injuries from back strain or lifting (the apparent highest percentage of accidents in agriculture) to slips and falls (the second most frequently reported accident).

Respiratory-system protection is described in terms of toxic-gas accident potential and the use of emergency gas filter and absorption face mask equipment that is truly life saving.

GLOSSARY

Blush wine   Wine produced from red grape varieties at grape sugar concentrations of 19–20° Brix, where little or no skin contact during the crush/press process step or thereafter. Often finished with 2–3% residual sugar.

Carbonaceous waste   Waste constituents that contain the element carbon, generally in combination with hydrogen and oxygen (organic compounds).

Charmat process   A batch process for the production of sparkling wine using an external source of carbon dioxide to create the necessary effervescence. (See *methodé champenoise* for sparking wine production technique to create carbonation within the bottle.)

Compressed natural gas   A petroleum product that is generally used uncompressed as an industrial fuel and for residential utility purposes, worldwide. At pressures in excess of 200 atm (3000 psig), the gas (mostly methane) can be liquified and used as a clean-burning fuel for internal combustion engines.

Factory Mutual Corporation (FM)   An independent testing organization supported by fire protection equipment manufacturers seeking FM certification and member fire insurance companies who require compliance with FM fire protection guidelines in order to underwrite commercial and industrial risks.

Leachfield/leachlines   Perforated plastic pipe placed in a gravel or crushed-rock-filled trenches to distribute septic tank effluent over a large area of soil.

Liquified petroleum gas   A petroleum product (generally propane or butane) that is used by many U.S. wineries for boiler and forklift fuel because of its availability and low air-pollution emission loading potential.

Makeup water   Water that is added to a closed-cycle system to maintain the design volume (cooling water, boiler feed water) at the optimum level.

Methodé Champenoise   A process for producing sparking wine or French champagne in the bottle. (See *Charmat process* for bulk method of sparking wine production.)

National Fire Protection Association (NFPA)   An institute with representatives from the fire protection equipment manufacturers, insurance carriers, and fire officials, who maintain design and performance standards for fire protection.

Sewage   Municipal or industrial wastewater.

Sewers   Conduits or pipes for conveying sewage.

Sewerage   The system of sewers, manholes, pump stations, and treatment works used for the transport and disposal of sewage.

Underwriters Laboratories (UL)   An independent testing organization supported by fire equipment manufacturers seeking UL certification for their fire protection products. (See *Factory Mutual Corporation* for an institute with similar functions and objectives.)

Uniform Building Code   One of several codes which are adopted by governmental organizations to provide standards for building construction, earthquake, and fire safety.

REFERENCES

1. THUMANN, A. 1978. *Electrical Design, Safety and Energy Conservation.* Atlanta, GA: The Fairmont Press.

2. TRAISTER, J.E. 1985. *Electrical Design for Building Construction,* 2nd ed. New York: McGraw-Hill Book Co.

3. WINKLER, A.J. and AMERINE, M.A. 1962. *Classification of Grape Growing Regions in California.* Berkeley, University of California Press, pp. 504–576.

4. PITA, F.G. 1984. *Refrigeration Principles and Systems: An Energy Approach.* New York: John Wiley and Sons.

5. MARINI-BETTOLO, G.B. 1983. *Chemical Events in the Atmosphere and Their Impact on the Environment.* New York: Elsevier Science Publishers.

6. JOHNSON, R., BROCKETT, W.A., and BOCK, A.E. 1958. *Elements of Applied Thermodynamics.* Annapolis, MD: U.S. Naval Institute Press.

CHAPTER 2

ELECTRICAL SYSTEMS

ELECTRICAL LOAD PLANNING

Considerable lead time is required in giving formal notification to the local electrical utility to allow their engineering staff to plan and design the system with the capacity to serve the proposed winery. The winery's principal electrical design engineer generally coordinates the master electrical service plan and obtains from the other design professionals involved in the project relatively firm estimates of the electrical load that their respective subsystems will require. It is not uncommon to have separate electrical engineering design subcontractors for the main service and winery building, for the site utilities, and for the landscaping (irrigation control systems and lighting). As this request has to be made anywhere from 4 to 6 months before final plans and specifications are complete, some oversizing of the electrical service requirements is made to cover any design shortfall that may occur because of special electrical equipment for which current and voltage requirements could not have been anticipated. A typical format for an electrical service application is shown in Table 2.1.

Note that in the potential electrical load and natural gas load inquiry

Table 2.1.   Typical Electrical and Gas Service Application and Load Estimate

	Date _____
(Address)	Telephone _____
(City)	Fax _____

ATTENTION: DIRECTOR MARKETING DIVISION

Subject: _____

Location: (Street) _____   (City) _____

Owner: (Name) _____   (Telephone) _____   (Fax) _____

Architect, Engineer, or Contractor: (Name) _____

(Address) _____   (Telephone) _____   (Fax) _____

Enclosures: (Three copies plot plan, elevation drawing, and single line diagram)

Date Service Required: _____   Data Sub-Structures Required: _____

Bid Date: _____

ELECTRIC LOAD INFORMATION

	Initially	Future		Initially	Future
Lighting.	____ KW	____ KW	1φ Air Cond	____ HP	____ HP
Receptacles	____ KW	____ KW	3φ Air Cond	____ HP	____ HP
Water Heating	____ KW	____ KW	1φ Heat Pumps	____ HP	____ HP
Duct Air Htrs	____ KW	____ KW	3φ Heat Pumps	____ HP	____ HP
Unit Air Htrs	____ KW	____ KW	Aux. Strip Htrs	____ KW	____ KW
Cooking	____ KW	____ KW	1φ Misc. Motors	____ HP	____ HP
X-Ray (input)	____ KVA	____ KVA	3φ Misc. Motors	____ HP	____ HP
Welders (input)	____ KVA	____ KVA	_____	____	____
			Largest Motor	____ HP	
MAIN SWITCH SIZE ____ Amps			TOTAL BUILDING SQ. FOOTAGE	____	

Type Service Desired:

(Phase) _____   (Voltage) _____   (Wire) _____   Estimated Demand _____ KVA

GAS LOAD INFORMATION

	Initially	Future
	(Use INPUT ratings in Btu/Hr)	
Space Htg. Equip.	_____	_____
Water Htg. Equip.	_____	_____
(Tot. Wtr. Htrs _____)	_____	_____
Ranges	_____	_____
Other Cooking	_____	_____
Boilers (indicate use)	_____	_____
Air Conditioning	_____	_____
Dryers	_____	_____
Incinerators	_____	_____
Pool Heater	_____	_____
Other _____	_____	_____
	_____	_____
TOTALS	_____	_____

Please determine or confirm the proposed electric and/or gas service arrangements.

Sincerely,

section in Table 2.1, a separate column has been provided for "initial" and "future" values. This will permit a reasonable degree of freedom for the utility to oversize the service to meet anticipated future needs at a considerable potential cost savings both to the utility and the winery.

Another key electrical service decision that must be made, which is a direct function of how other site utilities are located on the gross winery parcel, is the distance of load centers for water supply, process wastewater, and fire protection pump systems from the winery's main electrical service panel. Very early in the master planning for the winery's electrical needs, an estimate needs to be made of the total numbers of separate services and meters that may become satellite systems to the main service. Electrically, this equates to branch runs of primary or high voltage to points of use which then require step-down transformers to match the special motor and equipment specifications at that particular location.

Note in Table 2.1 that the power service company requests that a value be given for "estimated demand" in KVA (kilovolt-amperes). KVA values for lighting and other non-motor-driven devices can be readily approximated, even if a winery architect has not been selected. Electrical motors, on the other hand, require specific knowledge of the relationship between motor rating (in horsepower), starter type, motor efficiency, and motor power factor (8). The motor load calculations should be developed by the principal electrical engineer.

Occasionally, electrical power and/or natural gas distribution facilities may not be in reasonably close proximity to the proposed winery site. Service extensions are often possible under several options for construction cost responsibility alternatives. If the winery is literally at "the end of the line," it is highly probable that winery will bear all the costs of extension. If there are other potential residential or industrial/commercial users along the preferred alignment, the construction cost may subdivided among the potential beneficiaries in proportion to the size of service and/or potential revenue stream to the utility company. A third possibility is that the size of the winery service may be large enough that the utility company may extend the service at no cost to the winery and recover its expenditure of construction monies through future electric and gas revenues.

POWER DISTRIBUTION SYSTEM ALTERNATIVES

There is no question that an underground primary power and telephone system has the least impact on natural or man-made landscape features and beauty. Regardless of how you view an overhead system of conductors and poles, there is very little that is intrinsically attractive. If all of the factors favoring underground primary service were arrayed in order of importance, the list would be headed by aesthetic values. System reliability would also be high on the list because buried high-voltage cables are, under normal utility design construction practices, "looped" or duplicated in separate conduits, so that if a cable is burned through or damaged from inadvertent excavation, primary power can be rerouted to the secondary cable without a lengthy disruption of service. Adding to the reliability quotient, in forested areas underground service is virtually immune to damage from falling limbs, trees, or other extraordinary weather events. Once the decision is made by winery management to place electrical service underground, it must automatically follow that wired telephone service must be buried also. Unfortunately, there is little that can be shared in the installation, and each utility service company has their discreet set of rules and regulations relating to the spacing of pull boxes, depth of trenching, and the final bedding and backfill material to protect the conduits. The factor that weighs most heavily against undergrounding of primary service is the initial capital cost. The owner will usually pay for the following:

- All trenching, conduit bedding materials, and backfill
- Transformer pads and protective barriers
- All electrical conduit, pull boxes, and vaults from the utility company's nearest service pole to the winery's main panel and meter location
- All of the above for any branch services to serve motors or lighting at locations too distant from the winery's main panel

The utility will pay for the following:

- Primary power cable and cost to install in the owner's conduits
- Power transformers
- Meters

VOLTAGE CONSIDERATIONS AND PREFERENCES

One of the first tasks of the winery's principal electrical engineer is to construct a one-line diagram, which depicts graphically where electrical loads are located and how they are to be serviced. This procedure would require that electrical loads from other design subcontractors be compiled and included in the master one-line diagram. An illustrative example of a one-line diagram is shown in Fig. 2.1.

The utility company will provide the electrical engineer with their available primary voltages at or near the winery site. A classification of total winery electrical load in KVA (kilovolt-amperes) or KW (kilowatts) for service utility voltages that are generally available throughout the United States are as shown in Table 2.2. Table 2.3 lists possible service voltages that are compatible and energy efficient for various motor sizes and ratings.

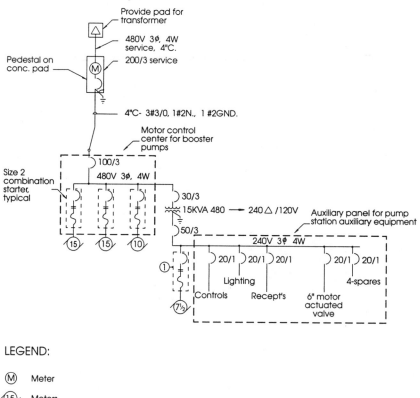

Fig. 2.1.   One-line diagram—potable water pump station.

Table 2.2. Winery Electrical Loads and Primary Voltage Preferences

Relative winery size	Estimated electrical load (KVA/KW)	Preferred utility supplied primary voltages (KV)
Small	≤10, 000 KVA/KW	2.4, 4.16, 12.0, or 17.2
Medium	10, 000–20, 000 KVA/KW	17.2
Large	>20, 000 KVA/KW	17.2 or 20, 780

Source: Adapted from Ref. 8.

Table 2.3. Motor Size/Rating and Recommended Service Voltages

Motor size/rating (horsepower)	Recommended electrical service voltages (V)	Frequency (Hz)	Power phase[a]
≤7½ hp	120/240	60	1
≥5 hp but <30 hp	208/240/480[b]	60	3
30–150 hp	208/240/480[c]	60	3

[a]For U.S. practice, electrical service can be provided as single phase (1φ) or three phase (3φ). The phases are the spacing(s) of alternating-current generator armature circuits. Three-phase generator armatures would be placed equally 120° apart; hence, the voltages generated are 120 electrical degrees apart.

[b]At option of customer, if available near winery site.

[c]At option of power utility.

The advantages of properly matching electrical load and voltage are as follows:

- Primary service and secondary service conductors and breakers can accommodate greater loads.
- For feeder systems to more distant load centers within the winery, voltage drops are not as prevalent on the higher secondary voltage system.

The other key consideration to be made by the electrical engineer in concert with winery management is the balance in cost and operational reliability of the system: (a) Installation of redundant transformers and other switch gear to allow load shifting in the event of a major electrical component failure and (b) to offer to winery management an option of a standby motor generator system and appropriate transfer switches to allow uninterrupted power supplies to certain critical winery subsystems. Figure 2.2 is a one-line diagram of the potable water system booster pump electrical arrangement previously depicted in Fig. 2.1, which for maximum water supply reliability shows an 80-KW emergency generator and transfer switch modification to achieve nearly fail-safe system reliability.

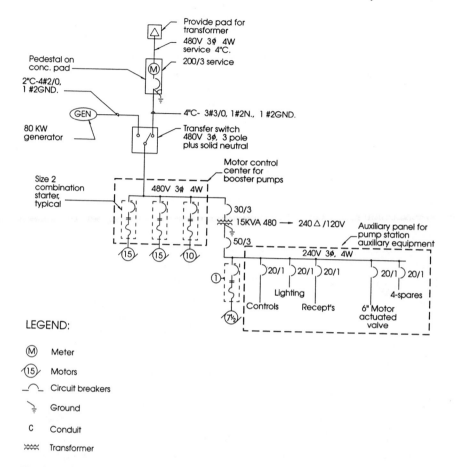

Fig. 2.2. One-line diagram—potable water pump station with standby emergency power.

Other electrical system auxiliary components and equipment can be offered as design options by the electrical engineer to

- Produce the greatest power efficiency to utility rate schedule advantage for the winery
- Suppress extraordinary peak loads and to stay within safe schedule limits and to produce optimum utility cost burdens by use of load control devices.

These two important features of long-term balance and conservation of winery electrical power consumption are described in more detail later in this chapter.

ELECTRICAL REQUIREMENTS FOR PROCESS EQUIPMENT

There are three principal factors to consider in selecting equipment for wine production from a purely electrical standpoint:

1. Personnel safety and electric hazards
2. Electrical to mechanical energy efficiency
3. Reliability

Both the electrical contractor and the electrical engineer/designer have a special responsibility to ensure that all electrical equipment used by winery personnel is carefully installed and grounded in accordance with the National Electrical Code (NEC); in particular, that portable pumps and similar equipment often used by a single employee in a remote sector of the winery (tank room or barrel-aging room) have ground-fault protection included on each motor for each piece of equipment. This electrical safety feature may, more appropriately, be the responsibility of the winery management, who make portable equipment selection purchases outside the electrical engineer's consulting service contract. Management, however, should seek the engineer's advice and recommendations on how to achieve the maximum electrical safety possible within the wet-floor environment that is the norm for most winery production areas.

Power cords and electrical convenience outlets should be clearly and permanently labeled, so that receptacles and power cord plugs cannot be inadvertently or purposely interchanged. A uniform system of power cord plug caps and winery receptacles should be coordinated with the winery's electrical engineer and made to conform with the recommendations of the National Electrical Manufacturer's Association (NEMA) (9). There are presently, at least 75 different plug cap and receptacle configurations which encompass 38 different voltage and current ratings (9).

Electric motors for both fixed equipment and portable equipment should be selected to provide the maximum efficiency while operating at ±5% of the motor's name plate voltage and with a power factor above 0.85. Motors can be made more efficient by reducing stator and motor core losses and preventing line current from passing through the stator winding (18). For the same motor, increasing the length of stator and motor (by increasing the mass of material in both) increases the motor's power factor and, hence, overall efficiency (9). Using energy-efficient motors is generally 30% more costly than their standard motor equivalents (9). The cost difference between standard and energy-efficient motors has decreased slightly since cost data were published in 1989 (8). The cost increment for energy-efficient motors over standard motors (1995 price levels) is about

Table 2.4. Comparison of Operating Characteristics and Cost for a Typical 30-horsepower (KW) "Standard" and "Energy-Efficient" Motor

	Standard	Energy efficient
Name plate data	230/460 V, 3φ, 60 Hz, 1800 rpm	230/460 V, 3φ, 60 Hz, 1800 rpm
Efficiency at full load	88.3%	92.4%
Power factor	0.86	0.89
List price: 1995 price levels ($US)	$1340	$1600

Source: Ref. 11.

20%. For illustrative purposes, specifications and approximate cost of two motors typically used for blowers, ventilators, or compressors are shown on Table 2.4.

A relatively uncomplicated life-cycle economic analysis could demonstrate the length of time to recover the additional capital cost of the more efficient motor in the form of reduced electrical energy costs and a more attractive power factor. The influence of power factor on a winery's operating electrical cost will be discussed in a subsequent section of this chapter entitled "Analysis of Electric Power Rate Structures and Service Contracts."

ELECTRICAL REQUIREMENTS FOR SITE UTILITY EQUIPMENT

Most of the electric-motor-driven equipment for wastewater treatment and disposal, sanitary wastewater treatment and disposal, fire protection water supply, and potable water supply systems can be classified as *critical equipment components*. Failure of these systems either from an electrical or mechanical malfunction can create a public health, stream pollution, or safety hazard if system reliability is not optimized physically, electrically, and economically.

One method of achieving high mechanical and electrical performance in water and wastewater pump stations is the use of duplex installations, where each pump and motor is duplicated. In an operational sequence, duplex motor controllers would actuate "Pump A" until a pumping cycle is completed. When the controller receives the next signal to operate a pump, "Pump B" would cycle so that both mechanical and electrical parts receive only 50% of the load that a single-pump installation would receive. The controller also can activate "Pump A" if "Pump B" malfunctions on its cycle in the rotation. Isolation valves and piping can be configured, so that the malfunctioning pump unit can be isolated and removed for inspection and repairs without interruption of the operating system.

Fire protection from water supply pumps and motors are critical components, but wear and use on a fire system generally consists of periodic pump test cycles, unless there is a winery structural fire or a wildland brush fire threatening the building complex. Thus, constancy of electrical and mechanical operation and load are a secondary concern to fire system performance when an actual emergency exists. Monitoring of the fire pump system performance parameters are of utmost importance. Fire pump system controllers are very sophisticated pieces of electronic equipment that periodically monitor and record the routine start-up and test cycles. The supercontroller and prudent maintenance are the best insurance to sustain fire system performance if and when it is ever needed.

Standby electrical power was discussed briefly in the section entitled "Voltage Considerations and Preferences" and shown in Fig. 2.2 as an option for a potable water supply booster pump system. The installed power for fire pumps is usually quite high to produce the necessary sprinkler and hydrant pressures and requisite rate of flow. For diesel-powered fire pumps, emergency electrical power to operate the pump controller and alarm systems is designed to draw direct-current (dc) power from wet cell storage battery banks in the event of a failure in the 120-V ac supply. This electrical backup feature gives the fire pump system a very high index of electrical reliability.

ELECTRIC POWER CONSIDERATIONS FOR NON-U.S. WINE-PRODUCING COUNTRIES

Both Italy and France are large exporters of winery equipment to the United States and other countries. European motors are generally rated for 200, 220, 400, or 440 V, and 10, 30, or 50 cycles per second (Hz) electric power (10). U.S.-based branch sales and distribution offices for Italian and French winery equipment are usually very accommodating and will often make U.S. motor substitutions at little or no cost to the purchaser. The equipment supplier should be able to provide to the winery a written copy of an electrical specification summary of all electrical motors, switches, insulation, grounding, and name plate ratings that clearly shows that the components equal or exceed the National Electrical Code (NEC) and National Electrical Safety Code (NESC) requirements.

A high percentage of the previous discussion on winery electric systems will apply to European winery practice as well. However, it is beyond the scope of this text to include and compare European and U.S. electrical engineering practices. Available services, voltages, phasing, and frequencies in Europe differ from U.S. electrical and utility norms, and professional winery planners and electrical engineers from outside the United

States who use this reference text for guidance in the design of expansion of European wineries will have to modify some of the conclusions and tabulations to fit electric source power characteristics for the particular country of interest. A number of references on European energy sources, practices, and electrical power distribution networks are available to contrast U.S. and foreign power networks (10, 12, 13).

ANALYSIS OF RATE STRUCTURES AND SERVICE CONTRACTS

One of the most important tasks for winery management is to have a thorough understanding of the long-term contract and agreement that the winery and service utility execute.

Previous sections have made reference to energy conservation as a means to lower monthly service charges by the following:

- Use of energy-efficient motors
- Use of energy-efficient lighting
- Use of automatic timers on noncritical cellar or tank space lighting that will deenergize lighting that is inadvertently left on at the end of a working day

The financial burden of bringing electric power to the site for which the winery will be responsible, has been previously identified in the section entitled "Power Distribution System Alternatives." During the time period when these costs are allocated between the winery and the electric utility, there also should be a disclosure by the utility company of the rate schedule or schedules that are appropriate to the winery's planned energy use. For example, if the winery's power is metered to well pumps, drip irrigation booster pumps, or vineyard drainage pumps that are discreetly part of the agricultural segment of the winery's operations, this latter class of energy use is generally charged under a purely agricultural rate schedule. Energy charges for such applications (in kilowatt-hours) are usually lower than comparable commercial or industrial energy costs. Also for agricultural use, "off-peak" power can be used to irrigate when the overall load on the utility's generating capacity is the lowest. Off-peak is usually defined as the period from 9:30 P.M. to 8:30 A.M. (Monday to Friday) and all day Saturday, Sunday, and holidays. Because vineyard irrigation is seasonal, the power service contract will also include a so-called "standby charge" for the reservation of the peak power (in KVA) that is required during the vineyard irrigation season, which is April or May through August in almost all of California's grape-growing regions.

Production equipment, winery lighting, and winery utility energy uses will be charged under a different rate schedule and framework than for purely agricultural purposes. Although most cost savings to wineries can result from a prudent energy conservation program, the electric utilities offer time-of-use incentives that, if carefully integrated into the winery's production schedule, can yield impressive savings.

The winery's monthly electric bill for production and site utility uses will generally be the sum of the following:

(a)   Customer charge—a flat fee which is assessed per meter per month.
(b)   Energy charge—sum of energy charges for use at the applicable rate for peak, partial-peak, and off-peak periods.
(c)   Demand charge—the demand charge is based upon percentage of an agreed-upon maximum power use for a specified interval of time (i.e., if the electric utility has a maximum of *x* kilowatts of power available at the winery's meter and only *y* kilowatts of power are used, the winery is assessed a "demand charge").
(d)   Power factor correction charge—power factors are printed on the name plate of each electric moter in the winery and ideally the value should be 1.0 or 100%. The name plate power factor is for the motor operating at its rated speed and power. The power factor varies with the load on the motor. Thus, if a wine transfer pump is pumping outside its rated performance standard (excessively long transfer hose and a lift higher than the lift for which the pump and compatible motor were designed), the power factor might, for example, fall to 0.81 (or 81%). Most power service contracts require that the user's net power factor be greater than 85%. As an example, for Pacific Gas and Electric Company's Northern California service area, the commercial/industrial rate schedule (for 1000 KW or more demand rate) covers power demand factor correction charges as follows:

> 6. "Power Factor Adjustments" The bill will be adjusted based upon the power factor. The power factor is computed from the ratio of lagging reactive kilovolt-ampere-hours to the kilowatt-hours consumed in the month. Power factors are rounded to the nearest whole percent.
>
> The rates in this schedule are based on a power factor of 85 percent. If the average power factor is greater than 85 percent, the total monthly bill (excluding any taxes) will be reduced by 0.06 percent for each percentage point above 85 percent. If the average power factor is below 85 percent, the total monthly bill (excluding any taxes) will be increased by 0.06 percent for each percentage point below 85 percent.

Fortunately, there are methods to control, within the winery's own electrical system, the extra costs for exceeding the boundary conditions for "demand" and "power factor correcting" charges.

Load demand controllers are solid-state electronic devices that compare the actual rate of energy usage to the ideal (i.e., no penalty) rate of usage for a preselected interval of time. Small, single-purpose computers can also be used for load demand control. As the actual usage rate approaches the ideal usage rate, the controller estimates if the demand limit will be exceeded. If the programmed demand curve information shows exceedance, then the controller will automatically begin to shed nonessential loads based on the winery's preset schedule of equipment use priorities. The load-shedding begins near the end of the service contract time interval and the controller automatically restores power when a new time interval begins.

Power factors for motors can be controlled by the use of capacitor banks to reduce the angle between the motor load's pure resistive power (watts) and the reactive power (volt-amperes reactive). Ideally, these are zero degrees apart, which gives a power factor of 1.0 or 100%. In practice, they are not zero degrees apart. The power factor is defined electrically as the cosine of the angle between the resistive power and the reactive power.

The capacitors have the capability of reducing the need for the reactive power component, which, in turn, reduces the electric utility company's need to provide additional generating capacity. If the winery wants to undertake a retrofit of a capacitor to improve the overall winery power factor, the local utility would probably provide technical assistance and advice, as the change is of considerable advantage to the power supplier as well as to the winery power user.

ENERGY CONSIDERATIONS IN WINERY LIGHTING

Area lighting within the winery working spaces is the joint task of the winery architect and electrical engineer with the engineer providing the technical input and the architect making the aesthetic decisions regarding lumanaire type and finish. Landscape lighting (generally all exterior lighting for plant material illumination and safety lighting) is generally the design responsibility of a landscape architect. Emergency lighting at remote utility system locations is an important part of the utility engineer's design effort.

The design of industrial space lighting systems is almost an electrical design subspecialty unto itself. Most electrical engineers, whose practice includes commercial and industrial building design, have the experience and technical knowledge to create the proper illumination for the unique activities which occur in winery structures. There are many excellent space lighting references which define the special terms used by illumination engineers and for which the winery's management should at least have a limited knowledge. (1, 2,9, 15, 16) It is beyond the scope of this book to encompass the details of lighting system design.

The important features of lighting systems for wineries in terms of overall electrical systems planning decisions and future operating costs should be given special attention during both the planning and final design phases.

Two very important factors deserve attention during the design review of lighting proposals as they relate directly to the long-term energy use commitments by the winery:

- The total lighting load expressed in KVA or KW
- The incremental heat load that is added to the winery structure from light fixtures

The lighting load will be the direct cost component of total winery electrical operating costs to be combined with other electrical power use and reported by the electric utility in their monthly bill to the winery. Thus, the selection of the lighting system with a balance in the greatest output in lumens per watt of power input and the lowest heat generation output would be the optimum design proposal. The indirect cost component of installed lighting is the incremental heat load produced by the lights that must be removed by the ventilating, and air-conditioning system to maintain the requisite ambient space temperatures for fermentation, aging, and case goods storage.

Emergency lighting is a safety necessity in the event that a major catastrophe produces a power outage. Employees and visitors must be able to exit the building and ascend or descend stairways and ramps without incurring injury. Battery-powered, wall-mounted lamps are the least costly and most effective illumination substitute. Power interruptions can be utilized to energize a mechanical transfer switch, so that periods of darkness are eliminated and only a perceptible dimming of the light source is experienced. Photoelectric cells on individual emergency lamps are not practical, as the sensors can not discriminate between a power outage and the end of a normal workday. Emergency lighting for site utilities is generally incorporated into the design of systems which are remote to the winery and on occasion may require repairs or special maintenance outside of normal daylight hours. The provision of a well-grounded, weather-proof convenience outlet on the emergency light standard which can be used for power tools is an inexpensive and useful adjunct to the utility system's emergency lighting arrangement.

CONTROLS, INSTRUMENTATION, UTILITY SYSTEM MONITORING STATUS, AND ALARMS

The fundamental requirements of controls and alarm systems were discussed previously in Chapter 1 under the appropriate section headings for each of the winery's utility systems. In practice, the utility engineer will make recommendations on the type and number of status or alarm signals that should be transmitted to the responsible staff person for utility operations. The terminal and status board (annunciator panel) for the alarm system should be located in an office where there will be a high probability of visual surveillance by one or more staff people during the working day, or for smaller wineries which cannot afford the luxury of full-time desk-bound managers, audio alarms can alert other nearby cellar personnel, who can convey the problem report to the proper winery staff manager for action.

For wineries with consolidated utility systems on a small parcel, where all equipment installations can be seen from some vantage point of the winery, the alarm systems can be simplified. Warning lights and an audio alarm can be placed on a mast giving a high probability that someone in the winery will see or hear the problem-alert signal. As most utility systems will involve piping systems for potable water or liquid waste, the signal cables, which must run from the signal source to signal terminus and annunciator, can be installed in conduits in a common trench with aqueous pipelines, provided that national and local building code requirements for multiuse trenches are observed. This can save substantial signal system installation costs. Other signals which are often transmitted from remote locations to the winery for monitoring and record keeping are flow meter readings from the following:

(a) Raw wastewater inflow
(b) Treated water (reclaimed water production) outflow
(c) Irrigation diversions from streams and reservoirs
(d) Potable water inflow to the winery

Hard-copy recorders can be installed to provide a continuous record of flows for reporting purposes to water-pollution control agencies (waste discharge permit compliance) and for monitoring performance of both water supply and wastewater systems. Figure 2–3 illustrates a possible alarm and signal monitoring terminus panel, which also shows the location of two possible hard-copy recorders and an automatic telephone dialer whose function was described previously in Chapter 1. Figure 2–4 is a portion of

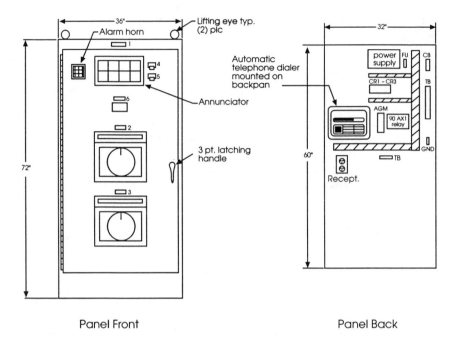

Panel Front   Panel Back

Fig. 2.3.   Typical utility system alarm/status annunciator panel.

a one-line diagram of winery utility system status and signals showing classifications of operational data of critical priority which are typically monitored, recorded, and alarmed.

One of the important secondary technological benefits of the Vietnam
War was the development of very sophisticated intrusion and security alarm
systems. The full electromagnetic spectrum was utilized for detecting infil-
trators by remote sensors along with very sensitive ground motion and
ultrasonic listening devices. Most wineries do not require ultra-high-tech-
nology systems which are installed routinely in banks and other financial
institutions, but newer winery developments and the retrofitting of older
wineries with simple intrusion alarms can prevent theft, vandalism, or even
eco-terrorism (17), the contemporary catchall term used in the security
industry for people or groups of people who are bent on causing economic
damage to selected industrial groups. The same commercial alarm signal
receiving and processing center that is the terminus for the winery's fire
alarm system can also become the after-working-hours monitor for the
integrity of the building envelope.

One additional set of winery parameters that could deserve a remote

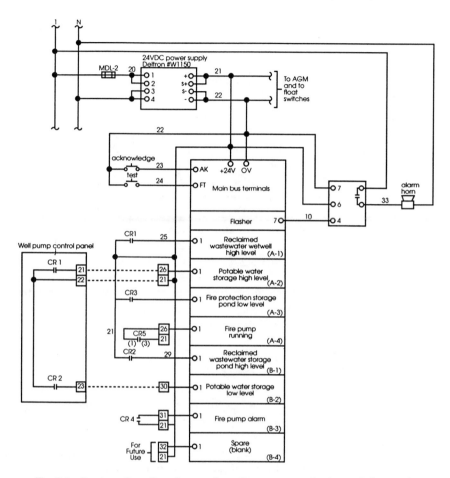

Fig. 2.4.   Portion of one-line diagram for utility system monitoring and alarm unit.

alarm status and monitoring are fermentation tank temperature status and emergency upper limits. Most refrigeration and chiller systems have a high degree of reliability, but the idea of having a forewarning of high fermentation tank temperature could prevent a stuck fermentation or at best an undesirable finished wine product. Precise-temperature probes, consisting of either thermocouples or resistance temperature detectors and their outputs, are easily converted to current outputs through simple signal processor/transmitters (19). These signals can be coupled to audio and visual alarms and to the winery's automatic dialer, a feature previously described for use in after-working-hours monitoring of critical utility system components.

POWER LOSS AND SURGE PROTECTION

The power utility company with which the winery contracts for electrical service may also help to provide protection for low- or high-voltage or low-frequency protection. The utility company may provide the relays and other equipment or automatically interrupt the supply. The winery would be required to provide, maintain, and pay for an open communication link to connect the terminal block at the winery's electrical service panel to the power utility's designated control center (14). The winery's participation in the utility company's interruptable power supply program does not preclude the need to protect electrically sensitive equipment components with built-in devices. The possible advantages to the winery are economic in accepting the utility's option of a nonfirm power supply. This program is less likely to fit a winery's unique pattern of electrical consumption, for example during the crush period, when power supply interruptions would not be acceptable. The potential for reduced energy and demand charges for curtailed or interruptable power at periods *other* than crush are probably the most worthy of serious consideration by a winery to reduce long-term overhead costs.

One very important aspect of planning for electrical system stability and for the protection of valuable power-surge-vulnerable equipment like digital computers and solid-state controllers is the installation of power-surge protectors.

Computers are sensitive to voltage fluctuations, and if the input to the unit varies by more than the manufacturer's stated limits, output errors and, in some cases, loss of stored information can result. Prior to 1987, computers were very vulnerable to so-called voltage spikes and undervoltages. International Business Machines (IBM) was the first major supplier of computer equipment to install built-in voltage regulators (20). All other computer manufacturers eventually added the built-in protection. Total interruptions of electrical power service are not acceptable for some sensitive pieces of computer and electronic equipment.

An "uninterruptable power supply" (UPS) must be part of a fail-safe electrical system, as, for example, a fire pump controller. Battery banks which can provide standby direct-current power to programmable controllers offer one method of avoiding total power loss. Emergency standby generators are another means of providing "UPS" for critical system components. This subject was covered in more detail in the section entitled "Voltage Considerations and Preference." One method of achieving this is by the use of an "inverter" to create an alternating current from a constant, rectified voltage, so that line source voltage fluctuations prior to rectification are filtered out.

OPERATION AND MAINTENANCE OF ELECTRICAL SYSTEMS

Small and medium-sized wineries (50,000 cases annual production and less) are the least likely category of wine production facility to have the trained personnel to accomplish electrical system maintenance tasks. Full-time cellar workers are often primarily responsible for maintaining production equipment, as well as site utility equipment. If the winery facility is recently constructed, the equipment manufacturer's suggested operation and maintenance instructions were provided to the winery after the start-up and test phase and before final payment is made to the contractor. These documents, together with supplemental instructions prepared by the design engineers and eventually the winery's management experience, provide the basis for maintenance, operations, and troubleshooting in the case of a breakdown or equipment malfunction.

Most maintenance for electrical system components is of the preventative type. Wineries have a distinct advantage in that much of the production line equipment is utilized seasonally with many opportunities for normal maintenance while the machinery is not in use.

The written safety orders should clearly outline the special precautions that should be taken for routine maintenance of electrical equipment. (See Chapter 12, "Winery Utilities and Health and Safety Programs.") For example, disconnect switches should be so located that maintenance personnel can have visual contact with the switch to prevent the inadvertent energizing of circuit while work is being performed. Brightly colored safety tape or tags should also be placed on the control panel to alert cellar staff that equipment is being worked on and that power should not be restored until the tape or tag is removed.

For larger wineries, which can employ a full-time maintenance electrician, the scheduling of work through a conventional repair or maintenance request system eliminates or at least balances the randomness of which jobs get needed attention and those which do not. All repair requests should be routed to the production supervisor for information and approval to ensure that winery production is not impaired and that major repair costs have been budgeted and purchase orders prepared and signed.

There are some large wineries with bottling lines operating year round on two 8-h shifts per day, with one shift for cleanup or to change the wine being bottled. Maintenance of electrical machinery and important equipment is a continuous effort for some wineries with unique operating schedules.

Good record-keeping is a theme that will be repeated often throughout this book. Whether it is performance records for wastewater treatment systems or monitoring wells for tracking the movement of septic tank leachate near an ecologically sensitive stream system, documenting maintenance and repair events is fundamental to keeping the winery in an optimum production mode and to create a continuous and unbroken record of operating history and performance. The training of new staff is simplified if well-kept records are available to familiarize personnel in their new assignments. Some examples of electrical equipment maintenance history, parts inventory records, and maintenance checklists are shown in Figs. 2.5 through 2.7.

The electric utility servicing your winery will have the responsibility for maintaining all power service components from the revenue meter, includ-

WINTERS WINERY — ELECTRIC MOTOR MAINTENANCE HISTORY NO. ______

LOCATION OF SPARE PARTS: ______

M.P. ______ MANUFACTURED BY ______ LOCAL MOTOR REP ______ TEL.: ______

SERIES	SHUNT	COMPOUND	SYNCHRONOUS	INDUCTION	
TYPE	FRAME	SPEED	VOLTS	AMPERES	
PHASE	CYCLES	TEMP FIRE	EXCITATION AMPS	MOTOR SERIAL NO.	
MODEL NO.	FORM NO.	STYLE	SERIAL NO.		
ORDER NO.	OUR ORDER NO.	DATE			
CONNECTION DIAGRAM MOTOR OR ARMATURE	STATOR				

SPECIFICATION	BEARINGS	SHAFT EXTENSION	PULLEY	GEAR	V BELT DRIVE
OPEN ☐ EXP. PROOF ☐ DRIP PROOF ☐ TOTALLY ENCL. ☐ VERTICAL ☐ ______ ☐	SLEEVE ☐ BALL ☐ ROLLER ☐	DIA ______ LENGTH ______ KEYWAY ______	DIA ______ FACE ______	TEETH ______ PITCH ______ FACE ______	NO. OF SHEAVES ______ PITCH. DIA ______ BELT SECTION ☐ A - 1/2 X 11/32 ☐ B - 21/32 X 7/16 ☐ C - 7/8 X 17/32 ☐ D - 1 1/4 X 3/4

MOTOR SERVICE RECORD

DATE INSTALLED	LOCATION	APPLICATION

DATE REPAIRED	REPAIRS OR PARTS REPLACED	CAUSE	REPAIRED BY	HOURS

Fig. 2.5. Typical electrical machinery history record.

WINTERS WINERY						MAINT. HISTORY CARD NO._____	
ELECTRIC MOTOR PREVENTATIVE MAINTENANCE							
H P	F.L. SPEED	TYPE		MAKE		LOCATION	
VOLTS		F.L. AMPS	FRAME NO.		PHASE	CYCLES	TEMP. RATING
SERIAL NO.			INSTALLATION DATE				

BEARINGS
☐ SLEEVE ☐ BALL ☐ ROLLER
FRONT END NO.
PULLEY END NO.
GREASE TYPE
GREASE CYCLE
COUPLING TYPE
BRUSH NO.
DATE OF PURCHASE   OUR ORDER NO.

DATE	REMARKS

AUTHORIZED SERVICE SHOP

TEL:

MONTHLY INSPECTION	J	F	M	A	M	J	J	A	S	O	N	D	BY
DATE CHECKED													
BEARINGS													
EXCESS NOISE													
EXCESS HEAT													
SPEED													
VOLTS													
AMPS													
CLEAN													
INSULATION													
BRUSHES													
COMMUTATOR													
EXCESS VIB.													
ALIGNMENT													
LUBRICATE													

Fig. 2.6.   Typical electrical motor preventative maintenance record.

ing the transformers, all conductors and conduits, and disconnect switches associated with the primary power supply.

Of great advantage would be to have all maintenance history, machinery history, and inventory records in computer files to simplify gathering data that are generated routinely, and which must periodically be compiled and analyzed for small wineries (50, 000 or less cases annual production). With uncomplicated production and utility system equipment, the hard-copy record card system is probably just as efficient and more cost-effective than using a data processing approach.

WINTERS WINERY		PARTS ORDER AND INVENTORY CONTROL CARD							
	PART DESCRIPTION:				PART NO.:				
MACHINE HISTORY FILE NO.:					PART STORAGE lOCATION				
PART SPECIFICATIONS:					ORDERED BY:		TERMS:		SUPPLIER CONTACT:
DATE ORDERED	ORDER NO.	QUANT.	SUPPLIER NAME AND ADDRESS	TEL. /FAX NO.S	PROMISED DELIVERY DATE	PHIPMENT DATE	DELIVERY DATE		REMARKS

Fig. 2.7.   Electrical parts order and inventory control card.

GLOSSARY

Alternating Current   Current that reverses direction at regular intervals, having a magnitude that varies continuously following a sine curve pattern.

Ammeter   An instrument for measuring the amount of electron flow in amperes.

Ampere   The basic unit of electrical current.

Annunciator panel   Lighted alpha-numeric alarm and status/performance array for electrical equipment.

Amplification   The process of increasing the strength (current, power, or voltage) of a signal.

Amplifier   A device used to increase the signal voltage, current, or power, generally composed of a vacuum tube and associated circuit called a stage. It may contain several stages in order to obtain a desired gain.

Amplitude   The maximum instantaneous value of an alternating voltage or current, measured in either the positive or negative direction.

Arc   A flash caused by an electric current ionizing a gas or vapor.

Armature   The rotating part of an electric motor or generator. The moving part of a relay or vibrator.

Attenuator   A network of resistors used to reduce voltage, current, or power delivered to a load.

Battery   Two or more primary or secondary cells connected together electrically. The terms does not apply to a single cell.

Breaker points   Metal contacts that open and close a circuit at timed intervals.

Brush   The conducting material, usually a block of carbon, bearing against the commutator or sliprings through which the current flows in or out.

Bus bar   A primary power distribution point connected to the main power source.

Capacitor   Two electrodes or sets of electrodes in the form of plates, separated from each other by an insulating material called the dielectric.

Circuit   The complete path of an electric current.

Circuit breaker   An electromagnetic or thermal device that opens a circuit when the current in the circuit exceeds a predetermined amount. Circuit breakers can be reset.

Circular mil   An area equal to that of a circle with a diameter of 0.001 in. It is used for measuring the cross section of wires.

Coaxial cable   A transmission line consisting of two conductors concentric with and insulated from each other.

Commutator   The copper segments on the armature of a motor or generator. It is cylindrical in shape and is used to pass power into or from the brushes. It is a switching device.

Conductance   The ability of a material to conduct or carry an electric current. It is the reciprocal of the resistance of the material and is expressed in mhos.

Conductivity   The ease with which a substance transmits electricity.

Conductor   Any material suitable for carrying electric current.

Core   A magnetic material that affords an easy path for magnetic flux lines in a coil.

Current limiter   A protective device similar to a fuse, usually used in high-amperage circuits.

Cycle   One complete positive and one complete negative alternation of a current or voltage.

Dielectric   An insulator; a term that refers to the insulating material between the plates of a capacitor.

Diode   Vacuum tube—a two-element tube that contains a cathode and plate; semiconductor—a material of either germanium or silicon that is manufactured to allow current to flow in only one direction. Diodes are used as rectifiers and detectors.

Direct current (dc)   An electric current that flows in one direction only.

Eddy current   Induced circulating currents in a conducting material that are caused by a varying magnetic field.

Efficiency   The ratio of output power to input power, generally expressed as a percentage.

Electrolyte   A solution of a substance which is capable of conducting electricity. An electrolyte may be in the form of either a liquid or a paste.

Electromagnet   A magnet made by passing current through a coil of wire wound on a soft iron core.

Electromotive force (emf)   The force that produces an electric current in a circuit.

Electron   A negatively charged particle of matter.

Energy   The ability or capacity to do work.

Farad   The unit of capacitance.

Feedback   A transfer of energy from the output circuit of a device back to its input.

Field   The space containing electric or magnetic lines of force.

Field winding   The coil used to provide the magnetizing force in motors and generators.

Flux field   All electric or magnetic lines of force in a given region.

Frequency   The number of complete cycles per second existing in any form of wave motion, such as the number of cycles per second of an alternating current.

Full-wave rectifier circuit   A circuit which utilizes both the positive and negative alterations of an alternating current to produce a direct current.

Fuse   A protective device inserted in series with a circuit. It contains a metal that will melt or break when current is increased beyond a specific value for a definite period of time.

Gain   The ratio of the output power, voltage, or current to the input power, voltage, or current, respectively.

Galvanometer   An instrument used to measure small dc currents.

Generator   A machine that converts mechanical energy into electrical energy.

Ground   A metallic connection with the earth to establish ground potential. Also, a common return to a point of zero potential.

Hertz   A unit of frequency equal to one cycle per second.

Henry   The basic unit of inductance.

Horsepower   The English unit of power, equal to work done at the rate of 550 foot-pounds per second. Equal to 746 watts of electrical power.

Hysteresis   A lagging of the magnetic flux in a magnetic material behind the magnetizing force which is producing it.

Impedance   The total opposition offered to the flow of an alternating current. It may consist of any combination of resistance, inductive reactance, and capacitive reactance.

Inductance   The property of a circuit which tends to oppose a change in the existing current.

Induction   The act or process of producing voltage by the relative motion of a magnetic field across a conductor.

Inductive reactance   The opposition to the flow of alternating or pulsating current caused by the inductance of a circuit. It is measured in ohms.

In-phase   Applies to the condition that exits when two waves of the same frequency pass through their maximum and minimum values of like polarity at the same instant.

Inversely   Inverted or reversed in position or relationship.

Joule   A unit of energy or work. A joule of energy is liberated by 1 ampere flowing for 1 second through a resistance of 1 ohm.

Kilovolts   1000 volts.

KVA   1000 volt-amperes or 1000 watts.

Lag   The amount one wave is behind another in time; expressed in electrical degrees.

Laminated core   A core built up from thin sheets of metal and used in transformers and relays.

Lead   The opposite of lag; also, a wire or connection.

Line of force   A line in an electric or magnetic field that shows the direction of the force.

Load   The power that is being delivered by any power-producing device. The equipment that uses the power from the power-producing device.

Luminaire   A lighting fixture.

Magnetic amplifier   A saturable reactor-type device that is used in a circuit to amplify or control.

Magnetic circuit   The complete path of magnetic lines of force.

Magnetic field   The space in which a magnetic force exists.

Magnetic flux   The total number of lines of force issuing form a pole of a magnet.

Magnetize   To convert a material into a magnet by causing the molecules to re-arrange.

Magneto   A generator which produces alternating current and has a permanent magnet as its field.

Megger   A test instrument used to measure insulation resistance and other high resistances. It is a portable hand-operated dc generator used as an ohmmeter.

Megohm   One million ohms.

Micro   A prefix meaning one-millionth.

Milli   A prefix meaning one-thousandth.

Milliammeter   An ammeter that measures current in thousandths of an ampere.

Motor-generator   A motor and a generator with a common shaft used to convert line voltages to other voltages or frequencies.

Mutual inductance   A circuit property existing when the relative position of two inductors causes the magnetic lines of force from one to link with the turns of the other.

NEC   National Electric Code.

Negative charge   The electrical charge carried by a body which has an excess of electrons.

NEMA   National Electrical Manufacturers Association.

Neutron   A particle having the weight of a proton but carrying no electric charge. It is located in the nucleus of an atom.

Nucleus   The central part of an atom that is mainly comprised of protons and neutrons. It is the part of the atom that has the most mass.

Null   Zero.

Ohm   The unit of electric resistance.

Ohmmeter   An instrument for directly measuring resistance in ohms.

Overload   A load greater than the rated load of an electrical device.

Permeability   A measure of the ease with which magnetic lines of force can flow through a material as compared to air.

Phase difference   The time in electrical degrees by which one wave leads or lags another.

PLC   Programmable logic controller (multifunctional microprocessor).

Polarity   The character of having magnetic poles, or electric charges.

Pole   The section of a magnet where the flux lines are concentrated, also where they enter and leave the magnet. An electrode of a battery.

Positive charge   The electrical charge carried by a body which has become deficient in electrons.

Potential   The amount of charge held by a body as compared to another point or body. Usually measured in volts.

Potentiometer   A variable-voltage divider; a resistor which has a variable contact so that any portion of the potential applied between its ends may be selected.

Power   The rate of doing work or the rate of expending energy. The unit of electrical power is the watt.

Power factor   The ratio of the actual power of an alternating or pulsating current, as measured by a wattmeter, to the apparent power, as indicated by ammeter and voltmeter readings. The power factor of an inductor, capacitor, or insulator, is an expression of their losses.

Prime mover   The source of mechanical power used to drive the rotor of a generator.

RAM   Random-access memory.

Reactance   The opposition offered to the flow of an alternating current by the inductance, capacitance, or both, in any circuit.

Rectifiers   Devices used to change alternating current to unidirectional current. These may be semiconductors such as germanium and silicon, and dry-disk rectifiers such as selenium and copper oxide.

Relay   An electromechanical switching device that can be used as a remote control.

Reluctance   A measure of the opposition that a material offers to magnetic lines of force.

Resistance   The opposition to the flow of current caused by the nature and physical dimensions of a conductor.

Resistor   A circuit element whose chief characteristic is resistance; used to oppose the flow of current.

Retentivity   The measure of the ability of a material to hold its magnetism.

Revenue meter   Meter used by electrical power service companies to measure power used for billing purposes.

ROM   Read-only memory.

Saturation   The condition existing in any circuit when an increase in the driving signal produces no further change in the resultant effect.

Series-wound   A motor or generator in which the armature is wired in series with the field winding.

Servo   A device used to convert a small movement into one of greater movement or force.

Servomechanism   A closed-loop system that produces a force to position an object in accordance with the information that originates at the input.

Single-phasing   Loss of one-phase of a three-phase power supply due to power company service malfunction.

Solenoid   An electromagnetic coil that contains a movable plunger.

Surge (power)   Delivery of source power supply at voltages greater than allowable variances.

Specific gravity The ratio between the density of a substance and that of pure water at a given temperature.

Synchroscope An instrument used to indicate a difference in frequency between two ac sources.

Synchro system An electrical system that gives remote indications or control by means of self-synchronizing motors.

Tachometer An instrument for indicating revolutions per minute.

Tertiary winding A third winding on a transformer or magnetic amplifier that is used as a second control winding.

Thermistor A resistor that is used to compensate for temperature variations in a circuit.

Thermocouple A junction of two dissimilar metals that produces a voltage when heated.

Torque The turning effort or twist which a shaft sustains when transmitting power.

Transformer A device composed of two or more coils, linked by magnetic lines of force, used to transfer energy from one circuit to another.

Transmission lines Any conductor or system of conductors used to carry electrical energy from its source to a load.

Vector A line used to represent both direction and magnitude.

Volt The unit of electrical potential.

Voltmeter An instrument designed to measure a difference in electrical potential in volts.

Watt The unit of electrical power.

Wattmeter An instrument for measuring electrical power in watts.

REFERENCES

7. _____ 1991. *Electric and Gas Service Requirements.* San Francisco: Pacific Gas & Electric Company.

8. _____ 1989. Energy efficiency in electric motors. *Consulting Specifying Engineer* 6(5). C. Publ. Co.

9. _____ 1981. *Successful electrical maintenance.* New York: McGraw-Hill.

10. DOLLINS, A.T. 1965. *Newnes Engineer's Reference Book.* London: Geo. Newnes Ltd.

11. _____ 1991. Personal Communication. Century Electric Motors Corp., Wentsville, MO.

12. BUCHAN, M.D. 1967. *Electricity Supply.* London: Arnold Publishers.

13. FRANKLIN, A.C. and FRANKLIN, D.P. 1983. *The J & P Transformer Book,* 11th ed. London: Butterworths.

14. _____ 1991. *Commercial/Industrial Schedule E-20; Service to Customers with Maximum Demands of 1000 Kilowatts or More.* San Francisco: Pacific Gas & Electric Co.

15. _____ 1972. *Illuminating Engineering Society Lighting Handbook.* New York: Illuminating Engineering Society.

16. KOHN, M.B. 1990. Task lighting; the key to a productive work place. *Consulting Specifying Engineer.*

17. YOUNG, S.S. 1991. Eco-terrorism. How safe is your plant? *Pollut. Control Eng.* 23(9).

18. FARRELL, G. and VALVODA, F. 1990. The art of protecting electrical systems: Understanding system ratings. *Consulting Specifying Engineer.*

19. SHENG, R.S. and NORDON, R.W. 1990. *Smart Temperature Transmitters Make Their Move.* Radnor, PA: Chilton Instrumentation Group. 6.

20. GALLERY, A. 1991. Personal communications. Industrial Applications Division, IBM Corporation, Chicago, IL.

CHAPTER 3

TELECOMMUNICATIONS

DEFINITION

Telecommunications, teleprocessing, and data communications are used interchangeably in this chapter and describe the movement of information from one point to another by any number of wired or cabled systems and devices (21).

BUSINESS ASPECTS AND WINERY COMMUNICATION NEEDS

Wineries are no different than any other commercial/industrial venture when it comes to a fundamental requirement for rapid, reliable, and cost-effective communications. For purposes of making a telecommunication needs analysis that is unique to wineries, six distinct classifications of activity can be made, each with different time-constrained objectives for sending and receiving information. The categories are as follows:

1. Grape production and vineyard management (all inclusive viticulture and grape purchases)

2. Production (from grape juice to packaged and finished cases of product)
3. Warehousing and shipping (storage, inventory, and transport)
4. Marketing and sales (on-premise and off-premise sales)
5. Special winery events
6. Business management and administration

Each of these winery activities can be considered a telecommunications user for which equipment hardware and systems must be matched. Although it is not the intent of this chapter to include detailed design of communication systems, it is important to highlight the special needs of wineries and how communications ideas and concepts can be prefiltered and judged by managers for merit, before soliciting vendor proposals for integrated systems. Table 3.1 lists winery functions, needs, and possible equipment choices.

TECHNOLOGY CHANGES INFLUENCING EQUIPMENT CHOICES

Since about 1985, some significant advances have been made in commercial communications. The inexpensive, but sometimes limited, FM two-way radio systems have all but been replaced with mobile cellular telephones. The addition of more microwave receivers to cellular phone transmission terminals and a broader geographic coverage have made the "dead spots" and "fade-out zones" almost nonexistent. The two-way radio systems were often plagued with interference from local taxi dispatcher frequencies that were sometimes close to the assigned carrier frequency. Message security was almost impossible, as anyone with a multifrequency scanner could monitor any transmission. (Message security is discussed in a subsequent section.) The radio connections were also limited by line-of-sight distance and the Federal Communications Commission (FCC) allowable wattage of the sending and receiving units.

Telex (or TWX typewriter exchange) systems have also been made nearly obsolete, in the United States and many other countries, by the widespread use and affordability of facsimile (or FAX) machines. As the name implies, telex systems consist of two typewriters that have the ability to both receive and send messages in real time and over telephone connections (21, 22). Computer-supported text message systems, such as electronic mail (computer to computer), have a much broader range of use, are faster, and can be combined with other information storage and retrieval functions of basic computers (22). Figure 3.1 depicts the sequence

Table 3.1. Winery Communication Needs and System Choices

Activity category	Most frequent communication link	System options
1. Production	(a) Winemaker/production supervisor to cellar and bottling line staff	Intercom plus paging system or PC network
	(b) Winemaker/production supervisor to vineyard manager during crush	FM two-way radio, mobile cellular telephone, and, pager
	(c) Winemaker/production supervisor to winery laboratory and off-premises commercial lab	Telephone, facsimile machine, and electronic mail
	(d) Winemaker/production supervisor to vendors	Telephone, facsimile machine, and electronic mail
	(e) Winemaker/production supervisor to utility systems foreman	FM two-way radio, mobile cellular telephone, pager, or PC network
2. Warehousing and shipping	(a) Winery sales manager to winery warehouse	Intercom plus paging system or PC network
	(b) winery warehouse to common carriers	Telephone, facsimile machine, and electronic mail
	(c) Winery warehouse to loading dock	Intercom plus paging system
	(d) Winery warehouse to commercial warehouse	Telephone, facsimile machine, and electronic mail
3. Marketing and sales	(a) Sales manager to brokers, reps. and distributions (U.S.)	Telephone, facsimile machine, and electronic mail
	(b) Sales manager to media reps	Telephone, facsimile machine, and electronic mail
	(c) Sales manager to distributors (overseas)	Telephone, facsimile machine, and electronic mail
4. Grape production and vineyard management	(a) Winemaker to vineyard manager in the field	Cellular telephone, or FM two-way radio or paging system, or network electronic mail
	(b) Vineyard manager to labor contractor (vehicle to vehicle or field to field)	Cellular telephone or paging system
	(c) Vineyard manager to mechanized harvester (vehicle to vehicle)	FM two-way radio
5. Special events	(a) Business manager to media reps	Telephone and facsimile machine
	(b) Business manager to special security and parking control (during event)	FM two-way radio and/or cellular telephone, or network electronic mail
6. Business management and administration	(a) General manager to division/dept. heads	Intercom plus paging or network electronic mail
	(b) General manager to corporate headquarters	Telephone and facsimile machine
7. Medical emergencies	(a) Winery or vineyard to "911" dispatcher and/or ambulance, emergency rescue vehicle, life-flight helicopter	Telephone or cellular phone from the field

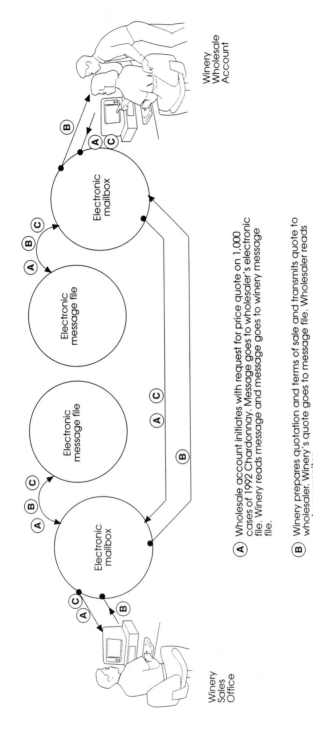

Fig. 3.1. Functional sequencing for typical electronic mail cycle.

Winery
Wholesale
Account

Electronic
mailbox

Electronic
message file

Electronic
message file

Electronic
mailbox

Winery
Sales
Office

(A) Wholesale account initiates with request for price quote on 1,000 cases of 1992 Chardonnay. Message goes to wholesaler's electronic file. Winery reads message and message goes to winery message file.

(B) Winery prepares quotation and terms of sale and transmits quote to wholesaler. Winery's quote goes to message file. Wholesaler reads quote on monitor.

(C) Wholesaler places order at quoted price and terms with shipping instruction. (Sequence of message handling same as (A)) of events and equipment functions in a typical winery electronic mail scenario. Nonetheless, Western Europe has clung steadfastly to the telex concept in part, and in 1983 several communication network organizations developed a service called "Teletex," which could link more sophisticated terminals, such as electronic storage typewriters and communication-capable word processing units. The character sets employed allow translation into all languages which use the Latin alphabet symbols (22). Thus, EU (European Union) members can avail themselves of this service and of the technology utilizing e-mail, and networking by means of World Wide Web access. Other factors which will continue to upgrade and improve telecommunication systems in the next decade are as follows (22):

(a) More widespread use of optical-fiber cable systems to replace aging coaxial cable systems.

(b) Replacement of silicon integrated circuits with gallium arsenide (GaAs)-based chips for production of integrated circuits (ICs) that are four to five times faster than contemporary ICs fabricated from silicon chips.

(c) Conversion of telephone center switches and other telecommunication system switches from electrical to digital (currently in progress) to so-called Josephson junctions (JJ), which are cryogenically cooled switches with zero electrical resistance. Switching speeds will increase from 10 to 100 times over current digital silicon and GaAs chip systems (22, 23). The widespread adoption of these devices could gain great market advantage for one telecommunication system or networking system over another. Wineries should be prepared to ask their communication equipment vendors specific questions regarding current and future plans for adoption of the apparently superior new technology and how it might beneficially impact the winery's system of choice.

(d) Availability of Integrated Service's Digital Network (ISDN) lines for more direct access to the World Wide Web. (faster than modem-telephone line connection)

(e) Use of artificial intelligence to design and develop the fifth generation of digital computers to design new devices and circuits to further advance telecommunication science.

The foregoing technological advances are presented to illustrate how quickly a winery's telecommunication system equipment and service company could change. The adoption of the new technology does not mean immediate obsolescence for a winery's chosen system, as some changes, such as faster switching, can be internalized in the telephone networking company's master switching terminals, which should mean better and more reliable service.

An analogy can be made to the situation developing in the "safe refrigerant'' field, where hydrochlorofluorocarbon (HCFC) compounds must eventually replace all of the ozone-depleting chlorofluorocarbon (CFC) refrigerants. (This conversion and its impact on wineries will be discussed in detail in Chapter 5.) Often, it is not just a matter of recharging the refrigeration unit with new gas, as the old compressor system seal materials were never designed to be compatible with HCFCs. In the telecommunications illustration, some technological changes and conversions can take place without great economic impact; some may not. Wineries will be less at the mercy of service company sales people and equipment vendors, if they remain informed on the telecommunication developments that are in the offing. This section is only meant to guide the study of that ongoing and never-ending learning process.

Some questions that should be asked by wineries to their system(s) vendors that will help in getting the proper mix of technologies for the optimum price to performance ratio are as follows:

(a) How has the particular technology changed over the last 10 years? How will it change over the life of the winery?
(b) Can the basic winery communication needs, as described generically in Table 3.1 and as modified to fit the particular winery's special situation, be met or exceeded with the vendor's proposed design package?
(c) What is the best estimate for the longevity of the system design?
(d) What new technologies should the winery be prepared to see arrive in the communications equipment/system marketplace in 2 years? 5 years? 10 years?
(e) How much flexibility can be built into the telecommunications system design to allow accommodation of new technology with a minimum of cost?

SYSTEM CHOICES, LIMITATIONS, AND ADVANTAGES

Mobile Communication Systems

Most frequency-modulated (FM) mobile radio systems transmit on frequencies between 100 MHz and 1 GHz (25). The so-called citizen's band mobile radios, made popular during the early 1970s, operated on a low frequency of 26 MHz. FM is utilized because of its immunity to noise (25). The limitations of the FM two-way radio system are the typically low-power output of the transmitters (5–50 W) and the fade-outs which occur because of topographic (hills) and structural (building) obstacles. Also, communication between vehicles equipped with transceivers is almost impossible unless the vehicles are very close to each other [500 yd (457m) and often in the line of sight]. Although the use of signal repeaters (higher-wattage transmitters with critically placed antenna to enhance signals) solve some of the system's problems, special band filters may be required to protect the signal's integrity. FM two-way radios work well in the marine and aircraft environment because obstacles are virtually nonexistent. There are mobile radio systems available in some areas of the United States which use geosynchronous satellites as the signal-repeater platform. Their performance can be imagined to approach ground to aircraft FM communication, which is usually trouble-free (25).

Cellular Telephones

These systems are becoming very commonplace in the wine industry and in commerce and industry in general (24). The system consists of a very-low-power transceiver which connects to an area repeater, which, in turn, connects the mobile telephone to a commercial telephone switching center and the familiar telephone system dial tone. The need for many repeaters to accommodate cellular phones required that urban areas, in particular, be divided into sectors or cells—hence the system name (25). A central computer tracks and switches the cellular telephone user from cell repeater to cell repeater as the vehicle traverses the service area. The central computer also controls the connection to the public telephone network and prepares billing information for the user's monthly statement of charges. The transmitting power of the mobile telephones range from 1.5 to 3 watts. Hand-held models are available with over-the-shoulder carrying cases for users who wish to leave their vehicle but remain in telephone contact without resorting to a pager (see description following). The initial cost for cellular telephones is in the "free" to $700 range (1996 price levels), depending on the manufacturer's discounts, optional features, and special offers for models to be replaced by a more current version. Some cellular telephones are given at no cost if the buyer selects a particular cellular telephone service company. For wineries (see Table 3.1), the cellular telephone for vineyard managers, sales and marketing directors, winemakers, and general or business managers would appear to be a must-have situation. The cellular telephones enable key personnel to conduct important business while in vehicular transit or away from their office.

Pagers

Pagers are devices that bridge the gap between the need for immediate voice-to-voice communication and learning via an audio or pager-to-body signal that someone is trying to communicate with you. The response to the message can be at your convenience. The advantages of pagers are that they are pocket-size; the receiving units are either rented or purchased. The monthly cost for equipment and air time is a modest $8 to $14 (1996 prices), depending on area coverage and the features and accessories desired (26). There is also an optional voice mail service that will allow the sender to not only leave their return telephone number but a message as well (27).

Coverage options range from local or statewide to regional, nationwide, and international. In the United States, there are paging service national dispatch centers, which connect the caller to a live operator, who will direct the short message to the receiver's pager. The receiver would have to be equipped with an alpha-numeric pager model to receive the message. The human-intervention link is more costly but offers several advantages to a receiver who must be reached in a timely fashion.

For the vineyard manager, a pager would appear to be an absolute necessity during the crush period. The need for the winery to know what is and what is not in the harvested grape pipeline can make the difference between a sane and a chaotic harvest.

Night-harvest operations are becoming commonplace in California's Central Valley and to a limited extent in other warm Region III and IV areas in the western United States. Night harvesting by mechanical pickers or conventional labor presents a greater risk for accidents than harvest during daylight hours (28). An almost fail-safe communication link with the winery–vineyard and nearest ambulance/public safety rescue team would appear to be a minimum requirement to ensure quick response for first aid and/or transport in the event of an accident. The advantage of night harvesting is significant, having the fruit arrive at the crush pad at 60°F (16°C) instead of at 78–85°F (26–30°C) and precludes artificial cooling of juice or must through heat exchangers, making the maintenance of the planned fermentation temperature profile much easier for the winemaker and cellar crew.

Facsimile Machines

As the heir apparent system to the less versatile telex, FAX machines are an essential piece of equipment in the fast-approaching end of the 20th century. Compact and carefully miniaturized, they have become a necessity in transmitting the written word for international trade and commerce. Whereas the first-generation machines employed thermal printing, several FAX manufacturers introduced plain-paper printers to eliminate one of the technical weaknesses of the earlier models. The fading of the thermal printed images with time, heat, humidity, and light have frustrated a host of users, much the same way as smudgy, weak carbon copies did before copy machines became more sophisticated. The advantages in the use of FAX machines are to be able to strategically time a transmission to another time zone, so that its arrival coincides with the beginning of the next business day, or overseas during non-working periods. None of the current machines (late 1992) have a preprogrammed timing device that will "send" at a prescribed time even though the operator may be absent. It is safer to monitor the transmission and be ready to intervene when papers jam or the paper feed runs two electrostatically bonded sheets simultaneously. Today, a fax modem from a personal computer (PC) avoids paper originals and printed copies unless they are needed. For all of their weaknesses (and there are not many), the advent of the FAX in business has made many legal and administrative functions cycle so rapidly between parties that executives are no longer afforded the luxury of sitting on a decision. The famous stall position of not taking any action until "I can see it in writing" no longer has meaning (unless, of course, you pull the plug on your FAX, which maybe is not such a bad idea when you have to stretch the timeline).

Other Devices

Some communications, which must be transmitted from the winery are in the emergency category and are of such importance that they will be briefly discussed as separate items.

Fire and intrusion alarms top the list of emergency signals which must reach the intended action terminal 100% of the time. The action terminal for fire is most often a regional public safety dispatcher, who initiates the call to the nearest fire station to get equipment to answer the emergency. In a winery with an automatic sprinkler system, at least two signals are generated: (1) a flow switch in the sprinkler line is activated when water in the system changes from hydrostatic to hydrodynamic and (2) if the winery is remote and has its own sprinkler/hydrant water supply pump, the master controller for the pump can also send a signal to an automatic dialer which can dial via a commercial telephone network to several preprogrammed contacts [i.e., winery general manager, winery utilities manager, local fire department and regional fire department (state's wildfire combatting agency such as California's Department of Forestry and Fire Protection)]. If the dialer fails to reach the specific telephone number, it automatically proceeds to the next number in programmed order until it completes the sequence. A DC power battery backup system is also available to permit the emergency notification system to operate in the event of a power failure. As suggested earlier in the introductory section of this book, other discrete signals of special winery importance can also be transmitted via the dialer, such as high fermentor temperatures, a tank leakage sensor alert, potable water storage low-level or wastewater storage high-level alarms, and intrusion and security-failure alarms.

At the present time, new communications devices are entering the marketplace. Dubbed PCSs (personal communication systems), the PCS will have a number of advantages over current cellular telephone systems (29):

- The first-generation models will allow short facsimile messages to be transmitted from the PCS to a table model FAX.
- They are earmarked as the wireless telecommunications of the near future.
- PCSs are a long-range extension of the cordless telephone concept.
- First-generation PCSs will have a maximum range of about 5 miles.
- Eventually, they will have greater range with the capability for use while traveling on commercial or private aircraft.
- The purchase price of first generation is expected to be five to six times that of current cellular phone cost, and maybe most important.
- An electronic scrambler may be offered to provide message security (see following section on telecommunications security).

Figure 3.2 graphically summarizes how an optimally equipped and fully automated winery office communication system could be configured.

TELECOMMUNICATIONS SECURITY

With any electronic transmission of financial, marketing, and sales strategy there is no sealed envelope. Electronic mail has been characterized as being similar to communicating with a postcard—anyone can read it. In the United States, there is currently no strictly enforced law to protect communication privacy. Many interactive computer networks have "watched gateways," with some monitoring done by federal agencies to apprehend wrong-doers, who routinely use electronic mail. For some large U.S. corporations with computer networks, employees' electronic mail can be monitored, not only to track administrative and operational functions but also to discourage the use of expensive computer installations for unauthorized use. Although the military has, of necessity, been in the vanguard for providing secure communication links to maintain absolute secrecy in matters of tactical and strategic importance, the private sector is only now in the last decade of the 20th century, moving toward message-encrypting devices. Message scramblers and unscramblers are currently available on a limited basis as a computer accessory. That still leaves cellular telephone message microwaves subject to electronic eavesdropping. Cellu-

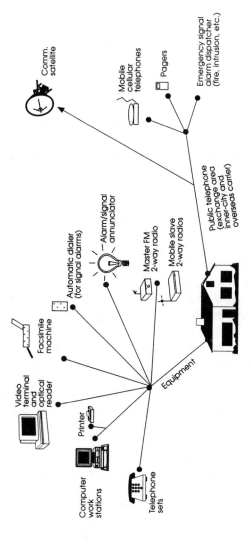

Comm.
satellite

Mobile
cellular
telephones

Pagers

Emergency signal
alarm dispatcher
(fire, intrusion, etc.)

Automatic dialer
(for signal alarms)

Alarm/signal
annunciator

Master FM
2-way radio

Mobile slave
2-way radios

Facsimile
machine

Video
terminal
and
optical
reader

Printer

Equipment

Public telephone
(exchange area
inner-city and
overseas carrier)

Computer
work
stations

Telephone
sets

WINERY COMMUNICATION SERVICE FUNCTIONS

- Word processing
- Electronic mail
- Data filing and retrieval
- Government and other periodic status and
  management reports
- Computer conferencing
- Security and alarms
- Warehouse and inventory status
- Vineyard status and operations
- Marketing and sales domestic and overseas
- Laboratory and quality control

Fig. 3.2. Ideal, optimally equipped, fully automated winery communication system and subsystems.

lar telephone scanners are unlawful, but the law is apparently not rigidly enforced (29). Black-market scanners can be purchased through computer-hacker underground magazines (30).

Although there is no need for wineries to become paranoid about telecommunications security, this section will have served its purpose if it is recognized that any unencrypted electronic message transmission, hard wire or radio wave, can be monitored without your knowledge using state-of-the-art equipment. A winery's information assets are just as important as any other commercial activity producing a commodity for sale in the marketplace. Thus, trust your most critical message traffic to a sealed envelope for viewing by only those eyes which you have preapproved as an information addressee.

COMPUTERS AND WINE-INDUSTRY-FOCUSED SOFTWARE

Computer systems and their various ancillary components constitute "equipment" and not a distinct subclass of winery utilities. Nonetheless, it is linked closely enough to the broad subject of telecommunications to permit its beneficial inclusion in this book. Wineries have embraced digital computers not only as work-saving devices but also to take advantage of their speed and simplicity in performing daily business tasks heretofore accomplished manually or not at all. Information retrieval, whether for winery machinery history, sales records, wine inventory, or equipment maintenance records, has streamlined business administration and operations for a relatively modest outlay in capital funds. At present, the Bureau of Alcohol, Tobacco, and Firearms (BATF) forms 2050 and 702 are easily compiled with inventory and sales records kept as current as updating input frequency requirements are ordered by winery management. In the word processing arena, the uses by wineries are almost endless, from routine correspondence to winery newsletters, brochures, price lists, and press releases.

Online services and use of the World Wide Web (WWW) by means of online service companies are commercial and industrial sector information retrieval and discrimination sources that are essential in the fast-moving, "knowledge–power" era of the late 20th century. Online services are relatively inexpensive, averaging about $25 per month for a 60-h base connection time each month (1996 price levels). Using a local online service and their network equipment makes good business sense, as the e-mail address will include the locale of the online service hub, which presumably will coincide with the winery's preferred geographical identifier. Cyberspace has also become an active arena for wine sales and barring any difficulties in the United States with interstate shipment of wine, the potential buying audience is enormous. The use of online services is, unlike the 10 A.M. to 5 P.M. wine tasting room, open 24 h, giving your current cellar offerings extraordinary exposure without regard to time zone, latitude, or longitude. As the commercial online service companies like to point out—"cyberspace is infinite."

Winery software packages are available from a number of different vendors, some tailored to a winery's specific needs, whereas others are standardized, off-the-shelf programs that are designed for PC-compatible equipment (31, 32). They are reported to perform but may not be limited to the following tasks:

"Wineries and Wine Warehousing for Single or Multiple Bonded Wineries:"

- Order processing
- Case goods inventory
- Warehouse billing
- Accounts receivable
- Accounts payable
- General ledger
- Payroll
- Transaction inventory
- Government reporting compliance
- Wholesale and retail sales orders
- Credit memos
- Discount structures
- Freight charge input
- Sales tax for retail sales
- Bills of lading
- Credit limits and terms
- Cash receipts (monthly)
- Payment history
- Delinquent accounts
- Mailing label writer
- Allocation of wine sales by customer and salesperson

In addition to the tasks that have been highlighted, wine production software programs are also available which can process, store, and manipulate, among other things, the following:

- Harvest forecast grape delivery data by grower
- Grower contract details, including weigh tag accounting
- Delivery position by varietal and grower versus harvest forecast data
- Projected grower payments
- Wine production history by lot
- Laboratory analyses tracking by lot
- Trial blending schemes

Finally, programmable linear controllers, PLCs which are in reality small computers, perform a variety of tasks in wineries. These extraordinary devices will be discussed under the appropriate utility category heading in subsequent chapters of this book.

MAINTENANCE AND OPERATION OF TELECOMMUNICATION SYSTEMS

Solid-state electronic devices and equipment have a high degree of reliability and are almost maintenance-free. Telephone network companies usually maintain their portion of the system. If the winery's telecommunication network choice includes, for example, a master TBX and intercom subsystem supplied by the same company, all troubleshooting is accomplished by the same repair technician. Since the antitrust breakup of AT&T, the lines of repair and maintenance responsibility have become more complex. Split responsibility for equipment and circuits is still workable but sometimes lacking in efficiency and is fraught with noncooperation among the equipment and network vendors.

The service companies which sell and provide network control centers for pagers and cellular telephones are generally very responsive to a user's request for technical assistance after damage, loss, or theft of an instrument occurs. Almost the same statement can be made for computer vendors. With the intense competition for corporate business, a user must be serviced quickly with highly qualified technicians or the vendor stands a very good chance of losing the account and incurring the wrath of not only winery management but other linked dependents of the system, such as warehouses, common carriers, and important retail accounts.

Large wineries may have the luxury of an in-house electronic technician, who can perform limited maintenance and repair tasks on some equipment. The latter capability is very rare and the risk of invalidating an equipment warranty or vendor maintenance contract is too great, even if the downtime seems impossible to cope with.

GLOSSARY

AM (amplitude modulation)   One of several ways of modifying a radio or high-frequency wave to carry information; the AM radio band is that portion of the radio spectrum allocated to AM radio services. (See *FM, frequency modulation.*)

Analog   The representation of information by continuous waveforms that vary as the source varies; a method of transmission by this method. Representations which bear some physical relationship to the original quantity, usually a continuous representation as compared to digital representation. (See *Digital*)

Antenna(e)   A device used to collect an/or radiate radio energy.

Artificial intelligence   Computer programs that perform functions, often by imitation, usually associated with human reasoning and learning.

Automatic dialer   A device that can produce either rotary dial pulses or DTMF tones automatically, to dial a previously stored sequence of digits. The PBX *speed dialing* feature is an example of this.

Band   A range of radio frequencies within prescribed limits of the radio-frequency spectrum.

Bandwidth   The difference in hertz (cycles per second) between the highest and lowest frequencies in a signal or needed for transmission of a given signal. The width of an electronic transmission path or circuit in terms of the range of frequencies it can pass. A voice channel typically has a bandwidth of 4000 cycles per second (or hertz); a TV channel requires about 6 million cycles per second (megahertz).

Baud   Bits per second (bps or BPS) in binary (two-stage) telecommunications transmission; after Emile Baudot, the inventor of the asynchronous telegraph printer.

Binary   A numbering system having only two numbers, typically 0 and 1.

Bit (binary digit)   The smallest part of information with equally likely values or states, "9" or "1," or "yes" or "no." In an electrical communication system, a bit is typically represented by the presence or absence of a pulse.

BPS   Bits per second.

Broadband   A term often used to describe a range of frequencies wider than that required for just voice communications. Also a term used to describe systems and equipment with wide bandwidth that can carry these ranges of frequency. (See *Narrowband Communications.*)

Broadband carriers   The term used to describe high-capacity transmission systems used to carry large blocks of telephone channels or one or more video channels. Such broadband systems may be provided by coaxial cable or microwave radio systems.

Broadband channels   Transmission channels used primarily for the delivery of video or for high-speed data transmission.

Broadband communication   A communications system with a bandwidth greater than voice bandwidth. Cable television is a broadband communication system with a bandwidth usually from 5 to 450 MHz.

Byte   A group of bits operating together. Bytes are often grouped into 8-bit groups but 7-, 9-, and 16-bit bytes are not uncommon. Today's fast computers often use 32-bit bytes.

Carrier   A signal with known characteristics—frequency, amplitude, and phase— that is altered or modulated in order to carry information. Changes in the carrier are interpreted as information.

Central office   The local switch for a telephone system sometimes referred to as the wire center or class 5 office.

Channel   A segment of bandwidth which is used to establish a communications link. A channel is defined by its bandwidth; a television channel, for example, has a bandwidth of 6 MHz, and a voice channel about 4000 Hz.

Channel capacity   The maximum number of channels that a cable system or some other delivery system can carry simultaneously.

Chip   The colloquial name for the semiconductor element that has become the heart of modern computing and increasingly of modern telecommunications. The silicon base on which the electronic components are implanted is smaller than a fingernail and looks like a chip. Increasingly, the chip refers to a single device made up of transistors, diodes, and other components interconnected by chemical processes and forming the basic element of microprocessors.

Circuit switching   The process by which a physical interconnection is made between two circuits or channels.

Coaxial cable   A metal cable consisting of a conductor surrounded by another conductor in the form of a tube which can carry broadband signals by guiding high-frequency electromagnetic radiation.

Common carrier   An entity that provides transmission services to the public at nondiscriminatory rates and which exercises no control of the message content. An organization licensed by the Federal Communications Commission (FCC) and/or by various state public utility commissions to provide communication services to all users at established and stated prices.

Comsat   Communications Satellite Corporation. A private corporation authorized by the Communications Satellite Act of 1962 to represent the United States in international satellite communications and to operate domestic and international satellites.

CPU   Central processing unit; the component in a stored-program digital computer which performs arithmetic, logic, and control functions.

CRT   Cathode-ray tube; a video-display vacuum tube used in television sets and display computer terminals. Today, a CRT is the display terminal for video games, word processors, personal computers, and a host of other electronic information products.

Dedicated lines   Telephone lines leased for a specific term between specific points on a network usually to provide certain special services not otherwise available on the regular or public-switched network.

Demodulate   A process by which information is recovered from a carrier. (See *Modulation.*)

Digital   A function which operates in discrete steps as contrasted to a continuous or analog function. Digital computers manipulate numbers encoded into binary (on/off) forms, whereas analog computers sum continuously varying forms. Digital communications is the transmission of information using discontinuous, discrete electrical or electromagnetic signals which change in frequency, polarity, or amplitude. Analog intelligence may be encoded for transmission on digital communication systems.

Direct broadcast satellite (DBS)   A system in which signals are transmitted directly to a receiving antenna on a subscriber's home or office via satellite rather than redistributed via cable or an intermediary terrestrial broadcast station.

Downlink   An antenna designed to receive signals from a communications satellite. (See *Uplink.*)

Earth station   A communications station on the surface of the earth used to communicate with a satellite.

Electronic mail   The delivery of correspondence including graphics by electronic means, usually involving the interconnection of computers, word processors, or facsimile equipment.

FCC (Federal Communications Commission)   The federal agency established in 1934 with broad regulatory power over electronic media, including radio, television, and cable television.

Facsimile   The transmission of images of original documents including print and graphics by electric means, usually via the telephone network but also by radio.

Final mile   The communication system required to get from the earth station to where the information or program is to be received and used. Terrestrial broadcasting from local stations, the public-switched telephone network, and/or cable television systems provide the final mile for today's satellite networks.

FM (frequency modulation)   One of several ways of modifying a radio or high-frequency wave to carry information. The FM radio band is that portion of the radio spectrum allocated to FM radio services.

Frequency   The number of recurrences of a phenomenon during a specified period of time. Electrical frequency is expressed in hertz, equivalent to cycles per second.

Frequency spectrum   A term describing a range of frequencies of electromagnetic waves in radio terms; the range of frequencies useful for radio communications, from about 10 kHz to 3000 GHz.

Geostationary satellite   A satellite with a circular orbit 22, 300 miles in space, which lies in the plane of the earth's equator and which has the same rotation period as that of the earth. Thus, the satellite appears to be stationary when viewed from the earth.

Gigahertz (GHz)   Billion cycles per second.

Hardware   The electrical and mechanical equipment used in telecommunications and computer systems.

Hertz (Hz)   The frequency of an electric or electromagnetic wave in cycles per second; named after Heinrich Hertz who detected such waves in 1883.

Kilohertz (kHz)   One thousand cycles per second.

Megahertz (MHz)   One million cycles per second.

Memory   One of the three basic components of the central processing unit (CPU) of a computer. It stores information for future use.

Message switching   A computer-based switching technique that transfers messages between points not directly connected. The system receives messages, stores them in queues or waiting lines for each destination point, and retransmits them when a receiving facility becomes available.

Microchip   Electronic circuit with multiple solid-state devices engraved through photolithographic or microbeam processes on one substrate.

Microprocessor   A microchip which performs the logic functions of a digital computer.

Microwave   The short wavelengths from 1–30 GHz used for radio, television, and satellite systems.

Modem (modulator–demodulator)   A device that converts digital pulses to analog tones and vice versa for transmission of data over telephone circuits.

Modulation   A process of modifying the characteristics of a propagating signal such as a carrier so that it represents the instantaneous changes of another signal. The carrier wave can change its amplitude (AM), its frequency (FM), its phase, or its duration (pulse code modulation), or combinations of these.

Multiplexing   A process of combining two or more signals from separate sources into a single signal for sending on a transmission system from which the original signals may be recovered.

Narrow-band communications   A communication system capable of carrying only voice or relatively slow-speed computer signals.

Optical fiber   A thin, flexible glass fiber the size of a human hair which will transmit light waves capable of carrying vast amounts of information.

PABX (Private Automatic Branch Exchange)   A private telephone switching system that provides access to and from the public telephone system.

Pager   A small receiver carried in a purse or worn on a belt that is part of a simplex communications system. It is used to receive a signal by telephone to a pager system which activates the number called and in some systems short alpha-numeric messages are displayed also.

PBX (Private Branch Exchange)   A private telephone switching system that is not automated.

Public-switched telephone network (PSTN)   A designation given to the nation's subscriber telephone services provided by the more than 1500 operating telephone companies, including the operating companies created by the 1984 divestiture of AT&T.

RAM (random-access memory)   A RAM provides access to any storage or memory location points directly by means of vertical and horizontal coordinates. It is erasable and reusable.

ROM (read-only memory)   A permanently stored memory which is read and not altered in the operation.

Software   The written instructions which direct a computer program.

Tariff   The published rate for service, equipment, or facility established by the communication common carrier.

Teleconferencing   The simultaneous visual and/or sound interconnection that allows individuals in two or more locations to see and talk to one another in a long-distance conference arrangement.

Telemetry   Communication of information generated by measuring devices.

Teletext   The generic name for a set of systems which transmits alpha-numeric and simple graphical information over the broadcast (or over a one-way cable) signal using spare-line capacity in the signal for display on a suitably modified TV receiver.

Telex   A dial-up telegraph service.

Terminal   A point at which a communication can either enter or leave a communications network.

Transponder   The electronic circuits of a satellite which receive a signal from the transmitting earth station, amplify it, and transmit it to earth at a different frequency.

"Twisted pair"   The term given to the two wires that connect local telephone circuits to the telephone central office.

UHF (ultrahigh frequency)   The frequency band from 300 to 3000 MHz.

Uplink   The communications link from the transmitting earth station to the satellite.

VHF (very high frequency)   The frequency band from 30 to 300 MHz.

WATS (Wide Area Telephone Service)   A service offered by telephone companies in the United States which permits customers to make dial calls to telephones on a specific area for a flat monthly charge or to receive calls ''collect'' at a flat monthly charge.

REFERENCES

21. FITZGERALD, J. 1984. *Business Data Communications*. New York: John Wiley and Sons.

22. CARNE, B.E. 1984. *Modern Telecommunications*. New York: Plenum Press.

23. BREWSTER, R.L. 1986. *Telecommunications Technology*. New York: Halsted Press.

24. _____ Cellular phones essential to wine industry management. *Pract. Winery Vineyard*, May–June.

25. WILSON, E.A. 1989. *Electronic Communications Technology*. Englewood Cliffs, NJ: Prentice-Hall.

26. OLSEN, L. 1992. Personal communication. Motorola Radio Engineers, San Francisco.

27. PHILSTOR, R. 1992. Personal communication. Metromedia Paging Services, Ashland, MD.

28. _____ 1989–1992. State of California Department of Labor. Summary of Agricultural Accident Statistics.

29. ROGERS, S. 1992. Personal communication. GTE MobileNet, San Francisco.

30. _____ 1992. Personal communication. Federal Communication Commission, District Legal Counsel, San Leandro, CA.

31. VAHL, J. 1990. *The Winemaker's Database*. St. Helena, CA.

32. GULSON, R.L. 1991. *WIMS and WIPS, Winery Information Systems for PCs*, Sonoma, CA.

CHAPTER 4

SANITATION, STEAM, AND HOT WATER

Although steam and hot water may not; in a true sense, stand a rigorous test of "utility" by definition, this subcategory of water supply is extremely important to winery operations for the production and maintenance of a sanitary and clean working environment for the long-term protection of quality wines.

The selection of the optimum steam/hot-water system involves a number of other issues that are connected in a complex matrix of fuel (energy), water quality (chemistry), water conservation, winery sanitation, winery size, air quality, and cost (capital and operating).

WINERY DESIGN CONCEPTS, WINERY AGE, AND SANITATION

The fundamental requirements for winery sanitation are generally firmly fixed in the minds of trained enologists. The formal study of spoilage organisms in a fermentation science curriculum is generally supplemented after graduation with continuing knowledge of the subject gained through exchange of wine spoilage horror stories between industry colleagues, enology journal articles, and professional society meetings and conferences. Industry and university research efforts are sometimes focused on microbiological problems of wineries and the results of the investigations can be found in the *Journal of the American Society of Enology and Viticulture* and in other scientific publications in Europe, Africa, and Asia.

Simply stated, the organisms that can produce spoilage in wine are never eradicated, as new populations arrive with each load of fall harvest grapes. Their spore-phase progeny have the advantage of dispersal by air transport, giving them a major advantage in maintaining their species in the preferred environment—wineries. Old wineries (pre-1970) are more difficult to keep clean (even with adequate steam/hot-water supplies) than are newer wineries, where thoughtful process design, space arrangement, and suspension of equipment, tanks, and barrels make cleaning tasks much easier. Troweled concrete floors and walls, which were the standard winery finish two decades ago, have, if left as they were constructed, become nicked, cracked, and otherwise abused by hard and continuous usage. Those cracks and recesses are difficult to clean and they become incubators and desirable habitat for the group of organisms with which the winery is constantly waging war. Old floors and walls can be given an overcoat of impervious acrylic, epoxy, or polymer paint that is, according to the several manufacturers, chip-proof and crack-proof (33). Figure 4.1 illustrates the main components for an old concrete surface replacement. EPA-approved antimicrobial agents can also be incorporated into the sealant coat to assist in controlling unwanted winery flora. There are many types of acrylic,

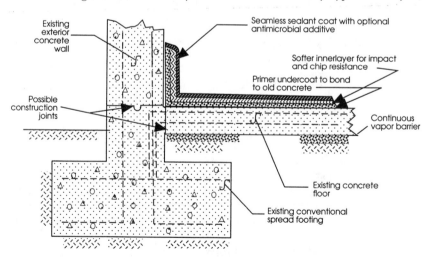

Fig. 4.1.   Concrete floor rehabilitation with three-coat polymer system. (Adapted from Ref. 33.)

epoxy, and polyurethane coatings that can be used to reclaim aged floors. The foregoing example is just one such system. Non-solvent-based coatings are also making their appearance in the marketplace in response to an eventual phase-out of solvent paints by the year 2000 as mandated by the Federal Clean Air Act Amendments of 1991.

There is little that can be done in a retrofit sense to old wineries, but if an inventory is made of potential "microbe incubators" on floors, walls, catwalks, and fixed equipment, often with the caulking and painting materials that are now available, some substantial improvements can be made over existing conditions. The following design tips can help sanitation in old and new wineries (34–38):

- A good rule of thumb is that equipment should not occupy more than 20% of the floor area in any production space to permit the ready movement of people, and product and to allow efficient cleaning of the space (37).
- Stainless steel should have a minimum polished finish of No. 4 (i.e., final polish with abrasives which correspond to 150-grit roughness) (38).
- Columns and supports should be tubular or circular in cross section (floor flanges should be avoided when attaching winery equipment).
- All fixed stainless-steel wine transfer piping should be mounted so that it will be self-draining. Pipe fittings should be of the sanitary type with long radius bends preferred over tight 90° elbows.
- If angle iron is used for equipment support mount the mounting should be as ⌐ ◿ not ◹ ; likewise with "Channel" ⊓, "Tee" ⊤ and "I" sections; mount upside down or in such a way that moisture cannot be trapped.
- All finished concrete ledges, sills, soffits, ducts, and exposed piping in process spaces should be sloped away from the wall to prevent dust accumulation and to prevent condensation or other moisture from remaining on the surface; concrete construction joints that form a recess should be sealed with a suitable, nonshrink, chemical grout. To facilitate cleaning, heating, ventilating, and air conditioning (HVAC) ducts of circular cross section are preferred over rectangular types.
- Multistory winery structures which permit the gravity flow of product are more difficult to sanitize than single-floor facilities (37).
- Openings in the outer winery building envelope must be screened with a minimum No. 150 mesh stainless-steel or fiberglass screen where practical to discourage fruit-fly incursion; creating a slightly higher than atmospheric pressure within the building spaces using air conditioning or two-way exhaust/night-air-cooling systems can also help control flying insects. Air curtains at large door openings can also be used at truck-loading docks and winery warehouse spaces.
- Slopes on all floors should have a minimum fall toward drains of ¼ in./ft; single floor drains with grated openings should be spaced no more than

20 ft apart. Trench drains with removable cast-iron or fiberglass grates are the best hydraulically and are the easiest to clean; stainless-steel basket strainers are also useful drainage accessories (basket strainers and their role in reducing solids loading in the process wastewater stream will be discussed in detail Chapter 8, "Wastewater Systems").

CLEANING SYSTEM OPTIONS

In small wineries (less than 25,000-case capacity), it is generally difficult to justify the installation (capital cost) of a fixed steam boiler, associated controls, and fixed piping to m,ixing-valve stations for a hot-water source. Portable units, which may include high-pressure washers and wheel-mounted steam/hot-water generators can serve a small facility well. For outside service, kerosene-, propane-, or diesel-fired boilers are available as standard stock items for portable steam or hot-water units. The petroleum-fueled systems reach operating temperature in a matter of minutes. Electrical resistance heater boilers are suitable for inside applications, as are propane-fueled boilers. However, the propane exhaust is not completely without odor or carbon monoxide emissions. One supplier of the electrically fired portable units is the Electro-Steam Generator Company of Alexandria, Virginia, whose equipment can generally be seen at wine industry conventions and often advertised in the several winery journals and magazines. The electric steam/hot-water generators are quiet, and they have a slightly higher first cost than their petroleum-fueled counterparts, and are slightly slower to reach operational temperature after the system is energized. Operating costs for electrically fired boilers are 30–40% higher than petroleum-fuel-fired units. For the budget-minded winery, the used market for reconditioned steam/hot-water generators should not be overlooked. First costs for a used, reconditioned unit are approximately 50–75% of new models, depending on age and condition. Even multiple portable units would likely be more economically advantageous if the winery's general arrangement of equipment and processing activity was spatially dispersed. Even large wineries in the 50,000–100,000-case annual capacity category have purchased small, portable high-pressure-water cleaning units to use both on vineyard equipment and for other difficult cleaning tasks that often occur away from the established cleaning stations. Figures 4.2 and 4.3 show examples of typical portable units available for industrial use that are commercially available.

For larger wineries, fixed boilers and permanent steam distribution piping to hot-water stations are generally preferred. The principal components of a clean-in-place (C.I.P.) system are shown in Fig. 4.4. In one

Fig. 4.2.   Portable cold-water pressure washer. (Courtesy of Direct Line, Inc., St. Paul, MN.)

Fig. 4.3.   Electrically fired, portable, hot-water pressure washer. (Courtesy of Direct Line, Inc., St. Paul, MN.)

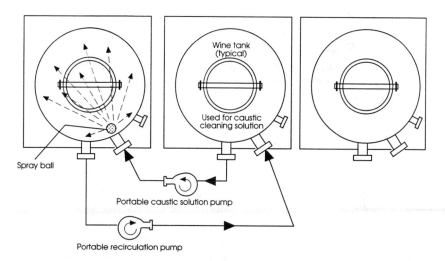

CHEMICAL CLEANING CYCLE

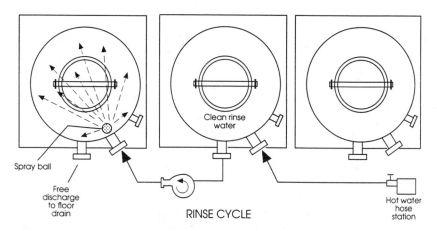

Fig. 4.4.   Schematic of typical winery clean-in-place system using portable hose and wine transfer pumps.

version of the C.I.P. arrangement, wineries utilize wine transfer pumps, hoses, and a small blending tank or another storage tank which serves as a cleaning-solution reservoir, and as a temporary component of the chemical-solution supply system. If the transfer pump is equipped with an internal timer, storage and fermentation tanks can be washed with a suitable caustic tank cleaner that generally contains a disinfectant such as chlorine or iodine. The time interval is set for the caustic cycle and hoses are interconnected with the solution tank, the pump, and the tank's spray ball as shown in Fig. 4.4. The drain valve on the tank is connected to a second pump which recirculates the cleaning/disinfecting solution back to the solution tank. During the rinse cycle, the recirculation pump can be disconnected and the drain valve allowed to discharge by gravity to the winery's floor drain system. Hot or cold water can be used for the rinse cycle and can be introduced to the chemical-solution tank to give a larger volume of water available at the suction side of the pump than might be available from just the 8–10-gpm (0.5–0.6 L/s) flow rate at the hot-water hose station. Somehow, the discharge rate of the pump must be balanced with the rinse water supply or pump damage could result.

The second type of C.I.P. system requires the installation of fixed-in-place, single-purpose pumps and solution tanks together with dual-purpose, fixed-in-place, stainless-steel valves and piping. The piping can be utilized for both product transfer and for cleaning. The process piping is valved so that individual tanks can be isolated. Figure 4.5 shows one possible arrangement for a fixed C.I.P. system. The most sophisticated systems can be operated from a single programmable controller, which can perform all the necessary cleaning or product transfer cycles. Air-operated sanitary valves (with a higher degree of reliability than solenoid or motor-operated) isolate tanks for cleaning and for product transfer. The layout of the fixed process/cleaning system should be undertaken by a plumbing or mechanical design engineer, who is familiar with plumbing codes, pipe supports, expansion joints, and sanitary fittings and valves. The amount of the solution can vary from as small as 300 gal (1136 L) to as large as 500 gal (1938 L). As suggested in the previous section, the tank volume-pump capacity must be in some sort of hydraulic balance, so that the pump will always have an adequate supply of solution or rinse water.

Because the pumps used for solution supply and recirculation may have variable rate of flow and head (lift) demands placed on them by their dual use in cleaning and product handling, a variable-speed motor or a variable-frequency drive (both of which add a substantial capital cost to the pump installation) would make the pump fully versatile through a fairly wide range of operating conditions. Multiple flow requirements can be met rather easily with parallel installation of pumps, all discharging to a common manifold but each having a different pump characteristic curve. Multipurpose pumps do not work well when the lift (total head) varies widely for the operational range. Obviously, the correct selection of pumps for the fixed C.I.P. system becomes even more critical for wineries with different floor levels. All of the foregoing physical and hydraulic constraints are highlighted to extrapolate the logic of the system concepts shown in Figs. 4.4 and 4.5.

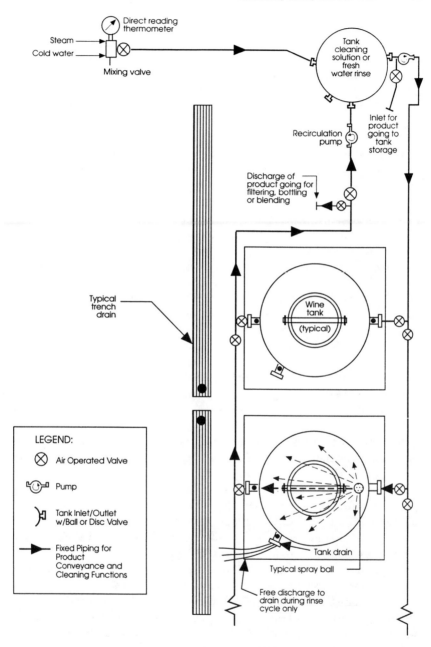

Fig. 4.5.   Schematic of a typical winery clean-in-place system using fixed piping and other equipment for product conveyance and tank/pipe cleaning.

Barrel-aging rooms are not readily adapted to an automated C.I.P. system because the wood cooperage is normally not subjected to heat (except during the toasting process at the time of manufacture or later, when shaving is undertaken). Soda ash, citric acid, and sulfur dioxide in small amounts are usually the only chemicals that are introduced internally to oak barrels. Thus, cold-water hose stations are generally all that is necessary for barrel rooms. If sanitizing of the floors with hot water is required, a portable unit such as that shown in Fig. 4.3 can always be used.

Wineries which utilize bored or driven caves for barrel aging have a special group of cleaning and logistical problems. Barrel racking is generally not accomplished in the caves; by the use of barrel racks and fork lifts, the racks are transported to a racking area in the tank room or other indoor or outdoor space of the winery proper, where the clean wine (hopefully not agitated in transit) is separated from the lees. Barrel washers rinse the sediment residuals from the cooperage and then the wine is returned, the barrel topped appropriately, placed again on the rack, and transported back to the cave. This racking protocol precludes extensive wash-down and cleanup of the caves. Plumbing is simplified (no ugly floor drains or water supply piping to detract from the raw geologic charm of the local native rock) and the cave environment remains a dry and relatively inhospitable site for harboring common spoilage organisms. If asked, most winemakers with caves would concur, saying that the nearly ideal, isothermic environment for aging, the tourist value, and, last but not least, the long-term energy conservation by precluding the artificial cooling and humidifying of a conventional aging room more than balances the extra expense of cave construction.

Water temperatures for cleaning, as produced at a hose station by means of a steam–water mixing valve, should probably not exceed 145°F (63°C) and will probably never be below 100°F (37°C). Higher temperatures can tend to "cook-on" some residues. Cleaning compounds and detergents perform better in the presence of warm water in the range reported above. Figure 4.6 is a typical wall-mounted, hot-water mixing valve.

Table 4.1 lists generally accepted temperatures for sanitizing using dry heat, moist heat, steam, and ultraviolet light.

For general sanitizing operations, the majority of wineries will use a combination of pressure rinsing, chemical washing, and rinsing. Live steam or hot water at 185–200°F (85–93°C) is frequently employed in bottling rooms, where a nearly antiseptic environment must be maintained. To give wine its best visual presentation and to meet acceptable public health requirements for sterile tasting room glassware, most wineries use machine dishwashers as opposed to double, deep, stainless-steel sinks for hand washing and rinsing, which is the other public health-acceptable glassware

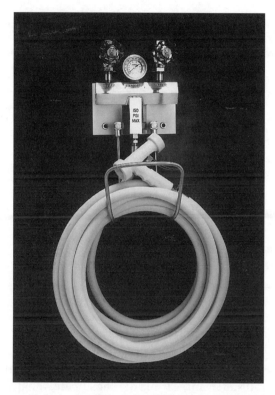

Fig. 4.6.   Wall-mounted, hot-water mixing valve. (Photograph furnished by Streamline Washdown Equipment, Inc. (800-926-3648), Carlsbad, CA.)

cleaning alternative. If the winery chooses the aesthetically preferable, thin-stem, crystal glasses for tasting, machine washing appears to be more destructive than hand washing. So-called "glass racks" can be purchased from restaurant supply houses to restrain wine glasses on the lower level of a standard dishwasher. The glass rack sizes are such that most all wine glass configurations can be accommodated. Dishwashers and autoclaves will also be found in winery laboratories, as they are efficient and effective devices for the small quantity of laboratory glassware required routinely for wine quality control analytical functions.

Finally, the chemicals that can be used in combination with warm, hot, and cold water are a satisfactory method for producing the level of microorganism mortality required in wineries. The most commonly used compounds are the halogens chlorine and iodine or the cationic quaternary ammonium compounds.

Table 4.1. Sanitizing Methods and Necessary Time–Temperature Parameters

Method	Energy level or temperature	Time for destruction of micro-organisms	Remarks
(1) Dry heat (autoclave)	180°F (82°C)	20 min	
(2) Ultraviolet light (surface treatment)	Wavelength = 254 nm	1 min	Surfaces only
(3) Moist heat (autoclave)	170°F (77°C)	30 s	Hand wash or hot dip
(4) Machine wash and rinse	150–160°F (66–71°C) (wash) & 180°F (82°C) (rinse)	10 min 10 s	Tasting room and laboratory glassware/equipment
(5) Live steam (surface application)	200–212°F (94–100°C)	2 min	Surface only; good for destruction of mold spores
(6) Boiling (immersion)	212°F (100°C)	10 min	
(7) Hot water (surface washing)	170°F (77°C)	15 min	

Source: Adapted from Refs. 35 and 36.

Cellar personnel should be thoroughly trained in the use of chemicals for personal safety and to provide assurance that the cleaning/sanitizing operations will have the desired end result. For example, the terms antiseptic, sanitizer, disinfectant, germicide, bactericide, fungicide, and bacteriostat may appear on chemical compound labels. The winery's safety and training officer should ensure through written and oral instructions that the correct dilutions of chemicals are prepared. The manufacturer's instructions/directions for use should be annotated by the safety officer and reprinted in boldface type in English and a second language common to the winery locale, for posting near the chemical storage area.

For purposes of reference, Table 4.2 contains a listing of commonly used winery cleaning chemicals and their concentrations for effective sanitizing.

Care should be taken to keep the suggested temperature regime during washing operations because both chlorine and iodine begin to volatize at solution temperatures above 120°F (49°C). Thus, an overzealous cellar worker could reduce the microbial killing power of a given solution by adjusting the mixing valve to give hot water over 75–80°F (24–27°C).

For all winery cleaning involving chemical agents, rinsing is, by far, the most important step. Hoses and fixed transfer pipe networks, for which the interior cannot be sniffed or visually checked, are the mostly likely locations for chemical residues to remain and cause off-colors, off-flavors, and

Table 4.2. Winery Cleaning Chemicals and Their Recommended Concentrations

Compound	pH	Concentration of cleaning solution[a] (mg/L)	Contact time for sanitizing (min)	Temperature (°F/°C)	Taste threshold after rinse (mg/L)
Chlorine	6.7	100–200	2	75/24	10
Iodine	7.0	12.5–25	2	75–100/24–40	7
Quaternary ammonia	6.8	200–400	1–2	75/24	15

[a]These concentrations should be compared to the manufacturer's written directions/instructions on the use of the product. If lower concentrations are suggested, ask the manufacturer to provide laboratory certification that the lower concentrations will give the desired microbe mortality at the suggested contact times and temperatures.

Source: Adapted from Ref. 35.

off-aromas in wines. Pipe openings and hose ends should be closed with something as simple as kitchen plastic wrap and rubber bands if they are to be left unused for any extended period of time. Stainless-steel blind flanges, attachable with sanitary clamps and washers, are the best closures. The industrywide practice of always rinsing all equipment intended for use (just prior to use), even though certified as being clean 2 weeks previously, is a good habit for cellar crews to adopt even though additional time and labor costs are required.

WATER USE PLANNING GUIDELINES (CLEANING)

Overall water use and water supply requirements for wineries will be covered in detail in Chapter 6, "Potable Water Supply Systems." Water use criteria for some cleaning and washing functions are based on historical records of a number of large and small wineries, where nonmetered water service at existing facilities required the synthesizing of data to provide reasonable design criteria for expansion or modification of both water supply and process wastewater treatment disposal systems. Table 4.3 summarizes the water use estimates for winery cleaning purposes.

BOILERS AND BOILER WATER QUALITY

Boilers are very sensitive pieces of equipment, and their rugged construction belies their requirement for thoughtful and delicate handling.

The mechanical engineering designer generally selects the initial boiler installation to meet both present and future needs. Boilers are generally

Table 4.3. Estimates of Water Use For Winery Cleaning Purposes

Item	Description	Estimated water volume/rate relationship	Remarks
General area or equipment cleaning	Washing floors, crush/press bays, equipment except bottling equipment	$V_{water\ volume}$ = cleaning time (min) × 7	Cleaning times highly variable for different types of crush and press equipment[a]
Bottle washing	For removal of fiberboard case residues and to ensure a sterile container	12 gal (48 L) per case of 750-mL bottles	
Hot-water hose stations	150–170°F hot water (60–70°C)	5–7 gpm (0.4–0.5L/s) at 50 psig (min) minimum design flow rate at each station at minimum pressure for general cleaning	
Tank cleaning (caustic wash and rinse)	C.I.P. (fixed) or C.I.P. (portable)	No. of spray balls × 7 gpm (28 L/min) × 30 min ≥ 210 gal (840 L)/tank	
Bottling lines	Filler, labeler/capsule spinner, corker, case sealer and conveyers	[7 gpm (0.5 L/s) × 45] × 2 times ≥ 630 gal (2500 L)/day minimum	Bottling lines washed 2 times per 8-h day or 3 times if wine variety changes during the shift

[a]Large belt presses used by some sparkling wineries may take as much as two to three times the time required to clean a bladder or screw press.

sized to about 200% of predicted normal steam demand. Most well-constructed boilers can operate at 150% of design capacity for short periods, so there is generally sufficient cushion in the sizing of the boiler (39). Thus, it is often oversized for a few years until the winery reaches full capacity. The capacity can often be split between two or more units so that if the timeline for wine production increase is rather long, the single boiler can meet steam and hot-water requirements for many years but precludes a major demolition and reconstruction process to add another boiler in the future.

Actual performance specifications will be downplayed in the short discussion of the hardware to highlight some special items that relate to boiler fueling and operation. Suffice it to say that there are boilers to fit any winery requirement, and it would be prudent of wineries in the design stage to have the mechanical designer provide a matrix of performance, life cycle, operation and maintenance history, and capital cost to highlight what might be the optimum choice for the winery. A professional engineer with mechanical design specialties and familiarity with winery facilities can provide sound guidance in the planning of steam and hot-water systems for new wineries and help filter out ideas for retrofitting older wineries with more energy-efficient and environmentally compatible boilers.

The winery's available water supply is an important parameter in the proper operation of an industrial boiler. Water hardness is usually expressed as milligrams per liter (mg/L) of total carbonate (calcium or magnesium) hardness. For boilers, when total carbonate hardness exceeds 85 mg/L or 5 grams of hardness per gallon (in boiler nomenclature), water softening must be included in the pretreatment of boiler feed water. The softening process, if it is performed on the entire winery water supply and not just the boiler feed water (usually an ion-exchange column, which trades calcium for sodium ions), also has a second benefit in that some cleaning tasks become easier, as cleaning compounds work more efficiently with soft water. Maintenance personnel for established wineries with hard, raw water supplies wage a constant battle to give the boilers the soft water they require.

Air-quality standards also enter the boiler selection process. The Clean Air Act Amendments of 1991 have set in motion new emission-control requirements for almost all industrial fixed sources, regardless of size. Nitrogen oxide and carbon monoxide have been made more stringent for both existing and proposed new or replacement boilers. Remote wineries for which natural gas service cannot be economically provided must rely on propane (LPG) or fuel oil for boiler fuel. The new rules are rather complex, with exemptions made only for very small installations. Contact the local Air Pollution Control District for the region to make sure that the proposed new or existing boiler is in compliance.

PRIMER ON BOILERS

Wineries which cannot get by on portable steam or hot-water generators should have someone in winery management or operations with at least a working knowledge of steam boilers.

A boiler is a steam generator with most units totally internalized and of either the fire-tube or water-tube type. The former type has the hot gases of combustion run through a network of tubes immersed in water, whereas in the latter, the water runs through the tubes. Fire-tube types are for small or highly variable-steam-load installations. Water-tube boilers are generally safer because of the smaller amount of water in the boiler and a water-tube fracture would be minor compared to the failure of the fire box. A sub-category of the water-tube boiler is the so-called steam generator. Only a small water capacity is provided, and from a cold start to operating temperature is usually very fast (in a matter of minutes) (39).

Boiler manufacturers are governed in construction safety standards by the American Society of Mechanical Engineers (ASME) Code for Boilers. All certified boilers have an "ASME Boiler Code" or similar symbol stamped on the boiler somewhere near the name plate.

Boilers are rated in terms of heat capacity generated, generally in BTUs (British Thermal Units) per hour. Also used interchangeably is the term boiler horsepower: 1 boiler horsepower = 33,472 BTU/h (39).

"Heating surface," which is another boiler performance parameter, differs from one boiler type to another.

Boiler Types Found in Wineries	Heating Surface Required [sq ft (m^2) per boiler horsepower]
Water tube	10 (1)
Automatic steam generator	2–5 (0.2–0.5)

The heating surface is defined as the outside circumference of the water tube times the net length of the tubes in the tube bundle. Relating energy requirements to heating surface for steam generation gives 240 BTU/h/ ft^2.

There are many boiler options such as the following:

- Indoor or outdoor installation
- Choice of fuel or fuels
- Alarms and lights for low water, flame failure, and pilot-light failure
- Separate heat exchanger in the steam boiler for hot-water production, if steam–water mixing values are not used

The winery's mechanical design engineer can assist the winemaker, the maintenance supervisor, and key owner/manager personnel in making the proper decisions regarding steam- and hot-water-generation equipment performance specifications, optional equipment, fuel choices, combustion emission control technology, and initial and annual operating cost optimization. Before the selection process of boilers is complete, the winery should be given a complete breakdown of the thermal efficiencies. Smaller boilers should have thermal efficiencies of 75–80%. Thermal losses should be identified and quantified by the manufacturer for several fuel choices and under excellent, good, and poor maintenance of heating surfaces (39). (See section following on boiler maintenance).

Just as many residential hot-water systems have their capacity exceeded, when three or four family members take long, closely grouped showers, so can wineries exceed their steam/hot-water capacity. Armed with the data provided in Table 4.3 and knowing the number of planned hot-water cleaning stations (or existing plumbing plan if the winery is operational), a comparison can be made to boiler performance and expected peak hourly demand with all hose stations in use. A factor of safety of 15–20% overcapacity would reveal that the winery has sized their system correctly, and if it is properly maintained, as will be described in the following section, long and trouble-free service will prevail.

ALTERNATIVE ENERGY SOURCES FOR HOT WATER

The high petroleum costs of the late 1970s and early 1980s have temporarily been reduced to the level where many forms of alternate energy available to wineries (direct solar heating of water and photovoltaic cells for dc electrical energy) became less cost-effective. Most wineries are in climatic zones where solar energy in some form could be considered for preheating boiler feed water or for augmenting scarce electrical power supplies in remote locations. A number of winery designs have seriously considered solar energy for two distinct but different reasons:

(a) A strong commitment to conserve fossil-fuel energy resources and to protect atmospheric resources.
(b) The proposed winery location was so remote and of such small scale that the cost for the line extension of electrical public utility service was inordinately expensive.

The winery's utility design engineer can clearly demonstrate, through a straightforward economic analysis, if solar energy is economically sensible to pursue as a viable alternative and how long of a cost-recovery period would be required to amortize capital expenditures. The best advice that can be given is for wineries to monitor electrical power costs and trends in fossil-fuel costs to get a feeling for the correct time to seriously look at alternative energy as a winery utility adjunct. Smaller wineries may be able to justify the installations based on purely benevolent environmental grounds if the stockholders are few and sympathetic to such a move and hard-hearted lending institutions can be convinced that it is a reasonably sound business choice. Rebates by power utilities to customers who wisely expend money for improvements that will improve energy use efficiency or exploit alternative energy sources were standard practice in the late 1970s and early 1980s. Encouraged by the U.S. government, such incentives may well be revived before the end of the 20th century.

MAINTENANCE AND OPERATION OF STEAM/ HOT-WATER SYSTEMS

Maintenance of steam/hot-water systems can be subdivided into four main classifications:

- Safety features
- Maintenance of heating surfaces
- Water quality
- Fuel quality

For a given piece of equipment, the manufacturer will provide detailed operation and maintenance instructions, which the winery staff should follow religiously. As in all manufacturer's warranties, failure by the user to follow the prescribed maintenance procedures voids any repair labor or replacement parts cost. Good maintenance history records are stressed throughout this book as an important element of a winery's utility management program. In the case of boiler malfunction or component failure, those records are the way that the manufacturer's representative can verify that the proper operational functions were, in fact, performed.

For portable steam-generator systems, the most likely failure will occur if the water supply is accidentally terminated while the heat source is still active. High-temperature inactivation switches can easily be installed in electrically fired units; however, the fail-safe fuel shut-off control for kerosene-fired units is more difficult. Because the operator is quite near the portable source while cleaning, flame-outs are readily observed and dealt with quickly. Steam and hot-water hoses and their fittings should be checked routinely, and age-cracked hoses replaced. It is a small expenditure to ensure cellar help safety and well-being. Heat-transfer surfaces in portable systems of the water-tube type are generally easily accessed for visual inspection by removing only the vent stack on the kerosene-, propane (LPG)-, or compressed natural gas (CNG)- fired units. If there is an accumulation of soot (more likely in the kerosene-fired units), the portable unit should be taken from the winery to an outside service and equipment maintenance area where the water tubes can be further exposed to permit wire brushing or cleaning with a portable, cold-water, high-pressure washer. During this cleaning operation, the exterior of the water tubes should be inspected for pits and corrosion, and if there is a convenient fitting that can be removed, without disassembling the entire boiler-tube bundle, the interior of the tubes should be examined to see if there are significant accretions of mineral salts (generally calcium and magnesium carbonate deposits resulting from their precipitation during the heating process; reference to this is made in the section entitled "Boilers and Boiler Water Quality," in which the hazards of hard-water use are described). The salt deposits can be removed by using the manufacturer's recommendation for a mild acid wash and thorough rinsing.

State-of-the-art steam–water mixing valves at hose stations are relatively simple and very reliable, but should be tested periodically to see that the mixture can never approach 100% steam (the probe-type throttling valves are made so that cold water cannot ever be entirely shut off unless the valve stem failed or some other low-probability mode of valve malfunction occurred). The direct reading thermometers on mixing valves should also be calibrated so that the cellar staff are confident that if 155°F (68°C) is registered at the mixing valve, the temperature of the water at the hose nozzle should be in the 135–140°F (57–60°C) range (deducting heat loss in hose transit).

Fixed boilers need slightly more attention than the portable units previously described, although some of the precautionary steps to ensure optimum thermal performance are the same.

The primary cause of boiler failures is due to water-tube leaks (40). All boilers are required by the ASME Boiler Code to have instrumentation and accessories that permit the safe and efficient operation of the boiler. Included in these requirements are the following (39):

(a) A steam gauge connected to the boiler's steam drum and which for operator safety is isolated from the live steam in the drum by an inverted siphon that is filled with water.

(b) A water-level sight glass, which is a glass tube which measures the level of water in the water side of a fire-tube type and with the sight glass connected to the top tube of a water-tube bundle; the operator can be assured that there is sufficient makeup water supply to keep the tube bundle filled at all times.

(c) A water injector pump or steam injector to force the makeup water supply into the higher-pressure boiler-tubing system. The steam injector is similar to the steam–hot-water mixing valve previously described, which allows the incoming water pressure to be elevated by means of a combining tube/chamber.

(d) A fusible plug is placed in the wall of a tube in the tube bundle, below which a reduced water supply level would cause damage. The lead filler in the fuse plug remains in its solid state as long as cooling water flows, but below the safe water level it melts and releases steam and water from the tube, which extinguishes the boiler fire and prevents extensive damage. [Thus, item (b), the maintenance of the proper water level, is a very important operational function to prevent costly boiler repairs.]

(e) A safety valve, which is spring-loaded, opens when a preset operating pressure is exceeded by 2 or 3 psig. The valve reseals itself after the pressure is stabilized.

Sight glasses should be located on equipment where they are readily accessible and easily viewed by the boiler operator. The glass should be removed periodically, when the boiler is inactive, and thoroughly cleaned so that rings of chemical deposits at or near the normal safe operational water level will not be mistaken for the actual water level.

Coal- and fuel-oil-fired boilers are less of an explosion hazard than propane (LPG)- or natural-gas-fueled systems. LPG fuels are heavier than air and could tend to settle into dry pits and other near-floor-level spaces, creating an explosion hazard. LPG gas lines from the storage tank to the winery boiler are maintained by the winery, whereas the pressure tank and its appurtenances are maintained by the propane gas vendor. Natural gas is lighter than air and, with a piping leak, can be dispersed throughout a closed mechanical room space very quickly. Included in the winery's maintenance equipment inventory should be a hand-held "gas sniffer" that can detect combustible gas concentrations down to a very low level. That device and periodic checks of fuel supply pipeline fittings with soapy water can reveal pinhole leaks and prevent a major catastrophe. Many design engineers provide double-walled gas piping as an added safety factor in preventing pipe failure. In seismically active regions, the design engineer will make provision for limited movement and flexibility of the fuel line, particularly where it passes through a rigid, concrete wall. Likewise, the aboveground LPG storage tanks will be protected from damage from winery vehicles and forklifts by being surrounded by protective pipe bollards (usually thick-walled galvanized steel pipe, 4–6 in. (10–15 cm) in diameter and filled with concrete).

Fuel oil storage tanks must conform to the EPAs underground storage tank regulations (double-walled fiberglass or steel) with leakage alarms placed in the void between the inner and outer shell. Aboveground fuel tanks must have full-perimeter containment and in the event of failure or spillage, drainage to a sump or tank that will retain the leakage and allow for removal and safe disposal. Petroleum discharges to streams or groundwater, accidental or otherwise, are no longer tolerated by United States, and state law and fines that can be levied are severe.

Monitoring the well-being of a boiler can be augmented by many devices that the winery or boiler service representative may routinely have in the maintenance toolbox. As described previously in the section entitled "Boiler and Boiler Water Quality," boiler stack emissions for some winery installations must be monitored to meet stringent air-quality management objectives. Combustion gas analyzers are available, and with a hand-held personal operator's terminal, adjustments can be made to the boilers combustion mix until the desired air-quality parameters are met (41). Here, the services of the boiler manufacturer's representative are a must, to help the winery's operator achieve optimum operating performance. Where dual-fuel systems are employed, for example, fuel oil and natural gas, the boiler's combustion variables must be radically adjusted each time a fuel type is changed. Seasonal changes are also possible in some U.S. locales. Sources for natural gas sometimes vary widely in quality from winter to summer, when lower quality gas from surge supply wells is introduced into the utility's gas network to meet higher winter demands. When this occurs, more attention to cleaning heat-transfer surfaces should be given, as the combustion residues that can be left on water tubes can increase.

Finally, once each year before peak boiler use, a major inspection of the boiler system and subsystems should be performed. Included but not limited to the inspection usually performed by the manufacturer's service personnel and assisted by winery staff or, for small units, by the winery alone are the following key work tasks:

- Check low water cutoff system and alarms and disassemble and overhaul all components of the boiler water supply devices.
- Burner components should be inspected and cleaned.
- Tube surfaces should be cleaned of all external deposits. Hand-held devices using ultrasonic technology can measure the tube-wall thickness (40). Tube walls should be checked for scale buildup and acid cleaned and rinsed if more than 1/16 in. (1.5 mm) has accumulated.
- Test the safety valve.
- Test and calibrate all gauges; clean the water level glass.
- Check fuel lines.
- Record results of tests and startup in winery's boiler maintenance history log.

Boiler downtime can be both expensive and frustrating for wineries. If failures occur during critical periods of steam need, the winery should plan on a downtime of 4–5 days, 1–2 days to drain and cool it down, disassemble the fire box, and gain access to the tube bundle. If a replacement tube is readily available, removal of the failed tube and installation of the new tube can take at least 1 day. Avoid retubing by approaching boiler maintenance with care and consistency.

GLOSSARY

Boilers Gas (natural or liquified petroleum)-, oil-, or coal-fired equipment which convert water to very hot (180°F) or saturated steam (212°F). Used in wineries for general cleaning and sanitation. Classed a low, medium, and high pressure.

Boiler code A set of standards prescribing requirements for the design, construction, testing, and installation of boilers and unfired pressure vessels; e.g., American Society of Mechanical Engineers (ASME) Boiler and Pressure Vessel Code, or federal, state, or local government codes.

Boiler rating Thermal capacity expressed in British Thermal Units (BTU) per hour or boiler horsepower.

C.I.P. Clean-in-place (portable) or (fixed) system of pipes, valves, pumps, and controls for winery tanks and product process equipment and piping. Fixed utilizes a permanent system of piping and valves to convey cleaning and disinfecting compounds and rinse water throughout the units to be cleaned.

Disinfectants Any number of halogenated chemicals that are employed in various concentrations to achieve microbial mortality (fluorine, chlorine, bromine, iodine, and astatine). Ozone and ultraviolet light are also capable of microbiological cell destruction under certain conditions.

Epoxy floor coatings Two-part paints that are used to seal porous concrete floors where frequent cleaning and sanitation is performed. Hard no-skid finishes are available with antimicrobial ablatives available as admixtures.

Insect vectors Insects capable of transmitting disease causing agents.

LPG (liquid petroleum gas) Petroleum by-products, propane and butane, are liquified at approximately 150 psig (pounds per square inch gauge) and used for boilers and forklift fuels at many remote winery locations.

Mixing valve A temperature-gauge-equipped valve that permits the adjustment of steam and water to achieve a desired hot-water temperature for effective winery cleaning and sanitation. Safety features prevent the discharge of steam only.

NOx (oxides of nitrogen) Clean-burning boiler fuels and stack gas emission scrubbers and screens prevent the discharge of nitrous oxides that exceed air-pollution-control emission guidelines.

Pathogen mortality Destruction of disease-causing organisms by means of high temperatures or disinfecting agents. (See *Disinfectants.*)

Pressure washers Cold = Portable, wheel-mounted, electric-motor-driven or internal-combustion-engine-driven multistage pumps provide cleaning nozzle pressures of 2000–4000 psig (pounds per square inch gauge) for removal of stubborn substances from floors, process equipment, or tanks. Hot = Same as cold except provided with electric resistance or LPG-fired boiler to give the pressure washer the added feature of sterilization as well as exceptional cleaning power.

Spray ball Available in many geometric configurations for use in the distribution of cleaning compounds and rinse water for winery tanks and barrels.

Structural shapes Fabricated steel or other metallic members used for framing and support for industrial buildings. Prudent orientation of the shapes by a winery building designer can prevent the creation of a surface for the accumulation of airborne dust, dirt and microbes.

Wand The long-handled nozzle device used at the end of the delivery hose from a hot or cold portable pressure washer to keep the operator a safe distance from the potentially injurious spray stream.

REFERENCES

33. _____ 1992. *Polymeric Floor and Wall Systems.* Cincinnati, OH: General Polymers Corp.

34. STORM, D.R. 1993. Sanitation, steam and hot water, Part II. Winery water and waste. *Pract. Winery Vineyard* May–June.

35. MINOR, L.J. 1983. *Sanitation, Safety and Environmental Standards.* Westport, CT: AVI Publishing Co.

36. BANERJEE, K. and CHEREMISINOFF, P.N. 1985. *Sterilization Systems.* Lancaster, PA: Technomic Publ. Co.

37. GOULD, W. 1990. *Food Plant Sanitation.* Baltimore, MD: CTI Publ.

38. BLOOMBERG, C.S. and MIZUNO, W.G. 1962. Cleaning metal surfaces. *J. Milk Prod.*

39. FARRALL, A.W. 1979. *Food Engineering Systems.* Westport, CT: AVI Publishing Co.

40. DUFFY, G. 1992. Talking to operators about boiler maintenance. *Engineered Systems* 9 (7).

41. _____ 1993. Personal communication. Quintox Combustion Gas Analyzer, Universal Enterprise, Inc., Beaverton, OR.

CHAPTER 5

REFRIGERATION, VENTILATION, AND AIR CONDITIONING

SCOPE

Probably more than any other aspect of winemaking, the maintenance of product temperature from fermentation through final warehouse storage is the most important process parameter in the vinification of high-quality wines. The rather restrictive ideal temperature regime of 55–65°F (13–18°C) can be achieved through a combination of means. What follows is an overview of the planning, design, and operation of the heating, ventilating, air conditioning, and refrigeration (HVAC&R) systems as seen from a winery perspective. Heating of winery spaces has been reduced in scope in this chapter because, except for administrative and laboratory spaces, heating is routinely handled by the mechanical designers and/or air-conditioning professionals. Heating is referred to incidentally in sections which discuss winery solar orientation and the Carnot refrigeration cycle operating as a heat pump. Insofar as possible, related topics that are or will produce a significant technical change or economic impact on HVAC&R systems now or in the near future will be highlighted. A selected list of references is given at the end of this chapter, including an excellent treatment of winery refrigeration technology, as practiced in Australia (42).

Where appropriate, cross-references will be used to other chapters in this book. For example, in Chapter 4, entitled "Sanitation, Steam, and Hot Water," the avoidance of rectangular cooling or ventilation ducts in winery production spaces was cited to help in preventing the accumulation of condensation, dust and dirt on flat, duct top surfaces, thus eliminating one potential substrate for spoilage organisms. As some of the largest electric motors will be part of the refrigeration–air-conditioning subsystems, the proper balancing of motor type, motor efficiency, and maintenance of an optimum power factor becomes extremely important in keeping electrical power services costs at their lowest possible levels. (See Chapter 2, Table 2.4 and accompanying text.)

BASIC PRINCIPLES

Refrigeration may be defined as the maintenance of the temperature of a body below that of the ambient temperature which surrounds it. One "ton" of refrigeration, the term commonly used in the industry to classify and measure capacity of refrigeration units, is defined by ASHRAE (American Society of Heating, Refrigeration and Air Conditioning Engineers), as the removal of heat from a body at the rate of 12,000 BTU (British Thermal Units) per hour, which is the equivalent of the production of 2000 lbs of ice in 24 h (47). The "body," of course, can be a volume of air, the contents of a wine fermentor, a cold storage room, or the small volume of wine and sediment in the neck of a bottle of sparkling wine that must be frozen before the classic *methodé champenoise* technique of disgorgement can take place. In practice, the term "refrigeration" includes both the energy required to lower the temperature and to maintain it at any given level.

In every refrigeration system there are elements that are common to all systems. Although there are options, substitutes, and alternatives for the principal features, the function of each part of the working system remains the same.

Because grape vines and, hence, wineries occur principally in temperate climate zones throughout the world, the ambient temperatures are much the same except for the seasonal reversal of crush periods from the Northern to Southern Hemisphere. Thus, the temperature gradients (heat flowing from a higher to a lower temperature) in must/juice heat exchangers and jacketed fermentation and storage tanks are quite similar for wineries throughout the world. There are, of course, exceptions. Southern England

and Germany, for example, have cooler growing seasons, and during harvest (September–October), most European wine-growing regions with latitudes north of France's famous Bordeaux are cold (43). For these high-latitude climatic zones, harvested grapes arriving at the winery might be at 40–50°F (4–10°C), certainly eliminating any need for artificial cooling of the must prior to the initiation of fermentation. Mechanical harvesting and hand picking at night in hotter wine-growing areas are becoming more commonplace. Taking advantage of the nearly 40°F (4°C) difference between nighttime lows and daily highs to have fruit arrive at the crush pad at 55–60°F (13–16°C) instead of 78–85°F (26–29°C) is a significant saving in time, energy resources, and money, with the potential quality benefits in the finished wine of special importance.

Climatic zone, winery size and construction, and the end product all influence the methods employed by the winery to achieve the desired level of cooling of wine, spaces, and people. For some, with abundant, cold-water supplies, simple immersion heat exchangers with "once-through" cooling water can control fermentation temperatures to desirable levels. Spray nozzles directed on uninsulated, outside fermentation tanks can be observed in several California wineries to augment heat transfer from the fermenting must to a cooler tank surface. Although these methods do accomplish the control of temperature, they require constant monitoring and surveillance by winery personnel. During drought periods, such as the 6-year dry period (1986–1992) experienced in California, single-purpose water use does not make good sense for resource management. Capture and reuse of the once-through cooling water for landscape or vineyard irrigation or other process use could make the scheme more environmentally compatible.

By far the most commonly used arrangement for refrigeration utilizes a so-called working substance (refrigerant) upon which phase changes are imposed (changes from liquid to vapor and back to liquid again) by the addition or rejection of heat energy. This cycle, identified by thermodynamicists as the reverse Carnot cycle, if designed for the proper winery heat loads, can offer the least expensive, most environmentally friendly refrigeration cooling apparatus.

From the smallest residential refrigerator to the largest industrial coolers and chillers required for the million-case-per-year wineries, the basic system elements are essentially the same. Figure 5.1 is a schematic diagram of a typical vapor compression refrigeration system. Historically, the most widely used application of this system in the wine industry is a two-part system consisting of a secondary coolant loop of brine or other high-specific-heat liquid. This feature, shown diagrammatically in Fig. 5.1, permits the "storage" of the cooling effect to accommodate large peak cool-

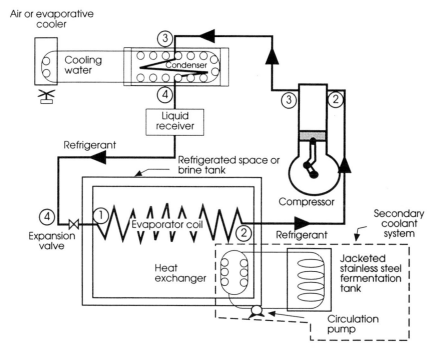

Fig. 5.1.   Schematic diagram of a typical vapor compression refrigeration system with secondary coolant loop for fermentor temperature control.

ing loads (precooling musts and juices and the control of fermentation temperatures). For example, if direct expansion of the refrigerant was the design option chosen to cool fermentation tanks [i.e., the typical dimple-jacket heat exchanger on the tank becomes the evaporator (see Fig. 5.1, refrigerant process steps 1–4)], the size of the refrigeration compressor and the volume of primary refrigerant would of necessity be much larger than a comparable system using a secondary coolant loop with a cold brine storage reservoir.

   Figure 5.2 is a graphical depiction of the work (hence, cooling effect) being done on the refrigerant in the reverse Carnot refrigeration cycle. The circled numbers shown along the refrigerant path in Fig. 5.1 correspond to the same numbers and thermodynamic state of the refrigerant, as work is done on and extracted from the refrigerant. Figure 5.2 coordinates are temperature (T) and entropy (S) with entropy defined as the change in heat in BTU per pound per degree change in temperature (°F absolute or degrees Kelvin). As the performance (efficiency) of any thermodynamic process depends on measurement of the heat energy changes which occur

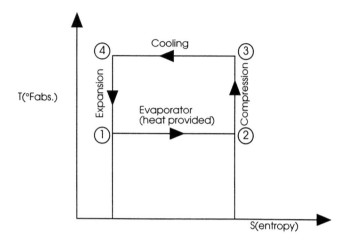

Fig. 5.2. Refrigerant temperature–entropy diagram for vapor compression refrigeration system operating as a reverse Carnot cycle. Circled numbers correspond to process steps in Fig. 5.1.

internally within the closed refrigerant cycle, the area inscribed under the *T–S* diagram becomes a measurement of the total work performed. Figure 5.1 is included to help in understanding an "ideal" cycle, for in practice there are heat losses in piping and internal friction. The foregoing thermodynamic discussion of the refrigeration cycle is necessary to describe one other possibility for this unique cyclic system to perform other useful work without radical changes in mechanical equipment or configuration. So-called "heat pumps" are refrigeration systems that can be reversed so that the heat that is rejected from the body to be cooled when the refrigerant receives the unwanted heat from the secondary loop brine or cooled space becomes useful heat for warming. Figures 5.3a and 5.3b illustrate how, with simple valving and possibly a dual evaporator installation to act as a heat sink for cold climate applications, the heat pump can serve an all-season role as cooler and heater. Heat pumps are not without operating problems, but it is an option for wineries that is worthy of consideration (4).

The heating, ventilation, and air conditioning, and public power industries have taken the thermodynamic storage concept (brine chillers and secondary coolant loops) a step further to reduce the operating cost and increase the net efficiency of commercial building space cooling. Stored ice made with low-cost, off-peak electrical energy is the key feature of the technology. The electrical energy savings are substantial, and as an incentive to install the more capital-intensive ice storage systems, which use the

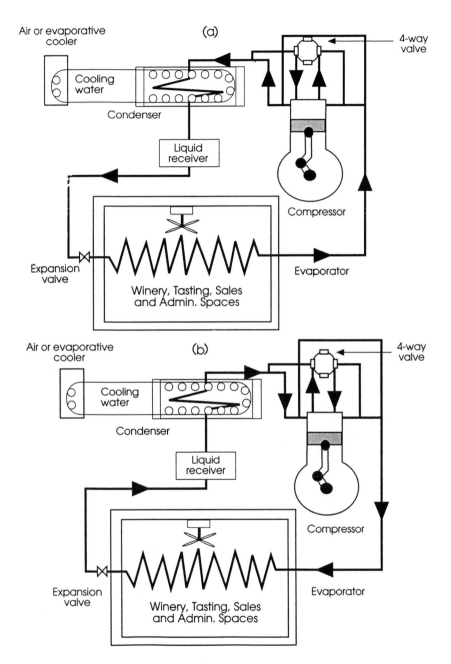

Fig. 5.3. (a) Vapor compression refrigeration system operating as a reverse Carnot cycle (cooling winery spaces). (b) Vapor compression refrigeration system operating as a conventional power Carnot cycle (warming winery spaces).

cheaper off-peaker power, electrical utilities have offered rebates of approximately one-third of the initial installation cost (45, 46). Although secondary cooling loops are commonplace in wine industry process cooling systems, space cooling loads are not large enough to generate the 20–60% savings in operating costs claimed by the heating, ventilating, and air conditioning (HVAC) industry literature for large buildings using ice storage in place of conventional refrigeration-based air-conditioning systems.

Before leaving the subject of refrigeration basics, one additional refrigeration system should be briefly mentioned. Only vapor compression systems have been discussed thus far in this chapter. A variant of the refrigeration cycle, called vapor absorption, can also produce useful cooling, and for some small-scale, nonwinery applications it is the *only* system of choice. The basic difference in vapor compression and vapor absorption systems is the source of the energy that is added to the refrigerant. Figure 5.2 illustrated the general heat energy relationships during the several discrete process steps of the reverse Carnot cycle. Whereas the compressor does mechanical work on the refrigerant to increase its level of heat energy, the vapor absorption system uses fuel combustion (flame) or steam as the energy source. Vapor absorption systems are not entirely without the need for electrically driven components. A very small (fractional horsepower) pump is required to circulate the refrigerant. Thus, the logical adaptation of vapor absorption refrigeration systems is for use in recreational-vehicle refrigerators using liquid petroleum gas (propane) and aboard steam-powered military and commercial ships which have a ready and continuous supply of low-pressure exhaust steam. Standard residential refrigerators of the vapor absorption type in the 18–20-ft^3 (0.5–0.6 m^3) size were produced historically for the limited, remote-mountain-cabin market where the only energy source was liquid petroleum gas. Research revealed that the largest, vapor absorption cycle, commercial refrigerator presently available (1996 data) was 10 ft^3 (0.3 m^3), and is produced in Sweden by Dometic, Inc.

Other methods of achieving a cooling effect such as air compression (reverse Joule cycle) or steam-jet–water-vapor refrigeration are more readily adapted to special situations where low-cooling loads can be managed with chilled air or water vapor at approximately 45°F (7°C). This low level of cooling effect would have scant application in the wine industry. Although refrigeration and cooling are important classes of utility service for wineries, the ice cream industry, in contrast, has brought low-temperature refrigeration (required for production and storage of product) to a very high degree of technical sophistication and system reliability. Whereas process temperature control plays an important role in the quality control of finished table and sparkling wines, without very low process temperatures, no ice cream product is created. Approximately 30 tons of refrigeration are required to produce 1000 gallons (3785 L.) of ice cream (39). The equivalent refrigeration required for 1000 gallons of table wine is approximately 0.67 tons. Beer production by contrast uses approximately 0.02 tons of refrigeration per 1000 gallons (3785 L.) of beer produced. Although factors of scale might tend to distort the unit values of refrigeration for the three food and beverage product industries selected for comparison, the installed refrigeration capacity for the wine industry would appear to be nominally low.

REFRIGERANTS

Most technical references on refrigeration that were published prior to 1984 gave little or no attention to refrigerants other than to recite their thermodynamic properties. The protective layer of stratospheric ozone which has protected the earth's surface from harmful ultraviolet solar radiation, has acquired a large hole in its otherwise continuous envelope, and shows very little promise for healing itself. Implicated as one of the culprits in the photochemical reactions with the ozone shield are a group of chemicals referred to as CFCs (chlorofluorocarbons). Also named as damaging to the ozone layer are so-called halocarbons. They, like their chemical cousins, CFCs, contain compounds of chlorine or fluorine. Historically used as propellants for hair spray, household insecticides, and shaving creme and as refrigerants, the halocarbons and fluorocarbons will soon be legislated out of existence in the United States and eventually worldwide under the provisions of an international treaty (Montreal Protocols) signed in 1987 by the United States and 57 other nations (49). The time schedule for the phase-out of undesirable refrigerants is contained in Title VI of the Clean Air Act, December 1990. In summary, the treaty specifies, among other things, the following:

- A goal of 100% recovery and recycling of atmospheric ozone-depleting substances during manufacture, testing, installation, servicing, repair, operation or disposal
- Reduce the 1986 level of manufacture and consumption of halocarbons and fluorocarbons by 50% by 1998
- The United States exceed Montreal Protocol requirements by federal legislation in 1989 to require complete phaseout of halocarbons and fluorocarbons by year 2000.

The purchase price of the eventually-to-be-banned refrigerants will increase dramatically until essentially all existing refrigeration systems have been converted to safe refrigerants. R-12, a commonly used "Freon" refrigerant in residential refrigerators and automobile air-conditioning units for nearly 50 years, was phased out in November of 1992 and its manufacture or import is prohibited in the United States. Federal law allows existing inventories to be consumed. The price of existing inventories of 30-lb (14 Kg) cylinders of R-12 went from $1 per pound in 1985 to $35 per pound in 1993. The increase in price is due to scarcity and also to an ozone-depletion tax that was imposed upon all halocarbon and fluorocarbon compounds to help offset federal costs for enforcing the law.

Wineries that have air conditioning and process cooling refrigeration units with halocarbon- or fluorocarbon-based refrigerants are faced with several choices, all of which should be thoroughly discussed with a competent HVAC engineer and an industrial refrigeration contractor. The choices are as follows:

- Delay refrigerant conversion on existing equipment until their use is totally forbidden
- Phase out old equipment and replace with more energy-efficient systems (motors, controls, and valves)
- Gamble that refrigerant manufacturers will develop inexpensive, environmentally safe refrigerants that are superior to the current offering of semisafe refrigerants (see Tables 5.1 and 5.2)
- Make the refrigerant conversion as early as possible and remove and recycle the potentially harmful material from the world's inventory.

Table 5.1 is a partial listing of the most promising substitutes for the hazardous refrigerants, that are commercially available as of the date of the preparation of this book. Although the compounds still contain fluorine and chlorine as part of their makeup, they are only 5% as damaging to the stratospheric ozone layer as their predecessors, R-12, R-22, R-114, R-500, and R-502.

Table 5.1. Partial List of Refrigerants That Meet or Exceed U.S. EPA Clean Air Act Standards

	Item-HCFC-22	HFC-134A	HCFC-123	HCFC-124	HFC-125
Ozone depletion potential[a]	0.05	0	0.01	0.02	0
Global warming potential[a]	0.4	0.31	0.25	0.10	0.84
Atmospheric lifetime (years), greatest = 240	19	18	19	19	19

[a]Both ODP and GWP are based on percentage of ODP and GWP for CFC-11, which has a value of 1.
Source: Adapted from Ref. 48.

By contrast, the soon-to-be-banned halocarbon- and fluorocarbon-linked refrigerants such as R-12, R-11, R-500, and R-502, have global warming potentials of 600–800% greater than the best apparent choices of safe refrigerants that can be used for conversion (i.e., HFC-134A and HCFC-123). In making the final decision on refrigerant conversion, a winery must decide, with the guidance from its mechanical engineering technical consultant, if the economics of conversion favor (a) retention of existing equipment with recharge using a compatible refrigerant or (b) substitution of existing equipment with new equipment designed for use with the safe refrigerants. As the useful service life of a refrigeration compressor and its ancillary equipment is somewhere between 30 and 40 years (50), the cost advantages for complete abandonment of existing refrigeration equipment must be very favorable because the salvage value of the old equipment will be almost negligible.

If all existing U.S. refrigeration systems were retrofitted with safe refrigerants simultaneously, supplies would be significantly less than the peak demand created by the hypothetical case. The availability of new refrigeration equipment is also expected to be difficult, if not impossible, as the deadline year 2000 approaches. The U.S. EPA estimates that as of 1992, some 63% of all CFC use has the potential for conversion to safe refrigerants. Table 5.2 summarizes CFC use by category.

Estimates have also been made of annual uncontrolled leakage of CFCs and CFC loss at 16% of the total original refrigerant charge during refrigeration equipment repair or servicing (52). Wineries should be fully aware that incidental discharge of CFC refrigerants is prohibited under the pro-

Table 5.2.   1992 Levels of CFC Use in the United States by Major Categories

Category	Percent of total CFC use (U.S. only)	Potential for refrigerant recovery and recycling
Automobile air conditioners	21.3	Yes
Commercial chillers (space air conditioners)	4.5	Yes
Commercial refrigeration (industrial process)	16.9	Yes
Solvent cleaning	14.2	Yes
Sterilization	5.1	Yes
Household refrigeration	1.2	Yes
Aerosol propellants	3.9	No
Foam insulation	20.3	No
Foam packaging	6.4	No
Flexible foam	6.2	No
	100.0	

Source: Adapted from Ref. 51.

visions of Title VI, Clean Air Act (1990 Amendments). For any repairs or servicing of a winery's refrigeration equipment, a CFC refrigerant recovery system (for both liquid and vapor phases) must be used and all material must be accounted for. The past practice of venting units to the atmosphere and then recharging are no longer possible system repair protocol. With new, safe-refrigerant prices at approximately $100 per pound (1996 price levels), recovering and recharging with the original refrigerant is a sensible maintenance and repair philosophy.

One final note on the use of safe refrigerants is necessary. If existing equipment is charged with one of the appropriate, substitute HFCs or HCFCs (temperature regime and compressor suction and discharge pressures are the major determinants of appropriateness), a loss in cooling capacity of approximately 15% should be anticipated (53). If prior to the decision to make the conversion the winery's system was operating at or near capacity, the anticipated 15% loss in cooling effect might swing the retrofit old versus purchase new equipment decision in favor of the new equipment option.

FERMENTATION HEAT BALANCE AND TANK THERMAL STRATIFICATION

During fermentation, the heat which is evolved is removed from the tank, vat, barrel, or open fermentor by several mechanisms, with each heat-transfer mode in a different proportion. The total amount of heat transferred from the fermenter will depend on the tank geometry (height to diameter ratio), materials of construction, ambient temperature, and the rate at which fermentation proceeds. For average conditions, approximately 10% (reds) and 5% (whites) of the fermentation heat will be transported with the carbon dioxide being evolved, as the yeast metabolizes the grape sugar (42, 44, 54). A reasonably accurate rule of thumb is that for each degree Brix of grape sugar, sufficient heat is generated to raise the temperature of the must or juice approximately 2.34°F (−16.47°C), if no heat is lost (42, 54). The other two modes of heat loss are through the walls, top and bottom of the container (if it is a closed fermentor), and radiation and direct transfer of heat if the fermentor is open and in direct contact with the surrounding atmosphere.

The most common practice in the United States and in most other wine-producing regions of the world is to double-jacket the walls of the stainless-steel tank. Generally, about 50% of the tank walls are chilled, as is current "common practice" in California. One-third of the 50% total jacketed wall is located near the top of the tank and the remaining two-thirds is located just above the racking valve and bottom manhole in most configurations. The remaining surfaces, which are not capable of lowered temperature through secondary glycol circulation in the jacketed areas (see the previous section of this chapter on "Basic Principles"), are subject to the vagaries of the fluctuation in ambient temperature surrounding the fermentation vessel, and the temperature gradient created between must temperature and the ambient air temperature. Thus, heat transfer from the must/juice is severely limited in the jacketed tank designs to approximately 40% of the total tank heat-transfer surface (unchilled top, bottom, and 50% of the sidewalls).

The two-jacket system is designed to take advantage of natural convection and mixing within the tank due to density differences in the wine adjacent to the chilled surface (heavier) and adjacent to the unchilled surface (lighter). The internal mixing in wine tanks is analogous to the seasonal "turnovers" of lakes and ponds, where cold overlying air in the fall chills the upper zone of the lake, increasing its density and allowing the heavier overlying water to replace the less dense, warmer bottom layers. After this event, the lake or pond is said to be thermally stratified. This phenomenon can also occur in wine tanks even with properly designed cooling systems. What often leads winemakers to erroneous conclusions regarding fermentation temperatures is the single thermometer or thermistor probe provided as part of a standard tank accessory package. For the most reliable quality control in the contemporary production of wines, other methods of fermenting must/juice mixing should be seriously considered. Wines which must meet stringent and consistent sensory criteria in order to be competitive in the marketplace cannot be left to chance with microfermentations at several temperatures occurring simultaneously in the same fermentor.

Artificial mixing is the option most often chosen by winemakers, who need to ensure the constancy of the fermentation process even though the grapes may sometimes vary from vintage to vintage. From the age-old method of breaking the floating must cap to sequentially programmed and controlled motor-driven mixers, thermal stratification can be destabilized. Sparging with nitrogen or carbon dioxide can also be used for thermal destratification and for other mixing tasks (liquid sulfur dioxide, bentonite slurries, and acid additions). Mechanical mixing is accomplished by means of specially designed equipment which can be attached to conventional tanks and other fermentation vessels by means of retrofitted valves. By means of a hollow-shaft motor, a stainless-steel impeller shaft can be inserted through one of the tank's sanitary 2-in.-diameter (51 mm) ball valves. A sanitary clamp fitting prevents the escape of the tank's contents. The impeller shaft can be fitted with one of several impeller configurations (i.e., two-blade, three-blade, or mixing head). The two-blade is hinged to allow insertion or withdrawal with ease as the mixing unit is moved from tank to tank. The hydrodynamics of the mixer are quite simple and the turning propeller creates a rolling kinetic wave within the tank which effectively destabilizes the thermal layering effect. An adjustable wheel-mounted stand allows the mixing unit to be used to accommodate different tank sizes and various ball valve vertical locations (55).

Rankine reported temperature stratification occurring not only in fermentation vessels, but in Australian wine cellars as well (42). Again, the wine quality impairment implications are obvious for both finished product, as well as for stored bulk wines, if storage temperatures are not kept in the narrow band of 55–65°F (13–18°C). Reversible ventilation systems, which will be discussed in the following section, can effectively disrupt thermal stratification in cellar, barrel rooms, and warehouse spaces.

VENTILATION AND NIGHT-AIR COOLING

The temperate climatic zones which prevail in most winemaking regions of the world provide an inexpensive option for using the space cooling effect that can be obtained from the wide fluctuations in the diurnal temperatures. Those same cool nighttime temperatures that have driven the development of night (or very early morning) grape harvest technology can be used to cool spaces and their contents.

As described in the following paragraphs, ventilation of enclosed winery spaces is often necessary to provide a safe working environment for cellar personnel. Many wineries have installed reversible-drive ventilation fans to take advantage of a ready supply of cold, outside air. The apparent single disadvantage of this dual-purpose air exhaust and supply arrangement is that during the usual 60–70-day crush period (California practice), the fans must be kept in the carbon dioxide exhaust mode (tanks located outside are the exception to this scheme); however, during the balance of the warm season, the fans can be programmed with a simple internal timer to charge the building with cool, outside air from, say, 10:00 P.M. to 6:00 A.M. (using the California typical summer-fall temperature profile as an example). Some wineries have also taken advantage of site topography and have buried or earth-bermed the building on three and as many as four sides to assist in the maintenance of a desirable interior heat balance. This building feature will be discussed in more detail in the section entitled "Insulation and Solar Energy Considerations."

The ventilation system in winery spaces with enclosed tank rooms must provide for the exhaust of the exogenous heat of fermentation and its associated volume of carbon dioxide gas. For example, 1000 gal of must of 22° Brix occupies approximately 134 ft^3 (3.8 m^3) of storage volume (slightly more in practice, as the foaming and cap rise can add 15–20% to the fermentation volume) and will evolve at 80°F (27°C) about 9400 ft^3 (266 m^3) of gas (54). Carbon dioxide can replace atmospheric oxygen in an enclosed space, creating a serious safety hazard for cellar workers. To illustrate, current California Department of Occupational Health and Safety regulations require that carbon dioxide (CO_2) concentrations not exceed 5000 ppm in industrial (read winery) working spaces (56). The CO_2 concentration in dry air at sea level is approximately 0.03% by volume (57). As CO_2 is heavier than air, it tends to accumulate at floor levels; thus, fermentation room ventilation systems should withdraw at floor level and introduce clean volumes of air above tank tops or catwalks. The winery's HVAC designers should formulate a ventilation system that will prevent CO_2 concentrations from approaching 0.5%, to comply with current versions of the OSHA standard.

Ventilation for winery spaces is governed by standards published in the American Society of Heating, Refrigerating, and Air Conditioning Engineers (ASHRAE) Standard 62-1989. (Current revision, ASHRAE Standard, 62R, not adopted as of June 1996 will be more restrictive). This standard is more applicable to office/administrative/laboratory spaces where at least 20 ft^3/min (0.6 m^3/min) of outdoor air per person is suggested for healthy work environments. Historically, space ventilation standards were designed to provide a specific number of air changes per hour, depending on the type of occupancy and use. For example, laboratories are required to have from 6 to 20 full-room volume outdoor air changes per hour, restrooms 10–30 air changes per hour, kitchens 4–60 air changes per hour, and mechanical machinery spaces 3–12 air changes per hour (58). Current building and mechanical code requirements should be referenced during ventilation system planning phases for new wineries, or for upgrading and retrofitting existing buildings.

Although CO_2 concentrations need to be closely monitored in enclosed fermentation spaces and exhausted, as previously described, sulfur dioxide (SO_2) is also a wine additive that is generally incorporated at some point during vinification in a saturated aqueous and sometimes gaseous form, to prevent color changes in white wines and spoilage in both red and white wines, and to denature spoilage organisms in barrels. Even in low concentrations, this gas can cause almost immediate respiratory difficulty in healthy individuals and harm or extreme distress in persons with asthma or seasonal allergies. Limiting concentrations of SO_2 of only 2 ppm are allowed in work spaces under California law (56). Areawide ventilation systems are generally not sufficient protection. All wineries must provide face masks, which have dual, acid–gas purification elements to protect cellar personnel. Safety goggles should also be worn, as eye-tearing is severe even at concentrations less than 2 ppm. The highest concentrations of SO_2 occur in aging cellars, when recently sulfured cooperage is being filled with wine and the SO_2 being forced out through the bung hole is in very high concentrations.

In the late 1980s and early 1990s, building and workspace interior environmental issues received considerable attention from federal, state, and local health and industrial hygiene agencies. Indoor air quality (IAQ) is an extremely important aspect of any winery structure. The so-called "sick building syndrome" is a problem to be carefully avoided in the planning and design of new facilities and in the upgrading and retrofitting of old wineries to meet new requirements of the HVAC industry and government regulations and codes.

INSULATION AND SOLAR ENERGY CONSIDERATIONS

The heat-transfer properties of the winery building envelope are an important, but passive adjunct to the maintenance of heat balance within. Further, preventing heat loss in tanks and piping can be balanced economically, and significantly reduce the winery's long-term commitment to energy consumption. Almost all power service utility companies have energy conservation teams trained to show power users how to conserve. The contemporary role of electrical utility companies in promoting the efficient use of energy resources was discussed previously in Chapter 2, "Electrical Systems."

Probably first and foremost in optimizing the major source of heat gains and losses is the careful solar orientation of the winery, within given site topographic and space limitations. Since the early 1970s when the first U.S. energy crisis emerged, the science of passive uses and management of solar energy was refined and tested to a degree that had not been seen in prior history or since. These findings and data are still appropriate in winery facility planning, and although the economic advantages of passive energy management and use change dramatically with the worldwide fluctuations in the price of fossil fuels, the benefits of including some features of solar energy should not be overlooked in winery planning.

Tunnels, underground cellars, and natural and man-made caves have provided an inexpensive, easily-temperature-regulated space for the production, aging, and long-term storage of wine. Contemporary caves can be of two types:

- cut and cover
- bored or dug

Cut and cover caves are structurally engineered, elliptical or semicircular spaces that are constructed in an open excavation (reinforced concrete, corrugated steel pipe arches, or combinations thereof), carefully water-proofed, and then backfilled with native soil. Bored caves are the most common method used in the wine-producing regions of the western United States to achieve a uniform cross-sectional tunnel. Either poured concrete or gunite (blown concrete) lining may be required for stability and seismic safety. The author has observed unlined tunnels in geologic formations where the soundness of the rock through which the bore traverses is both massive and unfractured. Boring machines capable of producing a 12-ft (4-m)-diameter tunnel are the most common dimensional choice of wineries and are compatible with the state-of-the-art boring machines used by cave contractors. A dug tunnel, on the other hand, would employ historic methods of excavation (i.e., drilling, blasting, and excavation). As the topographically flat sites in premium wine-growing areas dwindle, the less expensive, upslope sites eventually become more attractive. The sloping terrain and generally shallow soil cover presents a matrix of factors that can favor cave construction. To illustrate, a comparison is made of the unit cost of cave space to conventional building space (1993 price levels) for the two methods of creating typical barrel storage space.

	Estimated Unit Cost (1993 price levels) (59)
(a) Bored cave (hillside site with massive, sound, unfractured rock)	$700 per linear foot of 12-ft- (4 m) diameter bore
(b) Bored cave (hillside site with fractured, jointed rock, requiring lining)	$900 per linear foot of 12-ft- (4 m) diameter bore
(c) Conventional building (concrete or concrete block with wood frame roof; sloping, difficult site)	$200 per square foot
(a) Approximate cost per cubic foot of space (best site and geology conditions)	$12^2 (0.785) \times 1 = 113 \text{ ft}^3$, $\therefore 700/113 = \$6.20$
(b) Approximate cost per cubic foot of space (worst site and geologic conditions)	$12^2 (0.785) \times 1 = 113 \text{ ft}^3$, $\therefore 900/113 = \$7.97$
(c) Approximate cost per cubic foot of space (assume 20 ft high $\times$ 40 ft wide $\times$ 100 ft long, barrel storage)	4000 sq ft $\times$ $200/sq ft = $800,000 $\therefore 800,000/20 \times 40 \times 100 = \$10$

Unit prices for construction vary slightly from region to region in the United States and even more radically among winemaking countries outside the United States. If this illustrative cost analysis was expanded to include the annualized value of cooling energy expended for the conventional building, or foregone in the case of the caves, the choice (based purely on cost considerations) would favor the bored tunnel alternative. In a real-time winery planning situation, the combined counsel of architect, engineer, and contractor should be used in full measure to validate an underground versus a conventional aboveground structure, so that an objective array of options will evolve for winery management consideration.

The matrix of variables involved in situating a winery on a given site is extensive and complex.

- Large 18-wheel trucks, ingress and egress
- Drainage
- Visual esthetics
- Length of electrical and communication utility runs
- Soil stability
- Solar orientation
- Visitor safety
- Fire protection
- Security
- Relationship to other needed site utilities (water, wastewater, etc.)

Certainly not all of the variables can be satisfied for any given winery site, although it is the architect's professional obligation to optimize as many as possible, as influenced by the owner's strong feelings about particular site parameters.

Overlaying all of the considerations for optimizing solar energy resources and balancing, as nearly as possible, operating costs with heat balance in the winery building envelope are the statutory requirements that must be met under the broad national title of Energy Insulation Standards. With the federal Department of Energy preparing model regulations (ca. 1976), state governments followed with additions and modifications to their respective building codes and housing laws. It is sufficient to say, in covering this aspect of thermal energy management, that no winery will be constructed that does not meet the basic requirements for acceptable heat-transfer coefficients in windows, walls, ceilings, floors, and roof decks. The reader is referred to the special building code section which covers energy conservation for the particular governmental jurisdiction in question. For California, the standards are contained in California Administrative Code, Title 25, Chapter 1, Subchapter 1, Article 5, February 1975 (as amended) (63), and in annual updates published as the California Energy Code by the California Building Standards Commission.

Pipe insulation is a subcategory of state building and energy conservation codes, and for wineries, the possibilities for minimizing heat gain (air-conditioning ducts and secondary coolant lines) and heat loss (steam and hot-water lines) are boundless. Opportunities exist for older wineries with pipes wrapped with older, less efficient insulation material to retrofit accessible pipes with easily installed, molded, preformed one-half pipe, polyurethane foam sections. As a rule of thumb, flexible wrap-type insulation requires about two times the thickness to achieve the same heat-transfer properties of premolded foam sections. Older pipe insulation chiller lines also had a propensity to allow condensed moisture on the exterior of pipes to drip uncontrolled in and around production spaces, providing an ideal substrate for spoilage organisms to propagate (see Chapter 4, "Sanitation, Steam, and Hot Water"). On rare occasions, old steam line insulation containing asbestos fibers is a potential source of interior air contamination for both cellar personnel and as an airborne contaminant for wine. The industrial hygiene laws governing asbestos removal are so restrictive and onerous that sealing off the old insulation with new, efficient, premolded foam insulation is by far a better choice and an acceptable alternative to removal and replacement. The counsel of an industrial HVAC or insulation contractor should be sought before a winery undertakes the simplest of steampipe insulation retrofit projects.

Using standard values for pipe insulation that are contained in the most recent edition of the *ASHRAE Handbook of Fundamentals* (60) and temperature differences between the substance being conveyed and ambient air temperature, the winery's utility maintenance staff can verify that the insulating efficiency of the facility's existing pipe networks meet the recommended energy conservation code standards. The allowable heat loss/gain for various pipe sizes is shown in Table 5.3.

Tables of pipe insulation thickness for various materials and temperature gradients have been prepared for steel and copper pipe, making the assessment of existing pipe insulation relatively easy (60). If the ASHRAE Handbook tables fail to match pipe size, temperature gradient, or the insulation material's heat-transfer coefficient, the following equation can be solved for the dimension (thickness) of the insulation required (60):

**Table 5.3. Allowable Heat Loss/Gain in Chiller,
Steam/Hot-Water Piping as a Function of Pipe Size**

Pipe size	Heat loss/gain
Up to 2 in. diameter	50 BTU/h/linear foot
Greater than 2 in. diameter	100 BTU/h/linear foot

Source: Adapted from Ref. 61.

$$Q = \frac{(t_p - t_0)\left[\pi\left(\frac{r_2}{6}\right)\right]}{\left(\frac{r_2 \ell n(r_2/r_1)}{K} + \frac{1}{f}\right)^{-1}}$$

where

Q = BTU/h/linear foot or Kcal/hr/m
t_p = pipe temperature (°F or °C)
t_0 = ambient design air temperature (°F or °C)
r_1 = inside radius of insulation (in. or m)
r_2 = outside radius of insulation (in. or m)
K = thermal conductivity (BTU/h/°F/ft^2/in.) or Kcal/hr/°C/
    m^2/cm
f = air film coefficient ($\approx$ 1.65)

To simplify the equation, the variable r_1 can be assumed to equal the nominal diameter of the pipe. Again, the *ASHRAE Handbook of Fundamentals* (60) and the work of Tinsley and Kegel (64) provides the unit values of the variable parameters needed to solve the equation. A contemporary examination of the subject of the economics of pipe insulation with an emphasis on 1990 price levels (labor, materials, etc.) is contained in Kratowicz, 1993 (62).

The possible benefits of other forms of shading, solar reflection, and screens should be mentioned. Shades, drapes, and blinds are contemporaneously referred to as "window coverings" and should not be overlooked in managing the heat balance in exterior office and laboratory spaces. Blinds configured with either single, double, or quadruple cells of air-filled space-age fabric, make up one category of highly thermodynamically efficient window coverings. Glass tinting, roof overhangs, and near-building landscape shade should all enter the matrix of possible supplementary measures that can be employed. For the existing winery anxious for a "new look" or for passive ways to lower overhead costs, some of these latter items are affordable and have a readily demonstrated cost amortization in energy expenditures foregone, over a fairly short time frame. For new wineries in the planning stage, the winery's architect can facilitate the integration of all the possible interior and exterior structural heating/cooling economy treatments into a balanced blend of practicality and aesthetics.

OPERATION AND MAINTENANCE

The electrical and mechanical equipment associated with refrigeration systems requires the same periodic attention to routine maintenance and repair as any other complex of motors and mechanical parts in a winery.

The preferred rational approach in getting the longest life out of installed and mobile equipment for all industrial endeavors is "preventive maintenance." This methodology has been described previously in Chapter 2, "Electrical Systems," and the reader is referred to that chapter section on operation and maintenance for a summary of record keeping, parts inventory, and emergency repair procedures. No winery is going to be without some incidents of machinery malfunction, but the application of good maintenance engineering will reduce significantly the incidence of those critical downtimes.

Manufacturers' operation and maintenance manuals should be provided by equipment suppliers and the recommended daily, weekly, and annual procedures followed without compromise. Somewhere in the fine-print section of the warranty that accompanies the newly installed equipment is the exclusion clause that voids any repair or replacement responsibility if specified maintenance steps are not followed. The winery's maintenance supervisor should have in his office or workshop a master operations and maintenance (O&M) schedule that combines on a single calendar chronology, the maintenance events that should occur. Although it is also useful to have these data recorded on a computer hard-disk format, the wall chart presentation is much easier to serve as an ever-present reminder of who-does-what and when that is verified by a responsible initial when the task is completed. Sickness, vacation, or other unavoidable changes in staffing can be easily overcome, if everyone on the maintenance team takes part in and becomes familiar with the routine.

Whereas the routine work with refrigeration, air conditioning, and ventilation systems can be accomplished with winery personnel, the now strictly regulated task of recharging units with refrigerants must be done by certified HVAC personnel, who have in their tool and equipment inventory a refrigerant recovery and recycling unit. Venting refrigerants to the atmosphere has been prevented by the Federal Clean Air Act since July 1, 1992. Figures 5.4a and 5.4b show respectively a typical recovery system arrangement for capture of refrigerants in both the liquid and vapor stages.

If chiller units are only used in the winery during the crush period, the units should be thoroughly checked, operated, and monitored for performance several weeks before the first gondola of grapes arrives. This will allow ample time to rectify any electrical or mechanical problem prior to "C" Day. As part of this precrush program, all associated thermistors and other temperature probes which provide the input data for the more sophisticated tank temperature programmable controller need to be calibrated, adjusted, or replaced if defective. This is when the depth and breadth of the winery's technical manual data files and equipment operational history are tested. As-built drawings and shop and assembly drawings all become important in troubleshooting and problem solving.

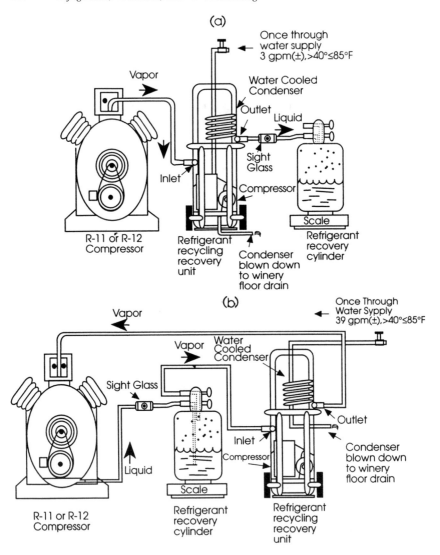

Fig. 5.4. Typical refrigerant recovery: (a) vapor phase and (b) liquid phase.

Personal computer software programs can be customized for each winery's special mix of equipment and controls that will satisfy management's interest in and priority ranking with operation and maintenance functions. With longer and longer operating times and maintenance/repair experience with each system, the additional data input each year further strengthens the informational base and permits such valuable outputs as O&M

manpower budgets, work orders, annual maintenance costs by system and subsystem, equipment useful life remaining, and replacement costs. Additionally, computer-generated vendor listings and their performance, and tools and parts inventory all help to streamline the O&M task and make it easier for justifying needed expenditures to management in a world of continually decreasing budget pies. The front office can also use information on the refrigeration equipment capability to, for example, take advantage of high-quality, low-priced, windfall grape purchases that were not originally in the crush plan. It is always impressive to provide answers to such technical questions quickly and be confident that your PC database will back you up. Such a software program is available from Ounce of Prevention Software, 1638 Pinehurst Court, Pittsburgh, PA 15327 (800-852-8075).

Methods of detecting refrigerant leaks are relatively simple (for large ones, at least) and the winery should have a refrigerant detector in its maintenance tool inventory.

Finally, the maintenance program for your particular winery must fit the manpower, technical skills, and budget limitation established by the size and nature of the operation. Refrigeration equipment is costly, and malfunctions cannot be tolerated, with loss or lowered quality of the finished wine, the end result. Give reasonable care and attention to the HVAC systems and they will provide the service for which they were designed.

GLOSSARY

AC/min Air changes per minute.

ACH Air changes per hour. A measurement of the ventilation of supply air rate calculated by dividing the volume of air delivered by the volume of the space receiving the air.

Air cleaning An IAQ control strategy to remove various airborne particulates and/or gases from the air.

AHU Air-handling unit. A component of an HVAC system that includes the fan or fans, filters, and coils to condition air.

ANSI American National Standards Institute.

ASHRAE American Society of Heating, Refrigerating, and Air Conditioning.

ASME American Society of Mechanical Engineers.

ASTM American Society for Testing Materials.

Axial fans Fans that produce pressure from air passing through the impeller, with no pressure being produced by centrifugal force.

Backward-curved fan Centrifugal fans with blades that curve backward, away from the direction of the rotation.

Balancing damper   A plate or adjustable vane installed in a duct branch to regulate the flow of air in the duct.

Blind flange   A flange used to close off the end of a pipe.

Breathing zone   Area of a room in which occupants breathe as they stand, sit, or lie down. Sometimes referred to as the portion of a space from 3 in. to 72 in. above the floor.

Building envelope   Elements of the building, including all external building materials, windows, and walls, that enclose the internal space.

Building sewer   The pipe running from the outside wall of the building drain to a public sewer or winery-owned treatment system.

Building storm drain   A drain for carrying rain, surface water, condensate, to the building drain or sewer.

Centrifugal chiller   A gas compressor in which the compression is obtained by means of centrifugal force, the force away from the center of a rapidly rotating impeller.

Chillers   Compression = Water is cooled by liquid refrigerant that vaporizes at a low pressure and is driven into a compressor. The compressor increases the gas pressure so that it may be condensed in the condenser. Absorption = Heat-operated refrigeration chillers that use an absorbent (lithium bromide) as a secondary fluid to absorb the primary fluid (water) which is a gaseous refrigerant in the evaporator. The evaporating process absorbs heat, cooling the refrigerant.

Compressor   The pump in a mechanical refrigeration system that compresses the refrigerant vapor into a smaller volume, thereby increasing the refrigerant pressure and the boiling temperature. The compressor is the separation between the refrigerant system's high and low side.

Condenser   The heat exchanger in a refrigeration system that removes heat from the hot, high-pressure, refrigerant gas and transforms it into a liquid.

Cooling tower   A device for cooling water by evaporation. A natural draft cooling tower is one in which the airflow through the tower is due to its natural chimney effect. A mechanical draft tower employs fans to force or induce a draft.

Couplings   Fittings for joining two pieces of pipe.

Direct expansion   Refrigeration systems that employ expansion valves or capillary tubes to meter liquid refrigerant into the evaporator.

Elbow   The angle is 90° unless another angle is specified.

Entropy   A quantity expressed as a function of the temperature, pressure, and density of a working substance that is equal to the potential work available.

Equivalent length   The resistance of a duct or pipe elbow, valve, damper, orifice, bend, fitting, or other obstruction to flow, expressed in the number of feet of straight duct or pipe of the same diameter which would have the same resistance.

Expansion joint   A joint whose primary purpose is to absorb the longitudinal expansion and contraction in the line due to temperature changes.

Flange   A ring-shaped plate at right angles to the end of the pipe. It is provided with holes for bolts to allow fastening of the pipe to a similar flange.

Gate valve   A valve utilizing a gate, usually wedge shaped, which allows fluid flow when the gate is lifted from the seat. Gate valves have less resistance to flow than globe valves and should always be used fully open or fully closed.

Globe valve   A valve with a rounded body utilizing a manually raised or lowered disk which, when closed, seats so as to prevent fluid flow. Globe valves are ideal for throttling in a semiclosed position.

Header   A large pipe or drum into which each of a group of boilers, chillers, or pumps are connected.

Horizontal branch   In plumbing, the horizontal line from the fixture drain to the waste stack.

ID   Inside diameter.

Manifold   A fitting with several branch outlets.

Nipple   A piece of pipe less than 12 in. long and threaded on both ends. Pipe over 12 in. long is regarded as a cut measure.

Nominal   Name given to standard pipe size designations through 12-in. nominal OD. For example, 2 in. nominal is 2⅜ in. OD.

OD   Outside diameter.

O.S. & Y.   Outside screw and yoke. A valve configuration where the valve stem, having exposed external threads supported by a yoke, indicates the open or closed position of the valve.

Plug valve   A valve containing a tapered plug through which a hole is drilled so that fluid can flow through when the holes line up with the inlet and the outlet, but when the plug is rotated 90°, the flow is stopped.

Plumbing fixtures   Devices which receive water, liquid, or waterborne wastes, and discharge the wastes into a drainage system.

Potable water   Water suitable for human consumption.

Reducer   A pipe coupling with a larger size at one end than the other. The larger size is designated first. Reducers are threaded, flanged, welded. Reducing couplings are available in either eccentric or concentric configurations.

Riser   A vertical pipe extending one or more floors.

Schedule number   American Standards Association designations for classifying the strength of pipe. Schedule 40 is the most common form of steel pipe used in the mechanical trades.

Screwed joint   A pipe joint consisting of threaded male and female parts joined together.

Service pipe   A pipe connecting water or gas mains into a building from the street.

Storm sewer   A sewer carrying surface or storm water from roofs or exterior surfaces of a building.

Tee   A pipe fitting that has a side port at right angles to the run.

Union   A fitting used to join pipes. It commonly consists of three pieces. Unions are extensively used, because they allow dismantling and reassembling of piping assemblies with ease and without distorting the assembly.

Vent system   Piping which provides a flow of air to or from a drainage system to protect trap seals from siphonage or back pressure.

Ventilation air   Outdoor air delivered to occupied spaces of a building.

VOC   Volatile organic compounds. Compounds that evaporate from the many housekeeping, maintenance, and building materials made with organic compounds. Biological organisms can also be a source of VOCs. In sufficient quantities, VOCs can cause eye, nose, and throat irritation, headaches, dizziness, visual disorders, and memory impairment.

REFERENCES

42. WHITE, R., ADAMSON, B., and RANKINE, B. 1989. *Refrigeration for Winemakers.* Adelaide, S. Australia: Hyde Park Press.

43. JOHNSON, H. 1980. *World Atlas of Wine.* London: Mitchell Beazley Publ.

44. BOULTON, R.B., SINGLETON, V.L., BISSON, L., KUNKEE, R.E. 1996. *Principles and Practices of Winemaking.* New York: Chapman & Hall.

45. DUFFY, G. 1992. Thermal storage emphasis shifts to saving energy. *Engineered Systems,* July–Aug.

46. _____ 1991. Cold air distribution with ice storage. 1991. *Proceedings of the Electrical Power Research Institute,* No. 2038.

47. McQUISTON, F.C. and PARKER, J.D. 1977. *Heating, Ventilating and Air Conditioning Analysis and Design.* New York: John Wiley and Sons.

48. BUKOWSKI, J. 1993. The many faces of industrial refrigeration. *Consulting Specifying Engineer* 14(4).

49. _____ 1990. Recycling programs for auto air conditioners and other industries using CFC's. *Pollut. Eng.* 56–57.

50. _____ 1991. *ASHRAE Handbook: HVAC Applications.* Atlanta, GA: American Society of Heating, Refrigerating and Air Conditioning Engineers, Table 3, p. 33.3.

51. LEE, D. 1992. *Potential for Recycling CFC's.* Washington, DC: USEPA Reg. and Analysis Branch; Global Change Division, Office of Atmospheric and Indoor Air Programs.

52. _____ 1989. *Refrigeration, Air Conditioning, and Heat Pumps—Technical Options Report.* United Nations Environment Programme, Paris, Tables 9.1 and 9.2, p. 80.

53. MECKLER, G. 1993. Integrated IAQ-CFC retrofit saves energy. *Consulting Specifying Engineer.*

54. AMERINE, M.A. and JOSLYN, M.A. 1973. *Table Wines: The Technology of Their Production.* Berkeley: University of California Press.

55. RIEGER, T. 1994. Tank mixers and pumpover devices. *Vineyard Winery Management.* 34–39.

56. _____ 1992. *Airborne Contaminants.* California General Industrial Safety Orders, Sect. 5155, Title 8. Cal. Code of Regulations.

57. BAUMEISTER, T. 1967. *Marks' Mechanical Engineers Handbook*. New York: McGraw Hill Book Co., p. 550.

58. KASHAB, A.M. 1978. *Heating, Ventilating and Air Conditioning*. New York: Mc-Graw-Hill Books.

59. _____ 1993. *Heavy Construction Cost Guide*. Leonard McMahon, Inc, Publisher, Quincy, MA.

60. _____ 1982. *ASHRAE Handbook of Fundamentals*, 5th ed. Atlanta, GA: American Society of Refrigeration and Air Conditioning Engineers.

61. _____ 1975. *Energy Insulation Standards*. Cal. Admin. Code, Title 25, Chapter 1, Subchapter 1, Article 5 (as amended).

62. KRATOWICZ, R. 1993. Which insulation thickness? An optimization principle. *Plant Services* 14(9):13–18.

63. _____ 1976. *Energy Design Manual*. State of California Department of Housing and Community Development, Div. of Codes and Standards, 1976 Ed.

64. TINSLEY, R.W. and KEGEL, R.A. 1993. Pipe insulation conserves energy and cuts costs. *Consulting Specifying Engineer*. 13(5):68–72.

CHAPTER 6

POTABLE WATER SUPPLY SYSTEMS

WINERY SITE WATER RESOURCE ASSESSMENTS

The availability of an abundant water supply of reasonably good chemical quality and that is microbiologically safe is as important to wine production as an artfully prepared and printed label is to the successful marketing of table wine.

Prepurchase Winery Site Analysis

In purchases of raw land for the establishment of a vineyard and/or winery, it is seldom only a buyer and seller involved in the negotiations and intricacies of the site acquisition. An "adequate" water supply is a key issue upon which purchase of the property is conditioned. Although a surface water supply is technically and economically feasible in some cases, preexisting water rights and environmental constraints might not allow diversion, storage, or other exploitation of the stream for industrial (winery) use. The doctrine of "riparian use" is a right to use water from streams which originate in or traverse a given parcel of land and the right attaches to the deed for the property. There is no permit or licensing protocol required for riparian water rights. The riparian water use can only be exercised by a landowner under a special set of restrictions that have their origins in part in English Common Law (65). Although this discussion uses California water law as a model, in other states there is a remarkable similarity in the statutes related to riparian use. These include the following:

- Water may be diverted for use only when sufficient flow exists in the stream to serve all riparian owners.
- Water diverted under riparian law may not be stored for later use.
- Water may be diverted for a variety of beneficial uses among which are

> Domestic (culinary) use
> Agricultural use
> Stock watering
> Mining
> Power production

With less than average stream flows existing during drought periods, the satisfying of all riparian rights and appropriative rights (those rights obtained through an administrative process to allow a broader development of the stream resource through licenses and permits to seasonally store water for later use) becomes difficult, if not impossible. Courts of law are then sought to allocate the limited supply, often with a "watermaster" being appointed by the court to arbitrate disputes among users and to police the diversion and metering of stream flows to match the court order for the particular stream. This brief excursion into water law is made to illustrate the importance of water rights as a sometimes overlooked element of the land–resource matrix. A prospective winery site developer should have an attorney and/or engineer who is versed in such matters, analyze carefully any and all riparian and appropriative water rights that are an intimate part of the real value of the parcel.

In Napa County, California, where some 250 wineries and vineyards occupy a land area less than 100 square miles, the competition for water resources is very intense. This scenario is repeated throughout many of the wine-grape-growing regions of the world. To illustrate how one governmental jurisdiction administrates the special water needs for wineries, the Napa County Development Code for Wineries requires that each prospective new winery demonstrate, prior to issuance of a Land Use Permit, that a raw water supply in an annual amount of at least 2.5 acre-feet per 100, 000 wine gallons of production can be produced. That is equivalent to about 815, 000 water gallons per annum. As will be reported later in this chapter, this unit water requirement is very close to values developed by the author from metered flow records of established wineries.

In the land use permit process, as practiced in Napa County, for example, a public forum for neighbors and interested public agencies to carefully examine the proposed winery project in minute detail, the water supply, noise, and traffic issues seem to receive the largest share of attention. Thus, the winery owner/developer should be adequately prepared to defend the proposed water supply plan for the winery. The plan should strongly support the precept that no adverse effects will be incurred by neighboring wells or surface supplies because of the increased draft on the local resource. What is required to quell the fears of falling groundwater levels and/or permanently diminished water supplies is a so-called pump test. (If a surface water supply is to be developed, the process is much more complicated. Also, the section on "Water Treatment" will show an added complexity for developed surface supplies.) The pump test is not costly if the proposed winery parcel already has a well penetrating the common aquifer (defined as the water-bearing portion of the underlying geologic structure) that near neighbors usually jointly. As a secondary benefit to the prospective winery developer, the pump test can also be extended over a 24–48-h period to verify (a) the safe yield of the well and (b) the groundwater level recovery characteristics after an extensive pumping cycle. If a new well must be developed, there is always a risk associated with bringing in a "dry hole" or a well with an inadequate yield to meet the winery's projected needs (a special section entitled "Water Demand Calculations" covers in detail just how winery water supply needs can be determined). To illustrate how the pump test can simulate postconstruction operating conditions and reveal adverse impacts on neighboring wells, a Well Log (Fig. 6.1) has been prepared, using the physical arrangement and geohydrologic conditions that prevailed during an actual winery development plan preparation. The simulation of the maximum pumping cycle and rate of groundwater withdrawal for the proposed winery during crush, as shown in Fig. 6.1, created a "cone of depression" in the groundwater level that left the small farm/residence pump on the adjacent parcel high and dry. With the mutual agreement of the adjacent landowner and the winery, the farm/residence well was deepened and the pump reset below the cone of depression at no cost to the adjacent property owner. This arrangement has worked to the satisfaction of both parties for some five

PLAN VIEW

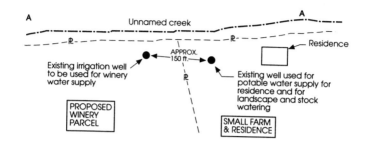

SECTION A - A

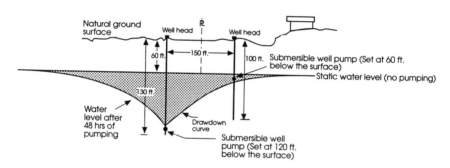

LEGEND:

Cone of Depression
℞ = Property line

NOTE: After 48 hours of pumping, well on parcel adjacent to proposed winery was rendered incapable of supplying water.

Fig. 6.1   Graphic results of pump test made during planning phase for a 100, 000-case table winery.

years and avoided a rancorous dispute that would have eventually erupted, if the winery well pump interference phenomenon had not been discovered.

Although less common than wells or surface water supplies, "developed springs" are, on occasion, used, sometimes with great success, to provide the only available water to small or very small wineries. For purposes of definition, a spring is groundwater that has risen to the surface because of a geologic structural anomaly. Because a properly developed spring is generally less expensive to develop than a well, a spring with a reasonably good historical record of sustained flow may be the water source of choice. The governmental entity responsible for approving the development plan for a spring will be for most of the wine-producing areas in the United States, at least, the local environmental health agency (local governmental subdivision of the state or county).

There is an imperative need for the winery developer to retain a professional (civil, environmental, or geohydrologic engineer who can demonstrate experience with natural spring development) to develop a preliminary report on the feasibility of using a spring for a winery water supply. This report should include the following:

- Site suitability survey to ensure that the spring's tributary watershed contains no current or future potential sources of contamination and that the geohydrology of the region is associated with spring flow of adequate quantity and mineral quality (see list of items following, related to water quality for wells),
- Dry season yield
- Zoning and development potential of geologically related areas which in the future might adversely affect the spring

If the preliminary report is positive, a spring development plan should be prepared by the design professional using the county's small water system development code as a minimum standard. The county's code will undoubtedly be at least equal to or, in some cases, exceed the state's Health and Safety Code which covers the use of springs for domestic water supply. Material upgrades or size increases may be suggested by the designer to add longer service life and/or system reliability without a large increase in capital cost. These recommendations should be carefully reviewed by winery management and given serious consideration for acceptance if the supporting technical data and economic justification appear reasonable. A simple checklist has been prepared to permit the winery staff to review a spring development plan with some degree of understanding, prior to giving approval to the design engineer for formal submission of the design documents to the county agency. An idealized plan (but which conforms to generally acceptable standards for spring box design) is shown in Fig. 6.2 to assist the reader in identifying spring development plan nomenclature and to show the general arrangement of critical system elements. A suggested spring development plan design review checklist follows:

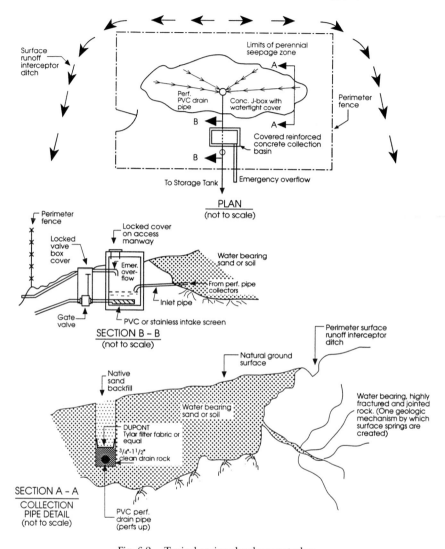

Fig. 6.2.   Typical spring development plan.

	Yes	No	Needs Clarification
1. Spring interceptor ditch to protect spring collection zone from contamination	———	———	———
2. Perimeter fence to restrict access by humans and indigenous fauna (locked access gate)	———	———	———

(*continued*)

3. Food-grade, cold-tar epoxy paint specified for inside surface of collection tank and waterproofing compound for exterior concrete surface   ＿＿＿ ＿＿＿ ＿＿＿

4. Locking hasp cast into watertight manway cover on top of collection basin   ＿＿＿ ＿＿＿ ＿＿＿

5. Warning signs to be provided and placed on all four sides of perimeter fence— "POTABLE WATER SUPPLY; DO NOT ENTER"   ＿＿＿ ＿＿＿ ＿＿＿

6. Emergency overflow pipe from collection basin provided with outlet plumbed to natural drainage course   ＿＿＿ ＿＿＿ ＿＿＿

7. Copies of spring dry season yield documented and attached   ＿＿＿ ＿＿＿ ＿＿＿

8. Copies of water chemical and microbiological profile provided and attached   ＿＿＿ ＿＿＿ ＿＿＿

9. Copies of water right permits or licenses provided and attached (if any)   ＿＿＿ ＿＿＿ ＿＿＿

In summary, controlling and reregulating the flow from a natural spring is an inexpensive alternative to a drilled well. Unless the chemical and microbiological testing program for the spring produces results that meet or exceed the safe limits provided by the now generally accepted Safe Drinking Water Standards, no treatment would be required of the spring water. Having the design engineer provide, as a minimum, a fully operational disinfection unit to serve on a standby basis is good insurance. Future contamination of the supply cannot be guaranteed and the inexpensive chlorine injection subsystem will allow the safe, uninterrupted operation of the winery, until the source of microbial contamination can be verified.

If natural springs do have any shortcomings, it is their fragile nature in active seismic zones. Ground accelerations associated with temblors, can modify rock structure to the point that water is either more efficiently conducted to the spring or hindered in its transmissibility through the rock substructure. The monitored discharge of six springs that are tributaries to Squaw Creek in the seismically active Geysers KGRA (Known Geothermal Resource Area) of California's Mayacamas Mountains from 1977 to 1980 showed both reductions and substantial increases in spring flow immediately after local groundshaking (Richter Scale events of approximately 3–5). Although a geohydrologist might shed some light on the probabilities of a change in spring flow with tectonic activity, without the observations made on spring flow and correlation with seismic events such as described for the Geysers KGRA, meaningful predictions would be difficult in the extreme. Taking a risk on the relative permanence of a spring is about equal in magnitude to the probabilities of economic success for the entire winery development venture.

Other critical elements of the winery site water availability analysis should include the following:

1. Assessment and determination of condition of any existing wells and their casing(s).
2. Geotechnical evaluation of well-drilling logs (records).
3. Water-quality assessment (chemical and microbiological).
4. Pump efficiency test(s).
5. If unused wells are discovered, check for documentation of well-abandonment proceedings.
6. Preparation of a "land-use-conversion-water equivalency-test."
7. When and how to utilize dowsers and geophysical surveys.

Each one of the elements will be considered together with their importance to the winery plan from an economic, operational, and regulatory perspective.

Well Casings

Since about 1983, "down-hole" video cameras have been used in the water well industry to assess the condition of wells and to estimate to what extent mineralization/encrustation of well screens or torch-cuts (the historical method of admitting the water from the aquifer to the well casing and pump) may influence the efficiency of the pump system (see item 4, pump efficiency test). The services of a competent well-drilling contractor are required to prepare a well casing for camera inspection. The well pump must be removed (as well as any discharge piping and power cable) if it is a submersible pump. If the well is an operating well, some provision has to be made (temporary storage tanks or temporary service from another well with surface pipe or hose) to provide an uninterrupted water supply to the ongoing operation. Before the camera can record a clear image of the casing and its condition, the water in the well must be allowed to clarify. This can occur naturally by sedimentation of any suspended material that might be loosened during the pump and pipe removal process. Settling and clarification can be augmented by using some type of settling aid such as aluminum sulfate (alum) or another common coagulant. Finally, the open well casing should have a security cover so that no foreign material or animal can enter the well. (The author's experience in reclaiming a winery's well into which a skunk had accidentally fallen is a water-quality problem with which no winery would want to cope.)

Well casing video records are relatively inexpensive and often reveal other information that is critical to the overall evaluation of the well system. For example, the seller of a parcel of land destined for winery and vineyard development had been billed and paid for some 350 ft of well casing by an unscrupulous well-drilling contractor. The video record recorded only 200 ft of installed casing and the uncased well sidewalls had collapsed, rendering the balance of the drilled well useless. Needless to say, the buyer of the winery site deducted from the purchase price the cost for cleaning and extending the casing through the balance of the water-bearing formation. Subsequent testing showed that the well yield increased about 20% over the historical safe pumping rate, which was certainly attributable in part to the repairs made to the well. Excessive corrosion and/or colonies of iron bacteria are easily seen on the walls and rough projections of the welded joints. Iron bacteria can be a troublesome organism in winery potable water systems, and on occasion, iron removal (in either the ferric or ferrous form) is required to eliminate the mineral which is necessary for the iron bacteria's survival and proliferation in pipelines, storage tanks, and wells. Rust stains on bathroom fixtures are one indication that iron bacteria may be propagating in the winery's water system. This water-quality problem and its optional solutions will be discussed in the section entitled "Water-Quality Assessment."

If the video camera inspection of the casing shows poor welding at joints, severe corrosion, or badly encrusted torch-cuts or encrustation of the well screen openings, the well contractor may suggest the following:

(a) Removal of the old casing and the installation of a new casing of like diameter [PVC pipe is often the current casing material of choice if the well is ≤ 1000 ft (≤ 305 m), as it is corrosion resistant and is easily jointed with solvent-weld or mechanical slip joints] (66). This also provides an opportunity to thoroughly clean mineral accretions from well screens or to replace them with noncorrosive fiberglass or stainless steel and to inspect the well's sanitary seal and verify its integrity.

(b) Cleaning of the well using acid treatment and rotary wire brushing. (It is presumed for purposes of this discussion that the well pump had been previously hydraulically tested as part of the prepurchase water system suitability assessment.) It is sensible to retest the well after the cleaning process to see, in fact, if any improvement in hydraulic performance occurred.

(c) Disinfection of the well prior to placing it back into service. After cleaning, the well is pumped (flushed) for several hours to remove any mineral or organic material that was dislodged during the active cleaning phase. This water is of sufficient purity that it can probably be released to a natural drainage course. The well-drilling contractor is responsible for obtaining any and all permits for the disposition of groundwater used during the testing and cleaning phases.

Well-Drilling Logs and Records

As part of the inventory of useful basic information concerning the water resources of a potential winery site, copies of any and all drilling logs should be obtained from the seller. They may also be useful to supplement other geological information that may be obtained for structural and foundation analyses for the winery building and gives a well-drilling contractor a good idea of water availability if another well needs to be developed. Although copies of the well-drilling log(s) are generally submitted to the state governmental jurisdiction responsible for construction, alteration, maintenance, and abandonment of wells, the logs are not, in California at least, public record. At the completion of the well construction process, the owner is given a copy of the log and he/she or an authorized agent only are permitted to release copies of the drilling records to other persons. The well driller submits a copy of the Well Completion Report to the state office having jurisdiction. The lore associated with the provisions of this rather unusual privacy statute is slightly obscure and indicates that because such information is extremely valuable to members of the drilling contractor's fraternity, for the general public to have access to such records could give an unfair economic advantage to one drilling contractor over another. Because of their confidential nature and because they are difficult to obtain, copies of drilling logs should be kept along with other important papers in a safety deposit box or fireproof safe. Figure 6.3 shows a copy of a typical well completion report (well log). This form is also used in California to report well abandonment (see "DESTROY WELL" under ACTIVITY, upper one-third near right-hand edge of form). The State of California is used as the model, although most states have adopted some sort of program to acquire data on newly drilled wells. Both the states of Nevada and Arizona have excellent administrative systems for monitoring and controlling groundwater and wells. This is probably in part because groundwater in both of those states is "appropriative" (i.e., use and development of groundwater is subject to a statutory permit and licensing process). The well completion report form also encourages the submittal of other well/water supply data that may be generated as part of the drilling operation (see "Attachments" section of Fig. 6.3). As there is often a geohydrologist present during the drilling process, the logging of the hole with more precise lithological description may be prepared by the geologist, using accepted mineralogical nomenclature. This does not relieve the well driller from his responsibility for preparing the "geologic log" on the form as shown on Fig. 6.3. A dowser (or "water witch") is sometimes part of the well development team. A brief discussion of dowsers, including the author's experiences with water witches, is contained in a following subsection.

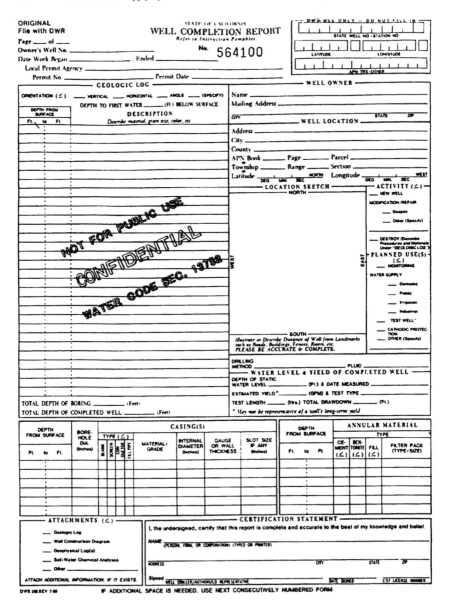

Fig. 6.3.   Typical well driller's confidential completion report.

Figure 6.4 is a copy of a visual well log form, which along with the video tape, forms the permanent data package for the well casing inspection. The utility of this film record is that if it becomes necessary in the future to locate a specific section of casing with questionable integrity (stress cracks,

P.O. Box 1326/275 County Road 98/Woodland, California 95695/Phone: 916/662-2825

VISUAL WELL LOG EVALUATION STUDY

Owner: _____ Date: _____

Location of Well: _____

CASING LOG   Wall Thickness: _____

Depth of Casing: _____ Casing Diameter: _____

Type of Perforation: _____

Liners and/or Tapers: _____

VISUAL LOG

Area Scanned: _____ SWL: _____

Purpose of Scan: _____ Logger: _____

Remarks: _____

DEPTH			CONDITION
From	To		

LW-317-1

Fig. 6.4.   Visual log form for video camera inspections of well casings. (Courtesy of Layne, Inc., Mission Woods, KS.)

areas of severe corrosion or deep pitting, and so forth), the tape can be reviewed on a VCR in fast-forward mode until the depth number (recorded on the film) corresponding to the area of interest is observed and compared with the subsequent video record for the same section of casing.

One final note concerning geophysical records for wells: In the petroleum and geothermal steam industries, complete borehole geophysical measurements are made including three-radiation logs, two-electric logs (E-logs), and a caliper log that records any and all variations in borehole diameter. All of the geophysical measurements made are for a more precise characterization of the nature of the materials through which the borehole passes and of the fluid contained in the subsurface formation. For water wells, the E-logs are used more frequently in deep wells and in subsurface formations where groundwater occurrence is confined to very narrow bands of fractured rock (not well-defined aquifers of alluvium, i.e., sand and gravel deposits). The interpretation of E-logs is the exclusive province of geophysicists, and is beyond the scope of this book. For further detailed information on the subject of E-logs and groundwater, the reader is referred to Keys, 1967 (67).

Water Quality Assessment

All "public" water supplies in the United States used for potable (drinking and culinary) purposes must meet or exceed the chemical and biological standards set forth in the Federal Safe Drinking Water Act. Presumably, a private water supply does not have to meet the federal standards, but more often than not, local county environmental health officials will strongly recommend that the water meet the standards, particularly in regard to so-called toxic heavy-metal concentrations. Table 6.1 is a copy of the U.S. EPA 1975 National Primary Drinking Water Regulations (as amended, 1988). The toxic heavy metals listed in Table 6.1 are cadmium, chromium, lead, mercury, selenium, and silver.

Table 6.2 lists the National Secondary Drinking Water Standards that were enacted to provide a benchmark for aesthetics (taste and characteristics). The so-called secondary standards include some of the important ions that are critical in the production of quality wine. If they are somehow incidentally introduced during the production process, wine color, taste and appearance can be affected (see Table 6.2).

Wineries without tours and tastings may escape the rigid water quality standard compliance required by the Safe Drinking Water statutes, and if water supplies are marginally potable, bottled water may be provided for employees (see the section entitled *"Bottled Water Service"*). From a purely winemaking standpoint, overly high concentrations of the chemical constituents shown in Table 6.3 have the potential to adversely affect wine quality by virtue of the rinse water residues left behind in transfer hoses,

Table 6.1. U.S. Primary Drinking Water Regulations

Constituent	Maximum contaminant level[a] (mg/L)
Inorganic chemicals	
Arsenic (As)	0.05
Barium (Ba)	1.0
Cadmium (Cd)	0.01
Chromium (Cr)	0.05
Fluoride (F)	1.4–2.4[b]
Lead (Pb)	0.05
Mercury (Hg)	0.002
Nitrate (as N)	10.0
Selenium (Se)	0.01
Silver (Ag)	0.05
Organic chemicals	
Chlorinated hydrocarbons	
Endrin	0.0002
Lindane	0.004
Methoxychlor	0.1
Toxaphene	0.005
Chlorophenoxys	
2, 4-D	0.1
2, 4,5-TP Silvex	0.01
Total trihalomethanes	0.1
Microbiological standards	
Membrane filter technique	Number of *coliform* bacteria shall not exceed 1 per 100 mL as the arithmetic mean of all samples[c]
Fermentation tube method (10 ml standard portion)	*Coliform* bacteria shall not be present in more than 10% of the tubes in any one month[d]
Turbidity	1 turbidity unit (TU, also known as NTU, JTU) as determined by a monthly averages

[a]Maximum contaminant levels are set according to health criteria.
[b]Limit depends on average air temperature of the region.
[c]These maximum contaminant levels may be modified depending on the system size and the number of samples taken.
[d]These levels are subject to further modifications.
Source: Ref. 68.

piping, tanks, pumps, and other equipment which comes into direct contact with the wine. This discussion has been simplified, as the damaging concentrations in wine may be impossible to achieve if the concentrations of the specific-ion in the rinse water are very low. It is highlighted here, because of its pertinence to the discussion of water resource assessments. Table 6.3 lists the chemical constituent and the nature of its potential adverse effect on wine.

Other researchers have expressed more concern for metal contamina-

Table 6.2. U.S. Secondary Drinking Water Regulations

Constituent	Recommended level[a]
Chloride (Cl)	250 mg/L
Color	15 color units
Copper (Cu)	1.0 mg/L
Corrosivity	Noncorrosive
Foaming agents	0.5 mg/L
Iron (Fe)	0.3 mg/L
Manganese (Mn)	0.05 mg/L
Odor	3 threshold odor number
pH	6.5–8.5
Sulfate (SO_4)	250 mg/L
Total dissolved solids (TDS)	500 mg/L
Zinc (Zn)	5.0 mg/L

[a]Maximum contaminant levels are set according to health criteria.
Source: Ref. 69.

Table 6.3. Selected Water-Quality Parameters With The Potential for Adversely Affecting Wine Quality

Constituent	Effect	Limiting conc. (if known) (mg/L)	Techniques for chemical alteration or removal from water supply
Iron	Taste	0.3 (Ref. 69)	Ion-exchange column
Bicarbonate	Taste	100 (Ref. 75)	Reverse osmosis
Sulfate (as potassium sulfate)	Taste and odor	250 (Ref. 69)	Reduce use of SO_2 in wine production
Copper	Haze and turbidity	0.2–0.4 (Ref. 75)	Ion-exchange column or reverse osmosis
Residual chlorine	Taste (with phenolics present)	0.2–0.4 (Ref. 75)	SO_2 rinse reduces residual chlorine to chloride ion
Manganese	Taste	0.05 (Ref. 69)	Ion-exchange column
Calcium carbonate	Taste[a]	70 (Ref. 75)	Ion-exchange softener

[a]See also comments in Chapt. 4 with reference to boiler feed water quality and carbonate hardness.
Source: Data from Refs. 73 and 74.

tion in wine manifested as adverse taste, odor, and turbidity similar to those identified in Table 6.3 as coming from wine-to-metal contact because of a poor choice in materials of construction for pumps, piping, and storage containers or from metal contamination from mineral dusts and soil picked up during grape harvest (73). The almost universal contemporary use of stainless steel and polyvinylchloride pipe and stainless steel pump housings with Buna-N rubber, brass, or stainless impellers throughout the wine industry has virtually eliminated wine–to–troublesome-metal contact.

Thus, the mineralization of grapes from dust/soil in the vineyard and rinse water residues appear to be the only serious sources for which wineries should be alert, if metal contamination problems begin to appear. The rinse water mineral residue component of potential wine contamination is self-mitigating because of the low solubilities of the metals in water and their propensity to attach themselves ionically to negatively charged particles. Thus, if the concentrations in the winery's water supply do not exceed the values in Table 6.2, the supply should be suitable for winery use.

Calcium or magnesium carbonate hardness in winery water supplies is also of concern because of its mineralizing effect on water tubes in steam and hot water boilers. If the boiler feed water is not softened in an ion-exchange column installed for that purpose, damage to the boiler can result. This was discussed previously in Chapter 4.

The chemical and biological laboratory testing for water supplies that are required to determine if the supply meets the Safe Drinking Water Act Standards are quite expensive. The cost for an analytical laboratory to search for microgram concentrations of pesticide residue is the same even if no residue is detected. Therefore, it is important that the water sample taken from the well or surface supply be truly representative of the source and that the samples be properly preserved, so that no chemical transformations can occur. (Chemical stability must be maintained and the sample must remain at the proper temperature to prevent any changes.) For microbiological testing, the container must be sterile and the point of sampling (faucet or valve) should be sterilized with a hand held propane torch. The water should be run for 5–10 min to flush any standing water from the system. A temperature of under $40°F$ ($4°C$) should be maintained in the sample by using an ice chest. If the samples are to be transported to the analytical laboratory, discuss the method and allowed time for safe shipping of the samples by a package express company that clearly understands the importance of a timely delivery.

Finally, in this era of "designer waters" that are produced from 300-year-old glaciers and magic bubbling springs, almost everyone has the trained water palate to decide on the subjective merits of water for purely drinking pleasure. Taste the potential winery water supply as if it was a medal-winning Chardonnay. If it is found lacking in some way, your negative taste score may be reinforced by the quality profile that will eventually be received from the analytical laboratory. Such a test performed by the author during a site suitability study for a winery, revealed a remarkably tasteful water supply that eventually received serious interest by a commercial bottled water company. The idea was left with the winery principals that if the wine venture failed, bottled water production might be a good fallback position.

Iron bacteria were mentioned in an earlier subsection entitled "Well

Casings". Both dissolved iron and manganese can be metabolized by several species of bacteria. Iron bacteria can proliferate with dissolved iron concentrations of as little as 1 mg/L, and at 25 mg/L they have completely plugged well screens with metabolic by-products of threadlike slimes (66). The mechanisms of iron microbe introduction into wells is not clearly understood. One theory suggests that there is cross-infection and that well drillers' rod, bits, pumps, and water tanks may be the source (66). Chlorine in the proper dosage can control the iron bacteria but not eliminate the nuisance entirely. If the concentration of dissolved iron is greater than 1.0 mg/L and the winery is concerned that iron bacteria might proliferate in the water storage and distribution system or that off-flavors and metallic aromas may occur in the wine from rinse water residues, ion-exchange columns are commercially available to strip low concentrations of iron in the ferrous form from the well water. Similar to domestic-type sodium zeolite resin charged columns for water softening, the iron removal medium is green sand that must be periodically regenerated with potassium permanganate. As a secondary benefit, reduction in hardness also occurs simultaneously with this method. For other methods of treating a winery water supply with high concentrations of iron (ferrous) >1.0 mg/L, or for iron in the oxidized state (ferric), a competent industrial water conditioning firm should be consulted.

Pump Efficiency Test

One final test should be made to validate the electrical–mechanical–hydraulic condition of existing well pumps and their appurtenant equipment. Every pump has a performance curve that graphically depicts the amount of water per unit time that can be delivered against a particular total dynamic head (TDH) (i.e., vertical lift plus friction loss). Also plotted is the efficiency for each set of head and flow rate combinations and the gross horsepower required to achieve the specified combination of TDH and flow rate. "Head" is defined for purposes of pump systems, as the total of "lift" (the elevation difference to which the fluid must be raised) plus energy losses from fluid friction in the system expressed in feet or meters of fluid. The pump efficiency shown on the performance curve for the pump is not the same efficiency that is determined from the in situ testing of the installed well pump. The pump performance curve efficiency is determined from manufacturer's "bench testing" of pumps under controlled laboratory conditions. If the winery is purchasing a parcel that has an existing well, then the performance curve for the well pump should be provided by the seller. If it is not in the seller's files, then the name of the pump manufacturer, the pump model, and horsepower must be transcribed from the pump nameplate (a brass or stainless steel plate attached to the pump with all the necessary identifying information). Armed with the pump data, the local manufacturer's representative can provide a duplicate copy of the pump performance curve. It is unfortunate if the well pump is of the submersible-type (i.e., both motor and pump are in the well) because in order to examine the name plate, the submersible pump must be removed in the manner described in the subsection entitled "Well Casings." A typical pump performance curve is depicted in Fig. 6.5.

Figure 6.5 is more easily understood if an illustrative example of a pump selection for a given flow and lift plus frictional loss (TDH) situation is presented. A TDH of 80 ft (24 m) and a desired flow rate of 1400 gpm (88 L/s) are assumed. The pump manufacturer has compiled in their catalog literally hundreds of pumps to accommodate different combinations of TDH and flow. Thus, how are the choices narrowed for the correct curve or group of curves, that might match your design requirements? The catalog usually also contains a pump selection chart which plots on the same coordinates of head and flow all of the pump characteristics for each model type. This "rough screening" of pumps that will meet the basic design parameters gets you at least to the right section of the catalog. Then, examine Fig. 6.5 and find that the 80-ft (24 m) TDH–1400 gpm (88 L/s) coordinate places the best pump combination at

- The B impeller curve (read 36 horsepower on the brake horsepower curve)
- A pump efficiency of 76%
- A minimum net positive suction head (NPSH) of approximately 11 ft (3.4 m)

The pump manufacturer specifies the NPSH to ensure that the pump will perform as designed at the given efficiency and not create suction vacuums that would approach the vapor pressure of the liquid being pumped (in this case water). If such a reduced pressure was allowed to prevail by hydraulically creating an excessive lift for the suction side of the pump, a phenomenon called cavitation might occur and imploding bubbles of water at the vapor pressure of water would eventually damage the pump impeller. A further examination of Fig. 6.5 (if, for example, this was the pump that was part of the real estate package purchased for the winery) would show other combinations of total head and flow rate which are possible for the given motor horsepower. Therefore, if it was the winery's intention to add more pipeline to service a production facility that would add the pipe friction loss of another 12 ft (3.7 m) of head [TDH now 92 ft, (28 m)] the corresponding flow rate would only be 1000 gpm (63 L/s) and the efficiency would be about 60%. The performance curve then indicates that the pump would be working outside a reasonable hydraulic envelope of flow and head.

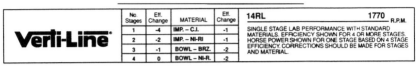

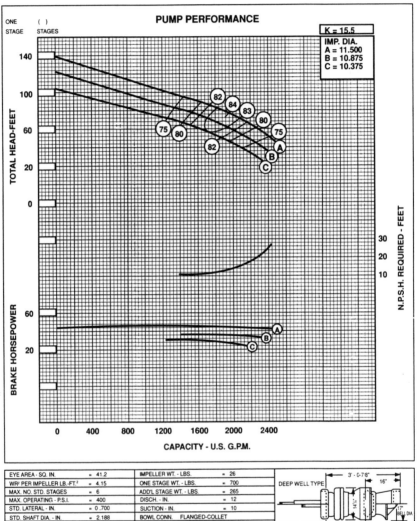

Fig. 6.5.   Typical pump performance curve. (Courtesy of Aurora Pump/Layne & Bowler, Aurora, IL.)

Electrically speaking, more flexibility can be integrated into a single pump motor by specifying a variable-speed motor or variable-frequency drive. Variable-speed motors (1200–1750 rpm or 900–1200 rpm) are respectively 2 and 2½ times more expensive than conventional single-speed models. Variable-frequency drives for conventional motors might add only $200 to $300 to the motor cost, but special insulation and an absolute necessity for motor and variable-frequency equipment compatibility require that the system designer write a flawless set of specifications. They would not serve a useful purpose for well pumps, but do have their place in a winery booster pump system that must respond to a variable-flow rate at constant pressure. This potential use will be described in the section entitled "Pressure Requirements."

Ideally, all pumps should operate in such a mode that the energy supplied to the pump (motor horsepower) is split about equally between flow and lift plus frictional losses (TDH). This is sometimes very difficult to achieve for a greatly unbalanced lift and flow rate requirement. In other words, if the lift is inordinately large (say 500 ft or 152 m) and the recharge rate of the well is very low (say 7 gpm or 0.4 L/s), pump manufacturers cannot build a pump that operates efficiently with such extremes of flow rate and lift. In this instance, the pump will operate at a very low efficiency, but because the pump discharge will not exceed the 7-gpm (0.4 L/s) recharge rate, it will not lower the water level in the well to a point where the pump is without a ready supply. As another precaution to prevent the well from running dry, a device called a "pump saver" can be installed. This relatively inexpensive insurance against a major equipment loss will be described later in this subsection.

The pump efficiency test is often a free service provided by electric utilities in response to consumer demand for more efficient use of electrical power supplies and lowered electric bills in an era of ever-increasing power rate increases. Small domestic wells (generally 5 horsepower and less) are usually excluded from the free testing service. The efficiency that this particular test provides differs from the pump performance curve efficiency previously described in that it reflects the overall electrical–mechanical efficiency of the pump and its appurtenant equipment and reflects both the age and condition of the system at the time of testing. Thus, it integrates the electric motor age/condition, wear, and pitting of the pump impeller and loss of hydraulic carrying capacity for riser pipes, valves, and fittings. Fig. 6.6 illustrates with a schematic diagram a possible testing arrangement and apparatus for a pump efficiency test and the application of the measured data to calculate overall pump efficiency. Efficiencies in the 70–80% range are not uncommon for new pumps that are working well within their range of flow rate and discharge head. Effi-

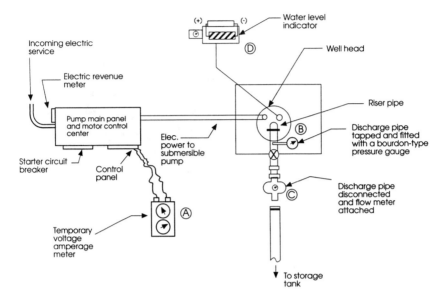

NOTES:

A) Input power to pump accurately measured throughout test period

B) Discharge pressure monitored throughout test period (≈TDH)

C) Flow meter monitored throughout test period

D) Water level in the well monitored to establish limit of drawdown during test and to provide data for termination of test when drawdown has stabilized

THEN:

$$\text{Overall efficiency} = \frac{\text{Output}}{\text{Input}}$$

$$= \frac{\text{H.P.} \cdot 550}{\gamma \cdot Q \cdot \text{TDH}}$$

WHERE:

H.P. = Voltage • Wattage • 0.746 Kw/H.P.

550 = 550 ft-lb/sec. H.P.

γ = Density of water 62.4 lb/ft^3

Q = Flow meter reading in cubic feet/sec.

TDH = Discharge pressure in pounds per sq. inch converted to feet of head.

Fig. 6.6.   Schematic of pump efficiency test apparatus and test protocol.

ciencies of 50–60% should be a basis for negotiating a presale owner repair or a discount in the winery parcel purchase price equal to the cost for pump replacement or repair.

The "pump saver" is a device to protect valuable well pumps and other hydraulic accessories that might otherwise have to be replaced or to undergo extensive repairs. It has had the greatest application in preventing submersible well pumps from running "dry." The device also detects electrical malfunctions such as power supply single-phasing (for a three-phase

Fig. 6.7.   Submersible pump protection device. (Courtesy of Hays Pumps, Anderson, CA.)

motor, this is fatal), low voltage, or complete power outage. The pump saver control box can be mounted (retrofitted) so that no great electrical panel modifications need to be made. In any of the aforementioned malfunction modes, the pump saver, upon receiving any of the warning signals, will inactivate the pump and provide an alarm (visual or audio at the well site) or remote with a compatible signal transmitter installed as a companion unit to energize an automatic dialer that will notify any or all of several preprogrammed telephone numbers that have been locked into the system—in other words, some responsible person (winery maintenance or commercial well contractor personnel) who can, with the greatest lead time possible, plan and execute a repair of the faulty well and anticipate its probable consequences upon winery operations and production. A typical pump saver device and its specifications are shown on Fig. 6.7.

Well Abandonment

Before you receive the keys for the gate to the parcel purchased for a new winery or vineyard, it is important that the area be thoroughly and carefully inspected by an engineer to look for wells that are no longer in use but that may not have been properly abandoned. Although not quite as serious as finding a leaking underground fuel tank after the close of escrow, failure to follow the proper procedures in the abandonment and sealing of a well can lead to hefty fines and criminal penalties under California's statutes (Section 24400 of the Health and Safety Code). The seller cannot be held totally responsible under real estate sales disclosure laws, if he or she can prove that the property was acquired without knowledge of the inactive wells. Abandoned wells are sometimes difficult to locate, often inadvertently covered with soil and encroached upon by native vegetation. Metal detectors and dowsers have been used successfully to pinpoint the location of old wells.

When the cogent reasons for sealing an abandoned well are revealed, the reader will clearly understand why such importance is placed on adherence to the statutory procedures. The reasons are as follows:

- To prevent pollution of groundwater from the surface through or around ᵗʰe annular space surrounding the well casing, if it has not been properly g₁ₒᵤted
- To eliminate a physical hazard (children and small animals are forever finding and falling into improperly abandoned wells)
- To conserve aquifer yield
- To preserve artesian conditions (if present)
- To prevent poor quality water from one aquifer from entering another.

Two reference sources are cited for well abandonment. (71, 72) Anon., 1984 (71) is a more generalized standard, whereas Anon., 1990 (72) is very detailed and provides illustrations and diagrams to assist the well drilling contractor in complying with the statute for a variety of well types and subsurface geologic conditions and formations. In California, the well completion log (Fig. 6.3) is also used to report the destruction of a well. See the form section labeled "Activity" with subcategory "Destroy."

Land Use Conversion Water Use Equivalency Test

In some areas of the United States, wineries are not embraced as welcome additions to the community and economy.

What often is reluctance on the part of governmental planning bodies to grant land use permits to winery facilities, appears to be related to the issue of water use. Somehow, if the winery project can be shown to require the same amount of water for its processing that the recent past activity on the same parcel utilized, the winery plan often becomes acceptable, waterwise, at least. The author was retained to assist in the conversion of a large horse-breeding facility to a 100,000-case annual production winery. This project clearly illustrates the importance of the water use equivalency test. Land use history showed that the adult horse population at any one time was about 70–90 and fluctuated some 10–15%, depending on births and sales of animals. As no metered records were available to help in estimating horse stock watering requirements, local horse boarding and training facilities were queried for their experience in unit water use by horses. The median value was 20 gallons per horse per day on dry feed and 12–15 gallons per horse per day on green pasture. A classic 1913 U.S. Calvary manual was also consulted, which validated the numbers obtained by telephone survey. Surprisingly, the comparative water use (horse breeding and wine production) were within about 2% of each other. This evidence was sufficient to sway the vote of the local planning commission in favor of the winery because in the governing board's opinion, postwinery and historical water use would not impact the area's water resources significantly.

Also of great importance in selling the concept of "reasonable water use" for a winery development proposal is the presentation to the planning agency of a water conservation plan that will become established policy for the winery. Because of its importance, the subject of water conservation in wineries will be dealt with in a subsequent section.

Water Dowsers

The history of water witches and divining rods is nearly 400 years in the making, and their rather mysterious art has prevailed to the present day. So it is with unusual golf putting stances or fish bait spray elixirs; if it works, use it. The author's rule of thumb is if the dowser is over age 50, there is a high probability that the person's reputation and water-finding skills have survived the test of time and high-tech geophysics.

If the winery is faced with the development of a new well and locating successful wells in the region is a hit-and-miss proposition, the well-drilling contractor may suggest a local dowser to assist him in fine-tuning the exact location for drilling. It is important to "bring in" a well on the first attempt, as the standard well-drilling contract provides for payment to the contractor, even if the well is dry or only marginally useful with a scant water yield. Thus, to increase the odds in favor of a productive supply well, the nominal fee charged by dowsers seems a small price to pay when compared to a $25–$30/ft unit cost to drill and develop the well. In very difficult locales with very low percentages in successful water well production, the services of a competent geophysicist or geohydrologist may also become necessary.

For further in-depth reading on water witching, U.S. Geological Survey Water Supply Paper No. 416, entitled "The Divining Rod, A History of Water Witching," by Arthur Ellis, 1938, is the complete anthology on the subject.

Water Demand Calculations

Water demand calculations can and should be developed early in the planning phase for the winery to

(a) match the winery site's potable water production capability with the winery's current and future process water demand schedule

(b) assess conjunctive water use for fire protection, vineyard irrigation, landscape irrigation, and water amenities (landscape, ponds, and other aesthetic water features)

(c) satisfy the local planning body's water availability requirements for partial fulfillment of a necessary condition for receiving a land use permit

(d) provide a database for an environmental impact assessment

As the feasibility of the winery often depends on a positive water availability report, it is often money well spent to commission a civil or environmental engineer to prepare a *preliminary water and wastewater engineering report*. This report can include the preliminary design phase for the wastewater systems (sanitary and process wastewater) because the water demand values that are created become the basis for sizing the components for the wastewater system.

If the winery operation is to include tours and wine tasting, the engineer must, in concert with the winery principals, synthesize maximum visitor-day estimates to which water use and wastewater discharge numbers may be assigned. One criterion is to use visitor numbers on the major travel holidays of Memorial Day, Fourth of July, and Labor Day to produce the desired peak visitor-day numbers that are sought. Established wineries or other tourist-based facilities in the immediate area are good sources of visitor data. State/county highway traffic surveys can also give the visitor potential for a given road or street. Often overlooked in this quest for visitor values is the "special event" that wineries frequently fail to include in their potential visitor traffic. For example, a "new wine release," "fall harvest," "wine writer's day," or "concert event in the vineyard" can produce more winery traffic than the peak travel holidays previously cited. Grand openings are also big on the list of winery attendance, which may far exceed expectations. From the utilities perspective, these ultrapeak loads can overtax systems that are designed for the "probable" peak load. The water system is generally not as vulnerable to overloading as the sanitary wastewater system. If the ultrapeak visitor events are wisely spaced, some system overloading can be tolerated, although it is prudent of winery management to arrange for portable toilets to prevent a hydraulic or solids overloading of the winery's on-site, septic tank and soil absorption (leachfield) systems. Table 6.4 shows a typical array of water supply design pa-

Table 6.4   Estimated Typical Winery Water Requirements by Category of Use (190,000-Case Annual Production Capacity)

	Gallons	Acre-feet	Remarks
(a) Process water use and wastewater discharge equivalent (annual)	3, 400, 00/year	10.4/year	16 gal water use per case of finished wine[a]
(b) Process water use (max. daily per Napa County Formula for crush/harvest period)	11, 400/day	—	Annual wine prod. (gal) × 1.5 ÷ 60
(c) Process water use (max. hourly flow)	190/min	—	Average for 10-h day × peaking factor
(d) Blow-down and makeup for water-cooled, evaporative refrig. condensers	10/min; 5, 256, 000/year	16.1	—
(e) Sanitary waste			
10 full-time employees (10 @ 25 gal/pers/day), plus 250 visitors/day (Apr.–Oct.) (250 @ 6 gal/pers/day)	1750/day; 373, 625/ season	1.2	7 summer/spring months
10 full-time employees (10 @ 25 gal/pers/day), plus 150 visitors/day (Nov.–Mar.) (150 @ 6 gal/pers/day)	1150/day; 175, 375/ season	0.54	5 winter months
(f) Landscape irrigation	346, 000/year	1.06	
(g) Vineyard irrigation	1, 800, 000/year	5.52	60 gpm for 4 h, 4 times per month for 6 months 18, 000 vines @ 100 gal/vine/yr

[a] See Table 6.6 for unit winery water use by production capacity and winery type.

rameters for a winery with an annual production capacity of about 190, 000 cases. The categories of use and estimated rates and total quantities should be considered a minimum compilation for land use permit administrative submittals. Each local governmental planning body will have its own peculiar set of requirements that must be met to satisfy the provisions of the special development code for the area.

There are several methods and approaches to developing water demand values for a winery that produce results that are both defensible and realistic. Most industries have developed water process unit demands based on unit of product created (i.e., gallons (volumes) of water per ton (mass) of steel, gallons of water per ton of paper, etc.). The wine industry is no different, and although there are appreciable differences from winery to winery on water consumption patterns, water use is more affected by the scale of the operation (i.e., large wineries use less water per unit of production than do small ones). Other factors can influence winery water demand values:

- Water scarcity
- Commitment to a water conservation ethic
- Special winery equipment that requires a nonstandard amount of water for cleaning and sanitation
- Methods of swelling and rinsing new cooperage for cellar use
- Method of refrigeration condenser cooling (once through water, water through evaporative cooler or air cooling)
- Greater or lesser use of dry methods of cleaning for general area cleaning (vacuum, compressed air, or squeegee)

If the winery plan can anticipate the presence or absence of the foregoing special elements, which influence unit water use, intelligent corrections can be made to reflect more precisely a more accurate water demand for the winery. For example, a well-conceived and implemented water conservation program can reduce overall usage by 15–25%.

The important role that water meters serve is worthy of mention at this point. How and where to meter will be discussed in a subsequent section. Water meter records not only verify design assumptions, they also provide a basis for measuring the efforts over time to conserve water. Further, they provide an excellent basis for planning winery expansions and for verifying the peak water demands that reflect the harvest schedule which represents not only a particular climatic zone but the varietal mix as well.

Table 6.5 is a listing of water uses that must be accounted for in the demand calculations. This table can serve as a planning checklist to ensure that no fixture or potential use is overlooked. It is a list similar to the one used by the plumbing engineer or mechanical designer to size the distribution piping within the winery. The listing of factors which may influence the unit water use values can be used in a corollary way to derive more precise estimates of flow.

Table 6.6 illustrates another method of synthesizing expected flow requirements during the planning phase for a particular winery process step, using the estimated number of cycles performed and water volumes per cycle. Such an analysis can be useful for an operating winery in analyzing where water use might be curtailed in the implementation of a water conservation program and for calculating the required pipe size for the winery's plumbing network.

Table 6.5.   Range of Water Use Rates for Various Winery Plumbing Fixtures and Equipment

Item	Approximate rate of water use (when in operation) gal/min (L/s) except where noted
(1) Hose connection	5.0–10.0 (0.3–0.6)
(2) Portable high-pressure water/steamer	5.0–10.0 (0.3–0.6)
(3) Landscape (impulse-type) sprinkler head	1.5–3.0 (0.1–0.2)
(4) Drip emitter (landscape)	0.5–1.0 gal/h
(5) Process refrigeration condenser cooling water (blow-down makeup)	5.0–6.0 (0.3–0.4)
(6) Spray cleaning bar in rotary-drum screen	2.0–4.0 (0.13–0.2)
(7) Miniflow toilet (water closet)	1.2–1.6 gal/flush
(8) Flushometer (flush valve operated toilet)	2.0–3.0 gal/flush
(9) Urinal (flush valve operated urinal)	0.5–1.0 gal/flush
(10) Miniflow shower	0.7–1.0 (0.05–0.06)
(11) Sink/lavatory	1.0–2.0 (0.06–0.13)
(12) Evaporative-type space coolers (3000–5000 cfm capacity)	0.5–1.0 (0.03–0.06)
(13) Boiler feed water	Highly variable, depending on horsepower rating

Table 6.7 can be used to estimate annual water requirements for wineries of different production capacities and different end products. The water use values were developed using metered flow records from a representative number of wineries in each product and production capacity category. The monthly distribution of water use varies considerably from facility to facility. The percentages of annual flow by month have to be adjusted for the time of harvest and whether the winery is in the Northern or Southern Hemisphere. The wine-producing areas of South Africa, Australia, New Zealand, Chile, and Argentina will all have grape harvest periods of about the same time of the year. Table 6.8 has been developed to illustrate a typical monthly distribution of annual water use as calculated from Table 6.7. A bonded cellar or a bottling facility would not show the monthly variation in flow that a typical table winery does. Likewise, wineries that receive finished or unfinished wine for special bottlings or second label wines as part of their product mix would have substantially different water use patterns than a winery that only provided estate-bottled vintages.

Peak water flows can be estimated using the Napa County (California) formula or by taking the maximum monthly distribution flows calculated from, for example, Table 6.8, reducing them to the typical crush week and day (generally 6-day work weeks with 1 day for serious cleanup and/or rest

Table 6.6.   Crush Period Winery Water Use Estimates by Process and Subprocess Step

	Cycles per day	Gallons (liters) per cycle	Total
Estimated Crusher-Stemmer Water Use			
(1) Gondola wash (5 gpm for 3 min)	30	15 (57)	450 (1710)
(2) Tank rinse (prior to filling; 5 gpm for 5 min)	2	25 (95)	50 (190)
(3) Crusher washdown (prior to use daily @ 10 gpm for 5 min)	1	50 (189)	50 (189)
(4) Crusher cleaning (end of day @ 10 gpm for 15 min)	1	150 (568)	150 (568)
(5) Tank cleaner cycle (sprayball for 15 min @ 10 gpm)	2	150 (568)	300 (1136)
(6) Tank rinse Cycle (sprayball for 15 min @ 10 gpm)	2	150 (568)	300 (1136)
(7) Crush pad washdown (15 min/day @ 10 gpm)	1	150 (568)	150 (568)
(8) Transfer hose washing (am & pm; 10 min @ 10 gpm)	2	100 (379)	200 (758)
Crush Subtotal:		1650 gal/day (6245 L/day)	
Estimated Bladder or Screw Press Water Use			
(1) Wash press (A.M. & P.M.; 10 gpm for 5 min)	2	50 (189)	100 (378)
(2) Wash hoses (A.M. & P.M.; 10 gpm for 10 min)	2	100 (379)	200 (758)
(3) Wash press pad (15 min @ 10 gpm)	1	150 (568)	150 (568)
(4) Wash pomace auger and hopper (5 gpm for 10 min)	1	50 (189)	50 (189)
Press Subtotal:		500 gal/day (1893 L/day)	
TOTAL CRUSH/PRESS CYCLE PER DAY:		2150 gal/day (8148 L/day)	

and equipment repair and work days that begin with the first grape gondola and end when the last load is delivered, usually 10–12-h days). For night harvest operations, the workday length is about 10–12 h but starts at 10 P.M. and ends at 8 or 9 A.M. A useful comparison can be made using the Napa County formula and values derived from Tables 6.7 and 6.8. The higher of the two peak flow values should be used, tempered with special factors of scale and operational technique that may be peculiar to the particular winery in question. Table 6.9 outlines two methods of peak water demand calculations.

Table 6.7. Estimated Range of Unit Water Use in Gallons (Liters) per Case of 750-mL Bottles

Production capacity (cases/year)	Table wineries	Bonded cellars or bottling facilities	Sparkling wines
>1, 000, 000	10–14 (38–53)	4 (15)	No data
200, 000–1, 000, 000	14–16 (53–61)	5 (19)	5–10 (19–38)
50, 000–200, 000	16–18 (61–68)	8 (31)	8–10 (31–38)
<50, 000	18–25 (68–95)	8 (31)	10 (38)

Source: Ref. 76.

Table 6.8. Typical Monthly Distribution of Annual Water Use for Northern Hemisphere Location

Month	Range of % of annual use	Remarks
January	6	
February	9	
March	7	
April	8	Bottling[a]
May	8	Bottling
June	7	Bottling
July	7	
August	10	Prep. for crush (table wineries); begin crush (sparkling wineries)
September	12	Crush
October	11	Crush
November	7	
December	8	
TOTAL	100	

[a]Many wineries in the 100, 000-case classification operate their bottling lines for 6–8 months.

WINERY WATER DEVELOPMENT AND ENVIRONMENTAL IMPACT ISSUES

The reality of the last three decades of the 20th century is that almost all projects (public, private, commercial, and industrial), big and small, must be examined in light of their impact on the environment. The National Environmental Policy Act (NEPA), and the 50 state versions of the same law to protect the valuable resources of the United States, must be complied for the construction of wineries. Agriculture for the most part is exempt from the provisions of NEPA. Thus, vineyard development does

Table 6.9. Comparing Two Methods for Peak Winery Water Demand Calculations for Hypothetical 40, 000-Case Facility

Typical Wine Industry Historical Water Use Data

- Winery production 40, 000 cases .... 96, 000 wine gal
- Annual water use (Table 6.7) .... $19 \times 40, 000 = 760, 000$ gal
- Water use for maximum month (Table 6.8) .... $0.12 \times 760, 000 = 91, 200$ gal
- For 26-day work-month (daily max. flow) .... $21, 200 \div 26 = 3507$ gal/day
- For 10-h workday (average hourly flow) .... 3507 gal $\div 10 = 351$ gal/h
- Average peak flow (gal/min) .... $351 \div 60 = 5.9$ gpm
- Peak hourly factor of 10 for maximum expected flow rate .... $5.9 \times 10 = 59$ gpm

Napa County Empirical Formula

- Napa County Formula (peak daily flow during crush)

$$= \frac{1.5 \text{ [ann. production in wine gal]}}{60}$$

$$= \frac{1.5 \, (96,000)}{60} = 2400 \text{ gal/day}$$

3507 gpd > 2400 gpd, $\therefore$ use more conservative estimate for predicted maximum daily winery process water use. The additional 1100-gpd cushion in the larger predicted peak daily flow is a good design parameter choice to give a little operational flexibility without a great increase in capital cost.

not generally require an environmental assessment, but a companion winery (industrial facility classification) might and often does.

The impact of water supply development for a winery is only one of many elements that must be examined to determine if a "significant impact" would, in fact, be created by the implementation of the winery project. In practice, one governmental jurisdiction assumes the responsibility for compliance with NEPA and/or its local state version. This so-called "lead agency" makes the determination whether an *environmental impact report* (EIR) must be prepared or that an environmental assessment would be sufficient for the planning body to make necessary findings. (An environmental assessment is a less rigorous administrative process whereby the issues are evaluated but with less supporting technical and original field data than the full EIR method.) The lead agency also has the option (and usually does) of requiring a "focused EIR" that deals with only those items identified as sensitive or having a marginally significant impact that require more technical support and scientific or engineering effort to characterize their probable impacts. If, after lead agency staff study, the winery project from the best information available, and sometimes the testimony of experts, during a public hearing, the lead agency can often declare that the project is apparently benign, or that there are no identifiable, significant, environmental impacts. The lead agency can subsequently issue a formal, so-called "negative declaration," sometimes with conditions or mitigating measures attached. The latter situation is the quickest and least painful of any of the consequences of the environmental review and compliance process. If a formal EIR protocol is followed, in addition to the time required to have a professional team prepare the report, there are several levels of review and other interested agency interactions that must occur with the proceedings. Finally, a public hearing is held where comments are received on the draft version of the EIR. The administrative time involved from the start until the public hearing can be as little as 3 months to as much as 1 year, if the issues are controversial.

After the public hearing, the EIR consultants compile all of the comments (presumably answering those questions that arose during the hearing) and perform those additional, supplemental technical studies that may be required at the request of the lead agency's governing body. Then the *final version* of the EIR is eventually issued and the conditions and mitigating measures put in a formal document, that must be complied with by the winery developers.

There are a number of precautionary planning measures that can be taken to avoid the winery's involvement in environmental water issues and to present a positive image and spirit of cooperation in dealing with the site's water resources and assets:

- Avoid encroachment upon any riparian corridors or habitat with pipelines, power or telephone utility lines, roads, or other infrastructure.
- Demonstrate, if possible, by presentation of hard numbers that postwinery water use and water use associated with previous land use will be the same or less than historical use.
- If surface water supplies are to be developed, seek opportunities to construct the necessary storage works off-stream, so that the aquatic resources of the system remain as undisturbed as possible.
- Look for opportunities to enhance or rehabilitate streams, which are downgrading severely (eroding in the vertical direction), by means of channel bank stabilization or other means.
- Look for opportunities to improve fishery resources by cooperating with the local fish and wildlife agency.
- With the guidance and approval of the local fish and wildlife staff, develop a riparian vegetation improvement plan (winery to install and maintain).

PRESSURE REQUIREMENTS

Several aspects of water pressure requirements for wineries have been previously mentioned in chapter 4. For winery sites with topographic relief, the most reliable system would have as a principal element, a storage tank strategically located, so that its vertical elevation above the working winery can produce the requisite dynamic pressure without variation. This arrangement could be loosely labeled the "gravity system." The simple relationship which relates vertical feet of fluid to static pressure [pounds per square inch (psi)] is

$$P = \gamma H$$

where P is the pressure (in pounds per square foot), γ is the density of fluid (62.4 pounds per cubic foot for water), and H is the vertical height of the fluid column. Thus, if the winery wants a constant pressure of a minimum of 50 psi throughout the winery:

$$P = 50 \text{ psi} \times 144 \text{ in}^2/\text{ft}^2$$
$$= 7200 \text{ lb/ft}^2$$
$$\gamma = 62.4 \text{ lb/ft}^3$$

$$\therefore \ H = P/\gamma = 7200/62.4 = 115.4 \text{ ft}$$

the tank would have to be placed at least 116 ft (40 m) vertically above the highest point in the winery at which the 50-psi pressure was required. The volume of the storage tank is discussed in the subsection entitled "Water Storage." The maximum rate of flow would also enter into the decision with the values for peak hourly rate calculated for Table 6.9 used to guide the selection of the most efficiently sized pipe. Once pipe size is determined, the pipe friction loss at maximum flow could be added to the 116-ft (40-m) elevation to give the net elevation of tank required to provide the minimum 50-psi working pressure at the winery. Other critical equipment items for this system would be a well pump or a booster pump from a spring box capable of lifting the water to the storage tank site and a water storage tank level control system to energize and deactivate the well or booster pump to keep the tank at the required volume of storage.

The more probable circumstance of winery location and water source location would involve a well pump, a booster pump, and the winery at nearly the same elevation, with the well pump discharging to a storage tank. A key feature of systems designed by the author that are similar is a so-called hydropneumatic tank. As the name implies, air is a functional component of the subsystem.

Figure 6.8 is a booster pump arrangement with a hydropneumatic tank that is best suited for a large winery complex that has a wide range of flow requirements. Rather than have a single large pump designed for the

Fig. 6.8. Triplex-type water booster pump with hydropneumatic tank. (Courtesy of PACCO Pumps, Inc., Brookshire, TX.)

probable maximum flow that upon a demand being placed upon the system [e.g., one, 1-in. (25-mm) hose connection at 5–8 gpm (0.33–0.5 L/s) is turned on for routine winery cleaning chores], the pump responds by pumping at the full rate of discharge and pressure for the pump. As the single-hose connection is for purposes of illustration, probably only 10% of the maximum winery water demand, all of the excess pump energy is either dissipated as heat in the volute (casing) of the pump or converted to pressure head, as was shown previously on the pump performance curve, Fig. 6.5. Thus, the designer has the option to install several pumps, each matched to some designated percentage of the winery water demand. The triplex system shown in Fig. 6.8 has a 20%, 40%, and 40% subdivision of total demand. In operation, the 20% lead pump operates alone, but if the winery's water demand exceeds 20% of the total system design capacity, a flow switch will energize the 40% pump and deenergize the lead pump. If the winery's demand exceeds 40%, the flow switch will start the second 40% capacity pump. If the 80% demand is exceeded, a differential pressure switch will restart the lead pump. The pumps are deenergized in reverse order as the flow demand decreases.

The 120-gal (455-L) hydropneumatic tank contains a rubber bladder that can be pressurized to a selected system value, say 65 psi. The tank operates virtually as a pressure cushion riding- on-the-line, eliminating pressure surges that might occur in the winery. It also maintains the fixed bladder pressure less friction losses throughout the system, thus protecting pressure-sensitive solenoid valves and other water system closures that respond adversely to pressure surges and cyclic fluctuations. Having this reserve at the proper pressure also precludes starting the lead pump for a very small demand and takes care of any fixture leakage in the winery in the same manner.

Variable flow can also be achieved with dual-wound motors (variable-speed motors) or variable-frequency drives that allow a variable flow rate from a single pump attached to a single electric motor. This electrical possibility was briefly discussed in the subsection entitled "Pump Efficiency Test." From the water system designer's perspective, the installation of a single booster pump and its unpredictable, future failure (electrical or mechanical) would place the winery in a no-water situation. This apparent shortcoming can be overcome by installing a duplex system with a duplicate pump and a duplex controller that alternates use from pump cycle to pump cycle for Pump "A" and Pump "B." The probability of simultaneous failure of both pumps is so remote that system reliability is in the 90–95th percentile category. Variable-speed motors are from 2 to 2½ times as expensive as conventional, single-wound motors. Variable-frequency drives are the product of the microchip age and bring about a change in motor speed by varying the line service frequency of the U.S. standard of 60 Hz. The response of the motor is reduced speed and the generation of heat energy. The motor manufacturer must be made aware that his product will be subject to variable-frequency power input, so that proper insulation and protective overheat relays can be incorporated into the factory specifications. In summary, the water system booster design must balance, for the winery, the optimum first cost, maximum operational reliability, and maximum system efficiency. Detailed methods of achieving the best electrical efficiency for the winery for all of its installed equipment was discussed in Chapter 2. Storm (77) also reveals ways of decreasing winery power bills.

Although it is not often a problem faced by wineries, the occurrence of operating pressures in the winery that are too high (say greater than 70 psi) is a possibility. The building designer for new wineries will balance pressures throughout the plumbing network to give 40–60-psi working pressures without zones of excessively high pressure. Sometimes, in multistory wineries or wineries built on hillside sites with significant topographic differences between processing areas, pressure balancing by judicious selection of pipe sizes is not possible. Retrofitting the high-pressure-zone primary piping with a pressure reducer can economically and rapidly eliminate the problem. These spring-loaded devices are designed for a particular range of upstream and reduced downstream pressures, so that a valve must be selected that will be compatible with the existing and desired pressure regime. For further reading on these specialty valves, see Storm, 1994 (78).

The author is heavily influenced, both by his experiences in operating a winery and for reasons of pure engineering economics, in favor of portable pressure washers rather than ultrahigh-pressure water subsystems that are hard-piped into the winery's plumbing network. Stubborn cleaning jobs, like removal of tartrate deposits from fermentors, presses, pomace augers, and bins are made easy by the use of portable high-pressure washing units. They are easily moved from cleaning site to cleaning site and their initial cost is low enough that multiple units are not beyond most wineries budgets.

WATER METERS

Good scientific method suggests, that if you cannot measure an event, a process, or a phenomenon, you do not know much about it. In a previous subsection entitled "Water Demand Calculations," the need and cogent reasons for keeping water use records were cited. A schematic diagram of a water system for a winery is shown in Fig. 6.9 with the preferred locations for and types of metering devices. Any arrangement is acceptable, as long as the subdivision of water use into its principal components can be determined. Wastewater production for process use should also be metered so that by subtracting process wastewater meter readings from the net volume delivered to the winery, the volume of water used for sanitary purposes can readily be determined. The necessity for developing flow rates for each of these categories of water use will be apparent when placed in the context of limiting capacity of wastewater treatment and disposal systems discussed in Chapter 8. The metering of pipes that are flowing full is the least complex of the flow measurement challenges that might face a winery. Table 6.10 lists the various types of metering systems and subsystems that are available for a full pipe flow condition. It is beyond the scope of this book to describe the characteristics of each metering category; however, the sonic, magnetic, and mass coriolis types have the distinct advantage of being non-flow-obstructing (79). Thus, they are nearly hydraulically perfect (insignificant energy loss from friction or turbulence) and are less prone to jamming or blockage. Well-stocked plumbing supply or waterworks supply companies generally have water meter offerings in each of the categories.

Metering open-channel flow, including natural streams, storm drains,

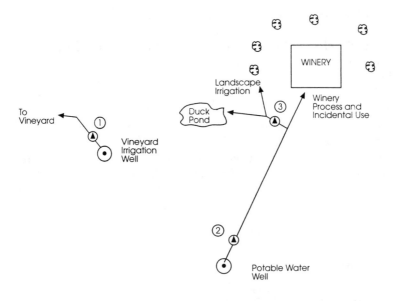

Meter	Description/Location	Type	Remark
①	Vineyard well discharge pipe	Turbine/ totalizer	Used by vineyard manager to verify total volume of water applied to the vineyard (manual read or data logged for PC input)
②	Winery potable water supply(process landscape and sanitary purposes)	Turbine/ totalizer	Records gross water use and well production (manual read or data logged for PC input)
③	Landscape irrigation and make-up water for small pond	Turbine/ totalizer	Meter ② minus meter ③ gives the net winery water use (manual read or data logged for PC input)

Fig. 6.9.   Winery potable and irrigation water metering plan.

other pipes flowing partly full, and irrigation canals or ditches, is a more difficult task. Some sort of stage (level) recorder is required to monitoring flow depth at a "control section" of the open-channel, pipe, or stream system. At the control section location, the geometry of the cross section must be precisely known and a "rating curve" must be prepared to correlate measured depth of flow to discharge.

**Table 6.10. Methods for
Metering Water Flow for
Pipes Flowing Full**

1. Differential Pressure
   - Orifice
   - Venturi tube
   - Flow nozzles and tubes
   - Pitot tubes
   - Elbows
2. Magnetic
3. Mass
   - Coriolis
4. Oscillatory
   - Vortex shedding
   - Fluidic
5. Positive displacement
6. Turbine
7. Ultrasonic
   - Transit time
   - Doppler
8. Variable Area
   - Rotameter

Source: Ref. 80.

Thus, for the stream example, a section of channel must be chosen that will not be altered over time by the action of the stream [i.e., an exposed rock stream channel that is scoured clean (no transient bedload materials present)]. The rating curve for the stream system must be prepared using a hand-held current meter to measure stream velocity at precisely measured intervals across the chosen control section. An accurate profile of the stream must also be obtained from bank top to bank top so that the cross-sectional area for each stream stage can be determined. This activity must be carried through for a full range of expected stream stages and stream discharges. During winter floods, the acquisition of current meter readings can be very hazardous and should be a task left to highly qualified water resource professionals.

Finally, a reliable method of transmitting stream level measurements to a stage recording instrument must be selected. The recording unit must be safely situated above the highest expected flood stage and protected from vandalism. The stage recorder is generally of the drum type with a spring-wound or electric-motor-driven clock mechanism connected to a float installed in a stilling well [a vertical length of pipe located out of the control cross-section whose base is connected to a submerged, perforated pipe to transmit stream stage (level) to the stilling well]. An ink or pressure sensitive paper stylus continually records the fluctuation in stream level. Canals, ditches, storm drains, and pipes flowing partly full can be metered in much the same manner as the natural stream example by the use of standard weirs and flume sections that are purchased with a factory-measured rating curve. The most likely applications for open-channel flow measurement in a winery are as follows:

(a) Compliance with water rights permit or license provisions for guaranteed downstream releases
(b) Storm drain flows for compliance with federal and state storm water-pollution-prevention permitting programs (81, 82)

For a winery to have a more sophisticated and expensive system, meters can be equipped with signal transmitters which, if the distance from meter to winery is not excessive [say less than 1000 ft/305 m], can be remotely read and recorded. The recorder's millivolt signal can be transmitted even greater distances if signal boosters are installed. From a control and instrumentation reliability standpoint, the more "black boxes" required in the circuit, the more likely there is to be a failure. Commercial telephone lines can also be leased to transmit control or remote recorder signals. This possibility will be discussed in detail in Chapter 8. The data should be stored on hard disk for eventual compilation and analysis on a PC. For ease of compiling and organizing this information, if the meters, are manually read and the numbers transcribed into a log book and then entered into the winery's computer, the reading should take place at the same time on the first of every month or daily, if a daily pattern of water use is sought. Thus, for the simple totalizing meter, performing the subtraction of the current reading from the previous reading for the time interval selected yields the measured volume of water.

Meters can also be equipped with microprocessors that can store the flow record data for several months. A data logger is periodically connected to the flow meter's microprocessor and the data are stored temporarily in the handheld device until it can be downloaded into the winery's computer. This method of data transfer and storage becomes economically practical when a large number of meters (wastewater, irrigation, and potable water) require servicing. The only disadvantage of this system is that if the return interval to the meters is several months and one of the meters malfunctions during that interval, then valuable flow data can be lost. Turbine meters are extremely reliable, with rare jamming of the impeller from inorganic detritus (rock or mineral scale fragments). Even this remote eventuality can be prevented with the installation of an in-line, manually cleaned trainer that prudently could be serviced at the same time as the meter is read. The strainers are standard plumbing items available in cast iron, brass, or polyethylene.

WATER STORAGE

If the winery is blessed with a large-capacity well, the relationship of maximum hourly flow, regulating storage, and optimum frequency for cycling of the well pump is less critical than a low-capacity well or spring, which is greatly out of phase with maximum winery water demand.

The crush and press, water use estimates, previously reported in this chapter, have been combined and incorporated into a typical 10-h working day during harvest to illustrate how a water system with reregulating storage can be balanced hydraulically to satisfy the limiting parameters of supply and demand. Figure 6.10 is a graphic depiction of daily water demand and of total cumulative daily demand. The latter curve can be used for establishing the *minimum* regulating storage that would permit the most efficient operation of the several system components involved and satisfy the maximum daily water demand.

The graphical format for Fig. 6.10 can be used for any winery water demand curve and the minimum operational storage volume determined. At this juncture, it must be noted that other considerations enter the decision on the size of storage facility to best serve the winery under all operating scenarios. For example, if the winery's well could only produce 10 gpm on a sustained yield basis, the approximately 11-h period (nonoperational period at night) for water to be accumulated in storage to meet the following day's demand would be inadequate and the well pump would be in operation constantly and still be unable to satisfy demand, or

Well production:	24 h × 60 min/h × 10 gpm =	14,000 gal	(52,990 L)
and	24-h winery demand =	24,700 gal	(93,490 L)
	water supply deficit =	(10,300) gal	(40,500 L)

Thus, even increasing the amount of regulated storage would not alleviate the water deficiency, as, clearly, water demand exceeds available supply. Although this example may seem simplistic, the author has analyzed existing, troubled, winery water systems and discovered this impossible situation on several occasions. In one instance where the development of an additional well was not possible, a well-structured water conservation program brought water demand more closely in line with water supply. Additional wells were the solution for the other unbalanced supply–demand water systems.

A good design philosophy regarding the optimum volume of storage to install for any given winery, is based on the premise that electrical/mechanical failures are going to occur in spite of careful specification and selection of the most reliable pieces of equipment commercially available.

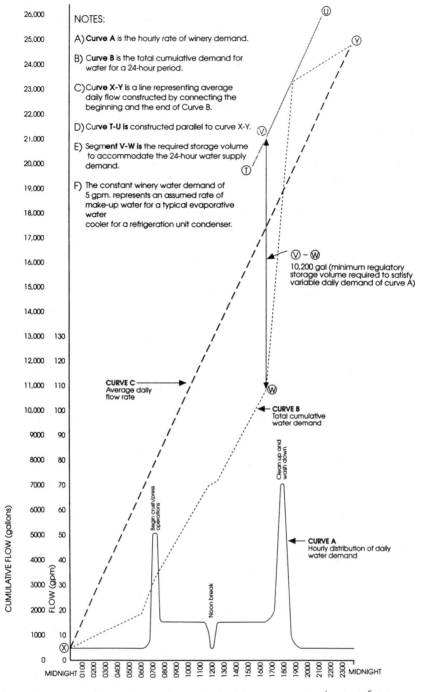

Fig. 6.10. Graphical method for determination of water storage requirements for a 120-ton per day crush capacity winery.

Therefore, as a minimum, the installed water storage volume should be a 2 days' supply at maximum winery demand. In most instances (except for complete and extended primary power failure) this will allow for continued winery operation, even though the well pump or other water source pumps may be inoperable. To give the highest probability of source pump failure, when the tank is nearly full with at least 2 full days' supply, the level controls for the storage tank should be set to activate the source pump when only 25% of the tank's volume has been depleted. The advantages of a gravity supply from a storage tank have been previously discussed in this chapter under the subsection entitled "Pressure Requirements." This gravity flow feature adds a high degree of system reliability.

A question often asked by winery developers is why the water storage tank cannot just be enlarged to accommodate the flow rate and duration specifications for fire protection? A basic and universal principle of all fire protection systems is that they must be *single purpose*. Thus, all pumps, storage vessels, pipes, valves, controls, and supply sources must be for the exclusive use of fire protection only. This theme will be repeated in Chapter 10, "Fire Protection Systems," but it is appropriate to report it in this chapter during this discussion of water storage facilities.

Standby emergency electrical generators can also be included in the achievement of an almost "fail-safe" water system. Although it is desirable to have a water system that is operable 100% of the time when there is total loss of primary power, the winery still cannot function unless the generator or generators can provide sufficient power to service the principal pieces of equipment and the lighting circuits as well. Therefore, it can be seldom justified on an economic basis because of the large first cost for the installed electrical load capacity needed to operate even a small winery (equal to or less than 40, 000-cases annual production).

Water tanks are manufactured in all sizes and shapes and are fabricated from a vast array of construction materials. Some wineries have made use of salvaged wine tanks and a generous coating of polyurethane foam to offer some insulation from temperature extremes and to soften the appearance of the highly reflective surface for a more visually integrated structure in a rural landscape environment. Although they are generally adequate for the purpose intended, their resistance to seismic forces should be examined by a structural engineer, including competent anchorage to the concrete base slab on which they are to be founded. Table 6.11 is listing of tank materials and their physical and operational advantages and disadvantages.

Table 6.12 is a handy reference for selecting the best geometry of a round tank to give the required water storage volume. The volume given for each diameter in U.S. gallons is for 1 foot of tank height. Obviously, structural

Table 6.11. Summary of Materials of Construction for Water Tanks

Tank material	Effects on water quality	Corrosion resistance	Structural integrity	Remarks
Mild steel	None, if properly painted and maintained	Good, if properly protected	Excellent	Can be bolted or welded; can be buried
Concrete	None, if properly painted and maintained	Good, if properly protected	Excellent	Usually cast in place; can be buried
Fiberglass	No coating required, if FDA approved resins used	Excellent	Excellent	Light and can be buried; up to 40,000 gal (151,400 L.); highway transportable
Wood	Slimes and algae tend to prefer wood substrate for growth	Excellent	Good	Erected in place on concrete slab base
Stainless-steel wine tanks	None	Excellent	Good	Usually come equipped with all fittings, valves, and vents; need to add emergency overflow outlet
Cross-link polyethylene	Must specify FDA-approved interior coating	Excellent	Good; some loss in properties from UV radiation over time	Limited to about 16,500 gal (62,453 L.) of storage for single tank; can specify multiples connected in series

problems begin to emerge if the tank height becomes too large in relationship to the base diameter. Providing base rings (lower sectors of the tank wall) with sufficient cross-sectional area to resist buckling during a seismic event becomes non-cost-effective. If you are in the planning stage for a winery water system, and you have determined the storage volume that will accommodate the winery's needs, select several combinations of tank construction material, height, and diameter, and obtain quotations from local tank fabricators. You may find just the tank you require at a bargain price, sitting in some supplier's storage yard, just waiting for a home.

Very strict federal and state laws are now in force, which require the guaranteed integrity and bottle-tightness of underground storage tanks for petroleum products and other substances, which present a pollution threat to groundwater supplies. On a national scale, the quantity of salvaged, used, steel storage tanks is impressive. Although some are for use only as

Table 6.12.   Cylindrical Tank Volumes per Foot of Height

Tank inside diameter (ft-in.)	Area of base in sq. ft or volume in cu. ft foot of tank height	Volume per foot of tank height (U.S. gal)
1-0	0.785	6.87
1-6	1.767	13.22
2-0	3.142	23.5
2-6	4.909	36.72
3-0	7.809	52.88
3-6	9.821	71.97
4-0	12.566	94.00
4-6	15.900	118.97
5-0	19.530	146.88
5-6	23.760	177.72
6-0	28.270	211.51
6-6	33.180	248.23
7-0	38.450	287.88
7-6	44.180	330.48
8-0	50.270	376.01
8-6	56.750	424.48
9-0	63.620	475.89
9-6	70.880	530.24
10-0	75.540	587.52
10-6	88.590	640.74
11-0	95.030	710.90
11-6	103.870	776.99
12-0	113.100	846.03
12-6	122.720	918.00
13-0	132.730	992.91
13-6	143.140	1070.80
14-0	153.940	1161.50
14-6	165.130	1235.30
15-0	176.710	1321.90
15-6	188.690	1411.50
16-0	201.060	1504.10
16-6	213.820	1599.50
17-0	226.930	1697.90
17-6	240.530	1799.30
18-0	254.470	1903.80
18-6	268.800	2010.80
19-0	283.530	2120.90
19-6	293.650	2234.00
20-0	314.000	2348.72

scrap metal, others which were installed within the last 5–8 years that were carefully painted and/or equipped with "sacrificial zincs" to prevent galvanic corrosion are available in some locales through tank fabricators. As a case in point, the author utilized two 40, 000-gal, salvaged, steel diesel fuel tanks for a winery's fire protection storage which had been in service for only 3 years. Sandblasting, painting with epoxy paint inside and out, the installation of larger-diameter inlets and outlets, air vents, and an access manhole for inspections, including transport to the winery site, all totaled considerably less than the cost for new steel or fiberglass tanks of equivalent size (1987 price levels). For older salvaged steel tanks that may be available in your particular area whose provenance is not so clear, third-party inspection of the tank by a competent welding–corrosion specialist should be performed. It would be wise for the winery to prepare a preinspection agreement to provide that the inspection cost can be shared between tank seller and the winery if the tank proves to be sound in all respects. The seller would absorb the cost, if it fails the examination.

One final case-study note on water tanks: During the winery utility planning phase on a steeply sloping parcel, topography would allow gravity flow from a storage tank to the winery if the tank was located above the winery in an environmentally fragile watershed. To preclude disturbance of the soils and destruction of a dense coverage of specimen, oak and maple trees with a 12-ft wide access road nearly 1000-ft long, a so-called bolted steel tank was specified, whose metal staves and other components could be skidded up the slope with a small, track-laid bulldozer. The amount of earth disturbance was minimized and no trees had to be removed. The well supply inlet and discharge piping to the winery and the level control signal cable were laid in a common trench excavated using a small, self-propelled trenching machine. Bolted steel tanks have their origins in the oil fields and fill a need for readily assembled crude oil stock tanks (American Petroleum Institute Specification 12B, Bolted Steel Tank). Their use as a component in potable water supply systems is also approved and covered under American Water Works Association Specification AWWA D103-87. There is one additional advantage to the bolted steel tank units, and that is, the method of factory painting of the stave sections. Field-welded steel tanks must have both primer and finished epoxy coatings applied in the field, where temperature, moisture, and fugitive dust are not easily controlled. Coatings for bolted tanks which receive shop application can receive warranties for up to 10 years (83).

Once you have retained a utility design professional, offer the tank options you have developed from the foregoing guidelines and seek an opinion on what will serve the winery best with your budget, foundation, and seismic zone status.

Beside the tank shell and the foundation, other important elements are required to make a complete storage system. Water storage tank level controls and alarms can be of several types:

1. float switch
   (a) Ball and rod
   (b) Rotating, floating ball
2. Flexible diaphragm
3. Open/closed electrode
4. Altitude valve

All of the devices that can be employed to regulate the storage volume in the reservoir have been designed and constructed with a high degree of simplicity and of operational reliability. There are some cautions that should be highlighted in approving one particular control system over the other:

1a. The float is nothing more than a float tethered to a mechanical arm that moves either in a radial direction for small tanks with small changes in volume or in the vertical direction with the rod passing through one or more guide rings as the attached float moves up and down. The attached rods in both cases are mechanically connected to a waterproof toggle switch that either energizes or disconnects the electrical power to the source pumps. In larger installations (in excess of 5 horsepower), the off–on command would be transmitted at a lower signal voltage to the motor starter/breaker in the motor control center panel close to the source pump, which, in turn, would activate or inactivate the line power to the system.

1b. The rotating ball differs in that although it depends on buoyancy of the float as in the two float–rod devices previously described, the rotating float type is connected to a flexible cable. As the water level rises, the float, as it is submerged, is inverted by buoyant forces through 180°, where a mercury switch contained within the watertight float opens. It disconnects the source pump power or provides a milliamp (mA) signal to deactivate the motor through the starter/breaker as in the previous example of a large power installation. If there is any explosion hazard associated with the stored liquid (sewage wetwell with chance of methane gas production or the like), the rotating float switch cannot be used.

2. The flexible diaphragm switching mechanism employs one or more preset pressure transducers or load cells that generate a 4–20-mA signal to a master control panel to either start or stop a pump supplying water to the storage tank. A third pressure cell is often added 6 in.–1 ft above the normal high water level cell to act as a high water alarm in case of an electrical malfunction in the control network.

3. The dual electrode has a shortcoming in that for water with a high total salt content or with individual cations or anions that were ionically unstable, the electrodes tended to become coated with mineral deposits over time, making them less than an ideal critical control component. As with the pressure cells, the electrodes were set at the desired "on" and "off" locations within the tank. As the water made or broke the connection between

the two contacts of the electrode, the source pump was either energized or inactivated. The third electrode was also generally added as a high water alarm, as described in item 2.

4.   The altitude valve is a time-tested waterworks system control device that, with the proper plumbing, pressure switches, and thermal protection, can be an inexpensive and reliable water control element. For wineries with storage tanks strategically placed for gravity flow to the production spaces, a multidirectional flow system is possible through a single pipeline to the tank from the source pump with a branch supply line to the winery. Figure 6.11 depicts a possible schematic arrangement of source pump, storage tank, and the necessary plumbing accessories to make the system operate properly.

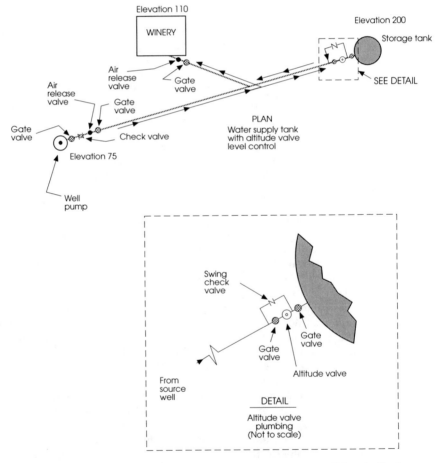

Fig. 6.11.   Concept plan for winery potable water system storage tank (gravity flow).

Examining the operating logic for Fig. 6.11, it can be seen that with the source pump not operating, the flow from the storage tank to the winery, bypassing the closed altitude valve, is possible. The check valve at the pump prevents the full static pressure at shutoff from damaging the pump, whereas the bypass permits flow to the winery on demand. When the pump receives the signal to energize because of the lowered static level in the tank, the altitude valve opens and the source pump fills the tank to the preset level, which, in turn, closes the altitude valve. During the pump operating cycle, the winery could still draw water from the system up to the capacity of the well pump. Thus, prudent design would have the capacity of the well pump exceed the maximum winery demand (assuming the safe yield of the well would permit), so that on each pump cycle with the winery in operation, some gain would be made in restoring the maximum operating volume of the tank. Note the placement of air release valves near the well pump and at the supply line entrance to the winery. Those are important plumbing fittings to prevent pneumatic blockage from entrained air.

The following potential operational problems associated with altitude valves should be noted:

(a) Altitude valves are susceptible to freezing because of the small water passages and tubing within and outside the valve body that are used to transmit pressure signals. An insulated valve box or a valve box with a resistance strip heater in very cold climates would solve the temperature problem.

(b) If the source well has a propensity to periodically discharge sand particles into the system, a strainer should be installed downstream of the altitude valve to entrap the solids.

High water level alarms were mentioned in connection with several of the choices for level control. (Tanks are always equipped with an emergency overflow pipe just above the maximum design water level). The alarm can be remotely transmitted to the winery or the automatic telephone dialer console previously described in the discussion of well-pump-savers, thus saving precious water. Low water level alarms are also sensible, if the tank is out-of-sight and no well-pump-saver alarm device is installed.

Most commercially available tanks (with the possible exception of the wine tank to water tank conversion previously discussed) have an emergency overflow. This screened outlet should be plumbed with suitable piping away from the tank base, hopefully to a natural drainage course. Some type of energy dissipation system should be installed at the outlet end of the emergency overflow pipe to prevent soil erosion and gullying of the stream channel. Rock aggregate or concrete rubble serves the latter purpose well. The tank should also be vented with a partial perimeter strip that is tightly covered with 150-mesh stainless-steel or copper screen to keep insects and other fauna from entering the clean-water holding tank.

Tank paints and coatings are essential for protecting all tank materials except possibly stainless steel and fiberglass (see Table 6.11). Even for those two materials, good landscape practice will suggest that if the storage tanks are in a prominent location, then they either be painted with a viewscape-compatible color or shielded from view with living plant materials as vegetative screens. Painting the winery's name and label logo on water tanks has also been tried, but such requests usually run counter to many current California signage ordinances. Sign regulations in other areas of the United States and outside the United States may be more lenient. Elevated water tanks (probably more frequently used in fire protection systems than potable water supply) have been observed throughout the winemaking world by the author. On their spindly, structural steel legs, the elevated tanks defy landscape-compatible painting and/or landscape screening, so that the use of the tank walls for advertising space somehow seems more appropriate.

The VOCs (volatile organic compounds), which are the drying and spreading agent in solvent-thinned or alkyl-based paints, will probably not be manufactured after 2000. The U.S. Clean Air Act and its 1991 amendments attempt to control fugitive volatile chemical emissions and to prevent further degradation of air quality. As identified earlier in this chapter, factory-applied epoxy coatings for tanks under tightly controlled conditions are warrantied for up to 10 years (83); thus, the repainting chore and attendant costs for tank maintenance touch-ups are widely spaced in time. To specify anything but an epoxy-based paint for a new steel water tank would be imprudent and economically disadvantageous. The principal variable in epoxy paints is the curing compound that is used. Whereas polyamide resins are the most widely used and produce the most durable finishes, water-based versions are now on the market. Some "chalking" or dulling of the epoxy paint occurs naturally and should be planned for in the finally desired architectural appearance of the tank's exterior. If a so-called high-build epoxy coating is specified [can be as much as a 40-mil (1 mil = 0.001 in.) dry film thickness applied in a single coat for flat and semigloss epoxy finishes only] (84) and under the worst conditions, chalking loss might approach only 4 mils in 12 years. Thus, the durability and life-cycle economy of epoxy coatings for steel tanks cannot be argued. As will be suggested in later chapters, equipment which may be located in other hostile environments within the winery utility complex can also benefit from reduced maintenance if polyamide epoxies and their closely related paint cousins, coal tar, ester, and polyester cured epoxy, are judiciously used. Added surface hardness, increased alkali acid resistance, increased adhesion, and variable drying time are some of the beneficial properties that can be reinforced with the proper class and concentration of the admixture. The winery's architect and utilities engineer will, as part of their design responsibilities, specify the paints and coatings for a new or expanded facility. For future maintenance and touch-up painting, the winery's master operation and maintenance manual should list the paints and primers applied at the time of initial construction so that no paint incompatibility will result (e.g., placing alkyd-based paint over water-based latex can produce a bubble-packed, chemically antagonistic coating that lasts in the best of circumstances for 1 or 2 years) (85). A competent industrial painting contractor can also guide the selection of proper paint for future exposed water system maintenance tasks. Paint technology continually advances, not only in lowered air emissions or in an environmentally friendly sense but also in offering to the industrial market longer-lasting paints to accommodate a variety of utility maintenance needs. Thoughtful winery maintenance supervisors are well advised to stay abreast of new and improved paint products.

WATER TREATMENT

Complete water treatment is an expensive undertaking for a winery that is dependent on a surface water source for potable purposes. "Complete" is defined as coagulation, sedimentation, filtration, and disinfection. Large municipal water treatment works contain the latter elements, as do the minitreatment systems that are factory assembled as a complete operating unit and shipped with the proper plumbing and electrical connections for rapid field installation. On the negative side of water treatment is the necessary commitment by the winery to provide skilled (hopefully certified) operators to monitor the performance of each of the critical subprocesses and through periodic sampling and analysis, to verify that the finished water meets or exceeds the purity standards set forth by the local environmental health and safety code. Less complex water treatment systems such as ion-exchange units for softening and/or iron removal can be serviced and maintained by equipment suppliers. Contract maintenance can also be obtained for complete demineralization systems such as reverse osmosis. The federal/state safe drinking water statutes, which specify benchmark purity levels, were previously discussed and presented as Tables 6.1 and 6.2. The professional engineer retained by the winery to design and/or specify the treatment equipment, will ensure that the equipment is capable of meeting or exceeding the required chemical and microbiological purity standards for potability. An additional problem that must be anticipated in the operation of the complete treatment system is that the treated water that is used to periodically flush the sand or mixed media filter (during a backwash cycle) is heavily laden with organic and inorganic solids and microorganisms and must be dealt with as a liquid waste product with the potential for stream and groundwater pollution (86). Softening columns and other ion-exchange units produce a brine discharge which may require special handling. Diverting this undesirable salt load from any water source that is earmarked for vineyard or other irrigation is the sensible thing to do.

Conventional versus Package Treatment System

For wineries, large and small "package" or factory-assembled treatment units are the system of choice, if the operational staffing, monitoring vigil, and management responsibility required of such units is accepted and followed throughout the life cycle of the system. During the period 1978–1983, package treatment systems (both wastewater and potable water) received serious criticism from state and local regulatory agencies in the western United States because of their spotty performance records. The problem was not so much with soundness of the electrical/mechanical/hydraulic components as it was the lack of attention given to the systems, once installed. Package treatment system suppliers have rectified this shortcoming and almost all major companies can now offer a service contract to provide the following:

1. Maintenance and periodic servicing or;
2. Complete operation and maintenance services

Culligan International Company has taken the monitoring/maintenance requirements for a demonstration package water treatment system several steps further by installing an automatic data sampling and analysis system, which monitors among other things, the following:

- Raw and finished water turbidity
- Chlorine residual
- Pressure drop through the filter
- Constant levels in chemical feed tanks
- Flow rate and total gallons produced

This control system would be extraordinarily expensive and an installation for a winery (even a well-capitalized winery) could not be justified. It is, however, an excellent illustration of the degree of sophistication to which a package water treatment control system can be raised. In the demonstration project, the data collected by the Culligan autocontroller are transmitted automatically by computer modem to the local water treatment equipment supplier (responsible for the operation of the Culligan treatment unit) and to the Culligan R&D laboratories in Northbrook, Illinois (70, 87). This feature has allowed the State of California to relax the statutory monitoring requirement for this demonstration system from daily to weekly. The manufacturer reported an average of about 3 h per week to maintain and operate the system for 1 year (87). The system was installed and operated as a demonstration project under the U.S. EPA's Small System Technology Initiative Program with the cooperation of the State of California Division of Environmental Health. Cost and space savings in the utilization of package water treatment systems at several locations throughout the United States have been reported (88). Wineries which have a serious interest in this treatment technology should work closely with their utility design engineer and have them obtain specific cost and performance information for the raw water supply that prevails at the winery location.

Problem Minerals

Problem minerals for wineries have been discussed earlier in this chapter. Iron bacteria (subsection entitled "Well Casings") and iron removal and specific ions or cations which could adversely affect wine quality (Table 6.3, subsection "Water Quality Assessment") were identified and, where appropriate, possible treatment methods were briefly discussed. Iron, which can encourage the growth of nuisance iron bacteria throughout the water system's conduits and storage tanks and create metallic tastes in wines, can be removed by an ion-exchange process which uses a manganese greensand matrix that exchanges the ferrous iron for sodium. Superchlorination and dechlorination with sulfur dioxide for iron removal is a method used historically in large municipal systems with chlorine contact chambers and liquid chlorine delivery systems to make the process work successfully. With this method of precipitation, ferric chloride must be removed downstream during one of the remaining treatment steps (coagulation, sedimentation, and filtration). Figure 6.12 shows a typical green-sand ion-exchange system that can be used to remove either iron or manganese with available flow rates and tank diameters noted.

Other problem water-quality constituents which may interfere with winery use (i.e., bicarbonate and copper) would require the use of a combined anion/cation-exchange system or a reverse osmosis (RO) system. These efficient demineralizers use a membrane that is selective for certain mo-

Fig. 6.12. Typical green-sand ion-exchange system for iron or manganese removal. (Courtesy of Culligan International.)

lecular weights. If raw water turbidity is maintained at a low level (≤ 0.5 JTU) by a prefilter, the RO units work efficiently and are able to totally demineralize up to water with ocean salinity or about 34, 000 mg/L total dissolved solids. They require a modicum of attention, as the pump pressures are high and the membranes can be blinded with suspended solids if the prefilter becomes clogged or is not performing properly. Production flow rates are very low, and the low dissolved oxygen, 5-6 pH, and total absence of minerals creates a finished product that is pure, but low in taste and aesthetics. Some RO units are configured with several stages or membranes, each designed to retain molecules of a particular size. Figure 6.13 depicts the general arrangement of a typical reverse osmosis unit. As suggested previously, periodic flushing of the system would require special handling for the strongly saline waste minerals discharge.

Hardness removal methods are familiar to almost everyone, because of

Fig. 6.13. Typical reverse osmosis unit for demineralization and specific ion removal. (Courtesy of Culligan International.)

Fig. 6.14. Typical softener, dealkalizer or nitrate reduction unit. (Courtesy of Culligan International.)

their widespread use in residential applications. Hardness in the winery's water supply is acceptable except possibly for the water required for boiler feed use and general cleaning (see Chapter 4). Thus, softening the entire water supply is not only expensive, but because the calcium or magnesium-ion is exchanged for the sodium-ion attached, which is ionically bonded to the zeolite resin in the exchange column, the additional sodium in the water supply adds another increment of unwanted sodium to the human body if the softened water is used for drinking and culinary purposes. Therefore, unless the total calcium/magnesium hardness is in excess of 250 mg/L, only the winery's boiler supply should be softened. For new wineries where potable water supply plumbing can be prudently designed, designated, softened-water, pipe runs can be installed and softener columns or tanks placed outside the building envelope for easy servicing and maintenance. Figure 6.14 shows an industrial-scale water softening unit that has the capability, with certain accessories, to also reduce excess alkalinity and nitrogen as nitrate.

Disinfection

Disinfection was discussed briefly in the subsection entitled "Conventional Versus Package Treatment Systems." Chlorine in some form is probably the most widely used of all the disinfecting agents available to make water supplies bacteriologically safe. The most attractive feature of chlorine disinfection chemistry is its ability to retain a degree of microbial killing power (so-called chlorine residual) for a period of time after the initial chlorine dose. The chemical reactions for chlorine in water can be seen in the following equation (89):

$$Cl_2 + H_2O \rightarrow HOCl + HCl$$
$$HOCl \rightarrow H^+ + OCl^-$$
$$HCL \rightarrow H^+ + Cl^-$$

To be effective, chlorination must be accomplished under a rather restrictive set of conditions:

(a) The water must be virtually free of turbidity [≤ 0.5 (J.TU) or Nephelometer Turbidity Units (N.T.U.)—see Table 6.1].

(b) pH must be as close to 7.0 (neutral) as possible.

(c) The other phases of the complete treatment process (flocculation, coagulation, sedimentation and filtration) must be performed in order to eliminate all potential pathogens.

(d) A chlorine contact tank or chamber must be provided, sufficient in size and volume, to provide a minimum of 0.5-h chlorine-treated water contact time at maximum flow for chlorine reaction products (i.e., hydrochlorous acid and chloramines to inactivate pathogens).

Chlorine gas can be fatal in concentrations as low as 1000 ppm for 5 min. The State of California allows a maximum concentration of 0.5 ppm in the work place for an 8-h exposure with a maximum of 1.0 ppm for any 15-min. period (56). Safety in the handling of chlorine gas must be emphasized by winery management, and all the proper precautions taken to prevent accidents. The winery's Accident Prevention, Health, and Safety Plan should contain a special section on procedures, that should be followed, should a gas cylinder leak occur, or if someone is overcome from a damaged head valve or a gaseous chlorine line is accidentally broken. Industrial safety orders for the winery's particular locale will specify the number and location of safety masks to be used for rescue or repair activities. (see Chapt. 12) One final shortcoming in the use of chlorine is that secondary organic compounds, called halogenated organics, are sometimes formed when minor concentrations of suspended organic matter are present in the treated water. Halogenated organics are suspected of being carcinogenic (90, 91). This might be a possible occurrence even downstream of the final filtration step if, for example, phytoplankton (algae) were allowed to grow in the water storage tank.

Other uses for chlorine in a winery become important when there is a sewer blockage, and overflow of sanitary waste in working spaces, roads, or walkways. Chlorine in a powdered form, high test hypochlorite (HTH), can be broadcast in the spill area to inactivate any pathogens that might be transported by foot traffic in and around the winery. Although most pathogens are subject to high mortality from desiccation and ultraviolet radiation, their immediate destruction with HTH is the safest choice.

The other options for potable water disinfection fall short of chlorine in providing the residual microbial killing power to prevent secondary contamination of the supply. The disinfecting agent options are, in order of preference, as follows:

	Disinfectant	*Shortcomings*
(a)	Ozone	One and one-half to two times as expensive as chlorine and no residual microbial killing power
(b)	Ultraviolet radiation	No residual microbial killing power
(c)	Gamma radiation	No residual microbial killing power
(d)	Chlorine dioxide	Produces "off" tastes and color
(e)	Pasteurization	High consumption of energy; no residual microbial killing power; not practical for water supplies
(f)	Ultrafiltration (membrane)	A filter membrane that would have a pore size sufficient to entrap bacteria and protozoans would allow viruses to pass

Disinfection will be discussed again in Chapter 8. The main differences between potable water and treated sanitary sewage effluent disinfection will be seen in the relative titers of potentially infectious organisms in each and the corresponding larger quantities of chlorine required to render sewage effluent benign enough for discharge to receiving waters or for recycling and irrigation reuse.

WATER CONSERVATION

The vagaries of the hydrologic cycle worldwide create water supply shortages from time to time. It is during these drought periods when water conservation practices force wineries to hone their water saving practices to a fine edge. Both surface and groundwater supplies are affected, although it is the surface system (rivers, creeks, and storage reservoirs) that exhibit the most dramatic, shrinking, visual impact. Groundwater basins and aquifers are hidden from view and the lay-viewpoint is often that these unseen, underground reserves have an infinite volume upon which users may draw. In an extended drought, such as California's 1987–1993 experience, groundwater levels in almost all basins fell at a precipitous rate when annual replenishment was in the order of 30–40% of normal (average) for nearly 7 years. Even though groundwater withdrawals were curtailed state-wide during the drought, the "mining" of water had a long-term impact on its chemical quality (less dilution with freshwater recharges). Concurrent and continuing lowering of water tables is the norm, as of the publication of this book, and most major California groundwater basins (Sacramento and San Joaquin) have not shown any signs of recovery (92, 93).

Forced water conservation that is adopted to cope with an imbalance in supply over demand, should not be the principal reason for wineries to embrace a well-structured program (76). The potential savings in operating costs and general overhead should be reason enough to get the attention of any winery manager, no matter how abundant the local water supply may be. When conservation plans have been seriously undertaken by wineries, water use savings of 20–25% are often easily attainable. The section of entitled "Water Meters" discussed their importance to sensible water system operations. Meter records are the only way that the success of a water conservation program can be evaluated. All of the successful programs to save water have a number of things in common:

(a) Water leak and water fixture inventory
(b) The assignment of a responsible, senior cellar person to head the program at the working end of the winery
(c) Monthly meetings of cellar staff to exchange ideas on methods of improving winery cleanup and wash-down protocol and to publish comparative (preconservation and postconservation program) water use figures from month to month for the same process step (bottling, racking, filtering, etc.)
(d) Use of posters and other graphic reminders near hose stations, lunch rooms, and restrooms to maintain the presence of the water conservation program at all locations where major water consumption takes place
(e) Cash offers or other incentives to employees to achieve preset goals of water conservation

Item (a), the water leak/fixture inventory, requires some explanatory guidelines on how a winery should proceed with that important first step. Wineries that were built 10–12 years ago did not have the choice of purchasing well-constructed and reliable low-water use or mini-water-use bathroom fixtures. Considerable progress has been made in the design and manufacture of plumbing fixtures that use far less water but, still perform their functions in a sanitary and hydraulically efficient manner. The poor performance of the early versions of low-water-use fixtures resulted from the need for filling the marketplace with fixtures that complied with federal and state laws mandating water conservation fixtures in new construction. Because of time constraints, the newly designed units were not rigidly tested by the generally accepted third-party testing institutes such as the National Sanitary Foundation (NSF) and ANSI, the American National Standards Institute. Thus, if winery management finds that its bathroom fixtures are high water consumers, they can be replaced over an extended budget period with low-consumption or water-saving units, so that the financial impact can be spread over several fiscal years. If this fixture replacement option did exist as a choice, a simple economic analysis could show that the savings in water supply and wastewater treatment costs would offset the capital cost of new fixtures in a very short time. (A 25% saving in pumping power for both well and booster pumps at 1996 Northern California energy rates of about 12¢ per kilowatt hour can balloon the water conservation benefits rather quickly.)

One final issue needs to be raised regarding the use of reclaimed and recycled wastewater as a method of reducing the overall winery demand for nonpotable water use. The water conservation plan should contain a section on the conjunctive use of all available water (primary drinking and sanitary use, vineyard and landscape irrigation, fire protection, and water amenities such as fountains and duck ponds). In the preparation of the winery's conservation plan, if all of the uses are arrayed in matrix form with quantity (both rate and total annual volume) and quality (chemical and microbiological purity, temperature, and appearance) listed as specific beneficial use standards, it will be readily apparent, if water exchanges can be made to reduce the total demand for the external water supply. For example, the almost continuous blow-down from the water-cooled, evaporative condenser (see Chapter 5) is often the least expensive method of condensing the refrigerant for the winery's process cooling equipment, but represents a fairly large volume of saline water discharge. In practice, once the salinity of the cooling water reaches a preset value, the water is discharged from the cooling unit and fresh makeup water is added. As suggested earlier in this chapter [see Table 6.5, item (5)], a medium-sized winery could produce 5–6 gpm (0.33–0.4 L/s) of saline water that might have reuse potential for the following:

(a) Fire protection
(b) Rotary drum screen spray-bar water supply
(c) Closed-cycle landscape fountain use

The foregoing reduction possibilities must be balanced against the feasibility and favorable economics of implementing the concept plan. The additional plumbing, controls, and regulating storage to achieve the conversion can become complex, so at this point in the conservation planning, professional engineering help should be sought to carefully document the costs and consequences of implementing the plan.

PIPING, PIPE MATERIALS, AND SECONDARY CONTAMINATION

After all of the care taken by the winery management to ensure a potable and safe drinking water supply, the possible contamination of water from the conduit materials and fittings within the winery should be considered. Thirty years ago, the analytical instruments, which can now detect harmful chemical constituents to the parts per trillion realm, existed only in a few university research laboratories. The battle against lead contamination in the human environment has been going on since the 1950s, when lead-based paints were phased out of the commercial market. Following closely behind were bans on the use of lead pipe in plumbing, leaded gasoline and lead–tin solder used for "sweating-on" copper pipe fittings. From 1985 to 1990, the focus of lead contamination reduction programs at the national level was on lead in public school water supplies (lead-lined drinking fountain cooler tanks and copper plumbing joints and fittings) (94, 95). Because the leaching potential of lead and other organic contaminants from plastic pipe joint compounds is a function of time, the short-term flushing of water lines before use will solve 99% of the problem, even though such a practice is counter to desirable water conservation behavior described in the previous subsection. After a weekend of inactivity in the winery, a simple procedure would have cellar personnel direct wash water to a floor drain for as little as 2 or 3 min to flush any leached metals and toxic organics (96).

Pipe distribution systems and storage vessels, that are to be installed for a new or expanded potable water supply, must be disinfected prior to use. The utility engineer's specifications will generally reference the American Water Works Association Standard Specification No. C196-87 (or its current version), which provides guidelines on the concentration of chlorine to be maintained in the storage tank (and pipe network) and the contact time required (generally 24 h). The disinfected system then needs to be thoroughly rinsed prior to being placed in service. Samples drawn from several locations in the winery must be submitted to a certified public health laboratory for routine microbiological analysis. If the sample results are within the narrow band of the allowable MPN (most probable number) for coliform group organisms, the local environmental health officer will certify the system. Thereafter, periodic testing of the water supply will be made part of the conditions attached to the winery's water system permit.

The federal Safe Drinking Water Act and its equal or superior state equivalent statutes have been cited previously in this chapter. The provisions of these laws give the U.S. public the best protection from waterborne disease of any country in the world.

BOTTLED WATER SERVICE

It is a simple task to rankle the conscientious and dedicated members of the community of municipal and special district water purveyors by the mere mention of bottled water. Using bottled water shows a certain lack of faith in what public water utilities produce at the faucet. Nonetheless, the popularity of bottled waters has grown dramatically in the last 10 years. It is less a lack of faith in public water supplies than it is the need to be able to:

(a) Drink water that has been designed not only to be safe but to be very palatable,

(b) Use water with a selected list of minerals that produces the best cup of coffee or tea that is possible to brew,

(c) Purchase water that is without minerals (deionized or distilled) or contains minerals in the precise amount (fluoride) to produce cavity resistant teeth in children.

Bottled water may have its place in a winery for boosting staff morale, such as hot- and cold-water dispensers near several work centers, where beverage breaks or just a cold drink can be made easily accessible to cellar hands, precluding extended walks to lunch rooms and loss of valuable production time. At approximately $5 per 5-gal carboy and $7 per month for hot/cold dispenser lease (1996 price levels), the potential benefits far outweigh the minuscule cost. The only negative aspect presented by this system is the rather enormous dead-weight lift required to center the inverted 45-lb carboy over the dispenser; however, for warehousemen at least, who are used to the 37-lb, 12–750-ml-bottle wine case, it should not be a problem.

STANDBY EMERGENCY POWER

Where an historical pattern of power outages, brownouts, and single phasing occurs, a standby power system should be considered to permit the operation of a portion of the winery's critical equipment components.

Included in that selected list are the winery's well pump and booster pump. Power outages are seldom extended (generally repairable damage to line transformers, transmission lines, or summer cooling/air-conditioning overloads) and usually last less than a working day. With a nongravity water supply, the booster pump is essential to winery operation particularly at crush time when the arrival of grapes for processing cannot be easily controlled nor delayed for many hours. In larger wineries, the feasibility of standby power becomes less practical. The built-in capacity in kilowatts required for operating the water system, crusher, press, and transfer pumps for a 200, 000-case annual production winery can become very expensive. Thus, for larger wineries, an emergency power loss plan should be prepared that would provide advance physical and financial arrangements for shipment of grapes to nearby wineries for preliminary processing with eventual transfer of product back to the winery upon restoration of power. Almost all other process steps in the annual table wine cycle, at least, could be interrupted for a short time without great harm to production schedules or to end product.

The major components of an emergency standby generator set are as follows:

1. Diesel, propane, or compressed natural gas (CNG) powered prime mover (engine)
2. Direct-coupled generator set
3. A transfer switch

An essential feature of the emergency power unit is a controller that starts and tests the generator at least weekly to ensure that the system will operate and pick up the emergency electrical load when, and if, it is required to do so. Artificial loads such as a resistance bank can allow the unit to come up to speed and actually produce power for a specified period of time before shutting down. The controllers also provide records of the performance of the unit on its periodic test run. Propane or CNG are the fuels of choice because of their stability over long periods of storage. Additionally, the federal and state requirements for the design and fabrication of underground storage tanks (USTs) for liquid petroleum fuels favor aboveground pressure vessels for propane or CNG.

In a retrofit situation, once the critical equipment motor capacities and their kilowatt equivalencies have been determined (possibly with a nominal additional capacity allowed for emergency lighting), the cost for the emergency standby power unit can be roughly estimated by contacting a portable generator set equipment supplier and asking for a quotation, indicating fuel preference, controller, and transfer switch. At this juncture in the planning process it would be prudent to discuss the idea with a professional electrical engineer, who can fine-tune the design concept and prepare bid documents, if a competitive bidding process is going to be undertaken.

There is one final point that needs to be raised on new winery site suitability studies and the need for emergency power, which was previously suggested in Chapter 2. The electrical utility that serves the area where winery site suitability studies are being made must, by law, keep a record of power outages, brownouts, and single phasing of three-phase power which are public record. If the utility is reluctant to provide such information, the Public Utilities Commission in the state in question may bring mild pressure to bear to have the data provided to the winery or the architect/ engineer team.

ECONOMICAL OVERSIZING OF WATER SYSTEM COMPONENTS FOR FUTURE EXPANSION

Although you may expect your design engineers to build in a small cushion of additional hydraulic capacity into your winery's water system, it is likely they will not do so without specific instructions from the winery principals.

We must go back to a previous section in this chapter entitled "Water Demand Calculations" to get the oversizing concept in its proper perspective. First, any oversizing of the system that would be counter to the approved use permit winery production capacity, might be difficult to discover in design review, even with a microscopic examination of the engineer's design criteria and calculations. Faced with a winery plan that suggests future expansion, but which has a land use permit (approved environmental impact report, approved public input on traffic, water use, and site plan), and winery capacity that falls short of management's future expectations, the winery's engineer should vigorously lobby the local design review agency for oversizing at least the major pipelines, so that at some unspecified time in the future that capacity can be economically utilized without resorting to expensive and environmentally disruptive post-winery-completion trenching and pipe laying.

Also, the design engineer should be instructed to offer to management the cost and incremental design effort to incorporate space, plumbing and electrical service features to permit, at some future time, additional units of booster pump, storage, or treatment capacity without major modification of the initial system layout. A perfect example of unplanned capacity is in electrical panels, where the electrical system designer by an act of good design sense, rather than intentional capacity increase leaves spares or blanks in breaker panels and motor control centers. In this same vein, electrical conduits which are tightly sized, only for the conductors required at the time the needed wires are pulled can be oversized at very little incremental cost to accommodate additional wires. The same logic applies as described previously with the major water transmission lines (i.e., future expansions which involve underground installation are going to be expensive and destructive to earth and hydrologic resource).

"Modular design" is an old concept that still carries a high degree of sensible engineering economics. Its basics involve the use of measured units of capacity, that can be enlarged and retrofitted to plumbing and electrical networks for which the future addition has been carefully thought-out in the initial design. Winery management must stay actively involved in the utility design process, so that their future needs will be met with the least expenditure of future design and construction dollars, governmental interfacing and administrative effort.

WATER SYSTEM OPERATION AND MAINTENANCE

Throughout this book, the basic theme of sound maintenance practices that involve good record keeping and unwavering adherence to manufacturers' recommended maintenance schedules is repeated. Such is the framework of good preventative maintenance, which, if carefully applied, can reduce to a very low statistical probability major equipment failures and costly production downtime.

Just as the mouth and nose of a good winemaker become the most important of tools in producing award-winning wines, so do the eyes, ears, and nose of the water system operations technician often give the clues that something is in the process of malfunctioning. Daily visits, (even more frequently with the system operating at full capacity), to booster pump stations, storage tanks, well heads, and treatment works are essential and good practice, if only to allow the senses to help anticipate a problem. The observer can eventually detect "off-noises" from worn or improperly lubricated bearings and from cavitating impellers. Touching a motor housing or bearing cover daily gives the maintenance person a temperature baseline that if warmer than the norm, signals a potential problem. The winery's maintenance supervisor can develop a checklist of simple observations to complement the manufacturer's periodic lubrication and troubleshooting schedule. Pump gauges are also useful instruments to monitor performance. If discharge pressure decreases appreciably over time, the likelihood of a worn or pitted impeller is predictable and provisions can be made to have the pump overhauled in a noncritical segment of the annual production cycle.

Wineries with package water treatment systems have the biggest maintenance burden of all and, as was suggested previously (in the section titled "Water Treatment"), the adoption of a subcontractual maintenance agreement with the treatment works equipment supplier may be the optimum choice for small and even large wineries, where a single maintenance technician's responsibilities are spread too thinly to permit the careful attention required of the complex treatment units.

Enlisting the help of someone from the winery's laboratory to be responsible for the routine water sampling relieves the maintenance technician for other tasks, and gives the job to someone, who is specially trained in sterile method procedures.

Spare parts lists and the machinery history cards and records can easily be integrated into the winery's computer system. The opportunities for acquiring winery utility system maintenance program software were previously described in Chapter 5 (see section entitled "Operation and Maintenance"). There are at least 25 separate programs that can perform the necessary operation and maintenance data storage, processing, and retrieval tasks.

GLOSSARY

Absorption (soil)   A process by which one substance is trapped throughout the volume of another, usually a liquid, by solution or chemical reaction.

Acre-feet   A unit used for expressing large quantities of water, as in conservation reservoirs, equivalent to the volume that would cover acre of land to a depth of 1 foot (approximately 326, 000 U.S. gal).

Activated carbon   A very porous material which after being subjected to intense heat to drive off impurities can then be used to adsorb pollutants from air or water.

Adsorption   A physical process involving the contact and trapping of water or air pollutants on the surface of a solid substance.

Algae   Microscopic single-cell plants suspended in water; phytoplankton.

Algal bloom   Visible overgrowth of algae in a lake, due to eutrophication.

Alkalinity   Capacity of neutralizing acid, usually due to presence of bicarbonate or carbonate ions. Hydroxide, borate, silicate, or phosphate ions may contribute to alkalinity in treated waters.

Alum   Aluminum sulfate, one of the most commonly used chemical coagulants used for water treatment.

Aquiclude   An underground layer of relatively impermeable soil or rock that does not yield appreciable quantities of groundwater. Often confines an aquifer above and below. (See *Aquifer* and *Artesian aquifer*.)

Aquifer   An underground layer of soil or rock that is permeable enough to yield significant amounts of groundwater.

Area-capacity curve   A plot of reservoir surface area versus cumulative volume and surface water elevation.

Artesian aquifer   An aquifer that is enclosed or sandwiched between two impermeable layers of soil or rock, also called a confined aquifer.

Backwash   The washing cycle for a sand filter used in water treatment in which clean water flows up through the filter, flushing away entrapped impurities.

Bacteria   Microscopic single-celled plants that do not contain chlorophyll and are not capable of photosynthesis.

Base flow   Dry weather stream discharge fed by groundwater seepage and/or springs.

Catchment area (watershed)   The area that contributes runoff to a stream; also synonymous with watershed or drainage basin.

Centrifugal pump   A mechanical device that adds energy to a liquid using a rapidly rotating impeller in a specially shaped casing; most common type of pump used for water treatment and distribution.

Chlorination   The process of adding chlorine to water, primarily for disinfection.

Chlorine residual   The small amount of chlorine compounds that remain in water after disinfection, providing continued protection in the distribution system.

Clarifier   A sedimentation basin or settling tank in which suspended solids settle to the bottom and the clarified water or wastewater is drawn off the top.

Coagulation   The addition of certain chemicals to water that augment the process of particle collision and coalescence to form larger particles which settle under the force of gravity.

Coliforms   A group of mostly harmless bacteria that live in the intestinal tract of warm-blooded animals, and which are used as a biological indicator of sewage contamination.

Colloid   Extremely small particles suspended in water or wastewater, which cannot be removed by plain sedimentation or filtration without coagulation.

Complete treatment   The several treatment processes that are essential to render surface water potable and safe for winery purposes (coagulation, flocculation, sedimentation, filtration, and disinfection).

Cone of depression   The shape of the groundwater table around a well from which water is being withdrawn.

Contact basin   A small tank or vault of sufficient volume to permit chlorine or other contact disinfectant adequate time to inactivate any and all waterborne pathogens.

Crown   The top inside wall of a pipe. (See *invert.*)

Deionization (DI)   Specially manufactured ion-exchange resins which remove ionized salts from water. Can theoretically remove 100% of salts. Deionization typically does not remove organics, viruses, or bacteria, except through "accidental" trappings in the resin and specially made strong base anion resins which will remove gram-negative bacteria.

Demineralization   The process of removing minerals from water, usually through deionization, reverse osmosis, or distillation.

Drawdown   The vertical distance between the static water level and the pumping water level in a well.

Drought   An extended period of below-average precipitation that causes low flows in streams and rivers and creates falling groundwater levels.

EPA   Environmental Protection Agency (United States): An organization that administers the Safe Drinking Water Act.

Ephemeral stream   A stream that becomes completely dry during some months; also called an intermittent stream.

Eutrophication   The natural aging of a lake, pond, or reservoir characterized by high nutrient levels, excessive plant growth, and accumulation of bottom sediments.

Evapotranspiration (ET)   A part of the hydrologic cycle involving the combined processes of evaporation and transpiration of water by vegetation. (See *Water balance.*)

Fecal coliforms   Members of the coliform group of bacteria. Their presence in a water supply is used to confirm sewage contamination.

Fecal strep   Fecal streptococcus bacteria are used along with fecal coliforms to differentiate animal or human sources of contamination.

Filtration   The removal of suspended particles from water or air using a porous material that allows the fluid to pass through, but traps and retains the particles.

Floc   A particle large enough to settle out of water or wastewater, formed during the coagulation–flocculation process; also, settleable particles of activated sludge. (See *Coagulation.*)

Flocculation   Gentle mixing of water after the addition of coagulation chemicals to form settleable flocs.

Floodplain   The land along a river that is inundated by water during flood events.

Freeboard   The vertical distance between the overflow level of a tank, pond, or reservoir and the maximum controlled level in the storage facility.

Frequency analysis   A statistical method to determine the recurrence interval of storms, floods, or droughts.

Hydraulic grade line   A graph of pressure head in a hydraulic system, usually comprising a series of sloping straight lines that show a drop in pressure in the direction of flow.

Hydrologic cycle   The cycle of water moving through the environment as rainfall, surface and subsurface flow, and vapor.

Infiltration   In hydrology, a term referring to the penetration of water from precipitation into the ground.

Inorganic   Mineral substances, usually not containing carbon.

Invert   The bottom inside wall surface of a pipe.

Ion   An electrically charged fragment of an atom or molecule.

Jar test   A procedure used to determine the optimum coagulant dose in a water treatment plant.

JTU (Jackson turbidity unit)   Turbidity tests results from a candle turbidimeter.

Lift station (pump station)   A pumping facility for lifting water from a low point and moving it in a conduit to a higher elevation, usually to a storage tank.

mg/L   Milligrams of an element per liter of water.

Microfiltration (MF)   Filtration designed to remove particles and bacteria in the range of 0.1–2.0 or 3.0 μm in diameter.

Micrometer   A metric unit of measurement equivalent to 10^{-6} m or 10^{-4} cm. Symbol: μm. British units: $1 \ \mu = 10^{-6}$ in.

Most probable number (MPN)   A statistical estimate of coliform bacteria concentration in a sample of water, based on the results of the multiple-tube fermentation test.

Multiple-tube fermentation test   A method for estimating the concentration of coliform bacteria in water or wastewater.

Multipurpose reservoir   A large reservoir built to satisfy two or more needs which can include flood control, water supply, power generation, irrigation, recreation, and aquaculture.

NTU   Nephalometer turbidity units ≃ to Jackson turbidity units (see JTU).

Perennial stream   A stream that flows continuously throughout he year.

pH   An expression of hydrogen ion concentration; specifically, the negative logarithm of the hydrogen ion concentration. The range 0–14, with 7 as neutral, 0–7 as acid, and 7–14 as alkaline (base).

Potable water   Fresh water that is chemically and microbiologically safe for all beneficial uses.

Pressure head   The height of a column of liquid, usually water, that a given hydrostatic pressure in a system could support.

Pump curve   A graph that shows the relationship between flow rate and pressure head on the discharge side of a pump.

Rainfall intensity   The rate of rainfall expressed in terms of inches per hour or millimeters per hour.

Rating curve   A graph that shows the relationship between the stage of a stream and its discharge; also, stage–discharge curve.

Rational method   An empirical formula for estimating peak storm water runoff rates for small watersheds.

Recharge area   A region where geologic and soil anomalies allow precipitation to infiltrate the ground surface and percolate deeply to the underlying aquifer.

Return period   The number of years between storms of specific intensities and durations, as determined by a probablistic analysis of historical storm events.

Reverse osmosis (RO)   The separation of one component of a solution from another component by flowing the feed stream under pressure across a semipermeable membrane. RO removes ionized salts, colloids, and organics down to 150 molecular weight. Also called hyperfiltration.

Shutoff head   The pressure head developed by a centrifugal pump that operates against a closed discharge valve.

Softening   A treatment process that reduces the hardness of water by removing much of the dissolved calcium and magnesium carbonate.

Stage   The depth or elevation of water in a stream or lake.

Standard Methods For The Analysis of Water and Wastewater   An important reference and standard for water and wastewater sampling and analysis, published jointly by the American Public Health Association (APHA), the American Water Works Association (AWWA), and the Water Pollution Federation (WPF).

Supernatant　The water that remains above the sludge layer in a settling tank or digester.

Total dissolved solids (TDS)　The residual material remaining after filtering the suspended material from water and evaporating the solution to a dry state.

Thermal stratification　The natural process by which separate layers form in lakes because of water temperature differences.

Total dynamic head (TDH)　The total energy that must be provided by a pump to overcome lift, pipe friction, and mechanical/electrical losses.

Water balance　A water accounting procedure that relates all gains and losses in a reservoir storage system to demonstrate that for any given set of hydrologic records (rainfall, runoff, evaporation, seepage, etc.), the net gain or loss in storage volume and the volume excess that is discharged as spill or overflow.

Water table　The top of the zone of saturation, where all soil voids are filled with water at atmospheric pressure. Generally, the free-water surface or highest elevation of an aquifer. Moves up or down depending on groundwater draft from wells and the annual recharge from precipitation in the tributary watershed.

REFERENCES

65. HUCHINS, W.A. 1956. *California Law of Water Rights*. Calif. State Engineer's Office, Ag. Res. Serv. USDA, July.

66. DRISCOLL, F.G. 1986. *Groundwater and Wells*, 2nd ed. St. Paul, MN: Johnson Filtration Systems.

67. KEYS, W.S. 1967. Borehole geophysics as applied to groundwater. *Proc. of Canadian Cent. Conf. on Mining Groundwater Geophysics*. Canadian Geol. Survey, Economic Geologic Rept. No. 26, pp. 598–612.

68. _____ 1975. *National Primary Drinking Water Regulations*. Federal Register Vol. 40, No. 248. Washington, DC: U.S. Government Printing Office.

69. _____ 1977. *National Secondary Drinking Water Regulations*. Federal Register 42: 17144, March 31. Washington, DC: U.S. Government Printing Office.

70. DICKSEN, R. 1989. Personal communication. Culligan, Inc., Napa, CA.

71. _____ 1984. *Abandonment of Test Holes, Partially Completed Wells, and Completed Wells*. AWWA Standard for Water Wells, AWWA A100-84.

72. _____ 1990. *California Well Standards*. State of California Department of Water Resources Bulletin, No. 74-90. January.

73. MRAK, E. 1937. Effect of certain metals and alloys on claret- and sauterne-type wines made with vinifera grapes. *Food Res.* 2:539–547.

74. _____ 1950. Saveur de Fer Dans Le Vin. *Mitt. Gibiente Lebensin Hyg.* 41(1/2): 56–57.

75. McKee, J.E. and Wolf, H.W. 1971. *Water Quality Criteria.* Publ. No. 3-A, State of California Water Resources Control Board.

76. Storm, D.R. 1984. Cutting winery overhead costs; from clean water to dirty water. *Pract. Winery Vineyard* 5(3):20–21.

77. Storm, D.R. 1992. Analyzing your electrical power bill; tightening the overhead cost belt. Part I and Part II. *Pract. Winery Vineyard* 12(6) and 13(1), XIII, No. 1.

78. Storm, D.R. 1994. The right valve for the right application. Part I and Part II. *Pract. Winery Vineyard.* 14(5) and 14(6).

79. Pomroy, J. 1994. Selecting flowmeters. *Instrum. Control Syst.* 67(3):61–68.

80. Storm, D.R. 1987. Flow measurement. *Pract. Winery Vineyard* 8(1):22–24.

81. Storm, D.R. 1993. Stormwater prevention plan update. *Pract. Winery Vineyard* 14(2).

82. Storm, D.R. 1991. New effluent control laws. *Pract. Winery Vineyard* 12(1).

83. Columbian Steel Tank Company. 1992. *Storage Tank Specifier's Guide.* Kansas City, KS: Columbian Steel Tank Company.

84. Banov, A. 1978. *Paints and Coatings Handbook.* Farmington, MI: Structures Publishing Co., pp. 26–28.

85. Storm, D.R. 1989. Paints and coatings; if it doesn't move, paint it. *Pract. Winery Vineyard.* 10(2).

86. Storm, D.R. 1994. Wastewater sludge; its characteristics and management. Part I. *Pract. Winery Vineyard* 15(2).

87. Brigano, F.A. 1993. *Low Cost Water Treatment System On-Line in Freestone, CA.* Culligan International Company.

88. Deshpande, S.S., Scanlan, J.M., and Winter, G. 1989. Small Community Compliance Using Cost Effective "Multi-Tech" Packaged Water Systems. AWWA Annual Conference, June 20.

89. Storm, D.R. 1988. Disinfection. *Pract. Winery Vineyard.* 8(5).

90. Bellar, T.A., Lichtenberg, J.J., and Kroner, A.D. 1974. The occurrence of organohalides in chlorinated drinking water. *J. AWWA* 66(12):703.

91. Jolley, R.L., Gorchev, H., and Hamilton, D.H. 1978. *Water Chlorination Environmental Impacts and Health Effects, Vol 2.* Ann Arbor, MI: Ann Arbor Sci. Publ.

92. Schuter, D. 1994. Personal communication. Groundwater Records and Wells, State Department of Water Resources, Sacramento, CA.

93. _____ 1993. *Water Resources Data, California, Vols. 1–4.* Department of the Interior, U.S. Geological Survey, Washington, DC: U.S. Government Printing Office.

94. _____ 1987. *Lead in School Drinking Water.* Supt. of Documents, Washington, DC: U.S. Government Printing Office. 20402. Stock No. 055-000-00281-9.

95. Ward, B. 1989. Lead poisoning; a status report. National Safety Council Environmental Health Center, *Pollut. Eng.* August.

96. Storm, D.R. 1991. Water contamination and plumbing materials, Part I and II. *Pract. Winery Vineyard* 11(5):7–8; 11(6):14–15.

CHAPTER 7

IRRIGATION WATER SUPPLY SYSTEMS AND RECLAIMED WASTEWATER

INTEGRATION OF WINERY AND VINEYARD FUNCTIONS

The principal reason for including vineyard irrigation in a winery utilities reference book is to detail the relationship between winery and vineyard water supplies. Chapter 8 will outline the real benefits and the problems which must be overcome to successfully recycle winery process wastewater for vineyard use. It is important in understanding the overall concept of water recycling and reuse to see vineyard irrigation from the utility engineer's perspective. It must be emphasized that the recycling concepts described in this chapter refer only to process wastewater, not to sanitary wastewater. The public health safety requirements are, for good reason, very conservative and restrictive for sanitary wastewater reuse and often extraordinarily expensive to achieve even on a large winery scale. Sanitary wastewater treatment and disposal will be discussed in detail in Chapter 8.

CONFLICTS IN RECYCLED WATER SUPPLY AND VINEYARD IRRIGATION DEMAND

The timing and occurrence of maximum vineyard demand and peak winery wastewater production are asynchronous in wine-grape-growing areas worldwide. Vine irrigation usually ceases for red varieties after veraison and about 1 month before harvest for white varieties, unless extraordinarily hot weather prevails in the 30 days prior to harvest. Thus, the production of recycled wastewater for May, June, and July, the irrigation months for the arid regions of the Northern Hemisphere, would be only about 22% of the annual wastewater production (see Table 6.8, Chapter 6). To implement the reuse concept for reclaimed process wastewater as supplemental irrigation for vineyards, a storage reservoir must be included in the system to regulate, blend, and retain reclaimed wastewater from the previous crush season until vineyard irrigation is required.

For areas that receive appreciable summer rainfall or have heavy soils that can store 16–20 in. of winter rainfall, supplemental vineyard irrigation is generally not required (97). For those grape-growing regions, the choices for wastewater reuse begin to narrow. Frost protection, using overhead, impact-type sprinklers and stored reclaimed water, is one possible use that requires very precise management of reclaimed wastewater production and the assured annual application of stored water for frost protection. California experience has shown that $5°F$ ($2.5°C$) of frost protection may be obtained from sprinklers set at every fourth vine in every fourth row, delivered at a rate of 0.1 in./h (2.5 mm/hr) (97). Frost protection in many California vineyards is often achieved by a combination of methods including smudge pots (kerosene-fired field heaters), the aforementioned sprinklers, and permanently installed motor-driven air fans to prevent cold-air stagnation.

The irrigation of the winery's landscape materials with reclaimed process wastewater is another feasible option. As with the vineyard irrigation scheme, in arid Mediterranean-type climates, plants become dormant in October and November and winter rains generally provide needed soil moisture from November through April. The landscape delivery system could consist of the following:

(a)   Conventional drip emitters
(b)   Overhead, impact-type sprinklers
(c)   Buried drip tape

Prior to selecting landscape irrigation as an option for reclaimed process wastewater disposal, two important questions must be answered in the affirmative:

1. Does the local governmental jurisdiction having responsibility for wastewater recycling and reuse permit landscape irrigation and under what conditions?
2. Has the landscape architect planted species that are tolerant to the highly variable mineral content of the wastewater?

If the local statutory requirements discovered under item 1 above, that are placed upon landscape irrigation use are excessively restrictive and costly, vineyard irrigation should be the reuse alternative of choice. During the planning phase for the winery or winery expansion or upgrade in waste treatment facilities is a good time to evaluate the potential volumes of treated wastewater that may be available for vineyard or landscape irrigation (see Chapter 6 for water requirements). Obviously, the scale of the winery operation dictates whether the wastewater recycling scheme will be a significant water supply benefit or just a convenience for permitting the confinement of the wastewater to the winery parcel. In reality, a well-designed and well-operated soil absorption system (leachfield or pressure distribution system of some type) is also returning the once-used water to the groundwater resources of the area and indirectly making it available for eventual reuse.

VINEYARD WATER REQUIREMENTS

Vine water use data are probably well known to anyone seriously involved in commercial viticulture and enology ventures, but they will be reported herein to have a common basis for estimating the sizes and capacities of major irrigation system components. Because irrigation requirements are based on

- Soil texture
- Soil depth
- Evapotranspiration rate
- Vineyard spacing and age
- Grape production objective (heavy or light cropping)

there will be a broad band of variance in both the timing and total volume of water to be applied to any given vineyard. As suggested previously in the introduction to this chapter, in grape-growing areas with appreciable summer rainfall, supplemental vineyard irrigation is not often required. Also, the unit water requirements reported, have been developed principally for California's viticultural and climatic data. Direct conversion of these reported data for other grape-growing regions of the world should be used with caution and strongly tempered with similar data from the local site, which clearly reflects those conditions. The basic wastewater recycling principles, however, are universally applicable.

A reasonably accurate rule-of-thumb for new vineyards (bench grafts or field grafting to established root stock) is applied water of 100 gal (400 L) per vine per year. This presumes a winter rainfall to replenish soil moisture during winter and spring months. Most irrigation requirements are universally expressed in feet or millimeters of water applied because historical irrigation practices involved broad, areawide applications of water, where depth was the measurement criterion. Field capacity, the term applied to the soil profile in the root zone of the particular crop being considered, is the amount of water that the soil can accommodate at saturation and is expressed in inches per foot (millimeters per meter) of soil depth. Overhead, impact-type sprinklers are rated in inches per hour, depending on orifice size and pressure at the sprinkler; thus, it is relatively easy to calculate the length of the irrigation set to achieve field capacity. Drip emitters, on the other hand, are rated in gallons (liters) per hour, so the simplest approach for drip irrigation practice is to convert historical areawide application values for vines to a unit volume of water to be applied per vine. Using applied water values as derived by Kasimatis (97, 98) for wine grapes grown in a Region I (cool) zone, only 4–6 in. (100–150 mm) of supplemental water are specified. For a 6 × 12-ft (1.8 × 3.7 m) spacing, there are 540 vines with an assumed root area at the 3–5-ft (0.9–1.5-m) depth of approximately 25 ft^2 (2.3 m^2). For illustrative purposes, the 6 in. (150 mm) of supplemental irrigation required for cool regions is assumed and converted as follows:

$$6 \text{ in.} \div 12 \text{ in./ft} \times 25 \text{ ft}^2/\text{vine}$$
$$= 12.5 \text{ ft}^3$$
$$= 94 \text{ gal/vine} \ (356 \text{ L/vine})$$

Corrected for deep percolation and evaporation losses, 25% or

$$94 \text{ gal} \times 1.25$$
$$= 118 \text{ gal/vine (total gallons of water applied per vine) } (447 \text{ L/vine})$$

For other wine-grape-growing regions and temperature groups, the following drip system emitter volumes in Table 7.1 have been calculated:

Table 7.1.   Calculated Supplemental Irrigation Requirements per Wine GrapeVine by Temperature Region

Region	Time–Temp. Group[a] (degree– days) [°F]	Assumed range of winter rainfall in inches (mm)	Areawide water requirement in inches (mm)	Applied emitter volume per vine for supplemental irrigation[b] (gal/vine/yr) [L/vine/year].
I	>2500	12–32 (30–80)	6 (15)	118 [447]
II	2500–3000	12–32 (30–80)	12 (30)	236 [894]
III	3001–3500	12–32 (30–80)	12 (30)	236 [894]
IV	3501–4000	12–18 (30–45)	18 (45)	354 [1340]
V	>4000	6–10 (15–25)	42 (105)	826 [3127]

[a]Average daily temperature in °F (°C) [measured at the vineyard site or from National Weather Service Climatological Data for nearest station minus 50°F (10°C) times the number of days for the months of April through October] summed for the period from bloom to grape harvest (not necessarily applicable to grape growing areas outside California).
[b]Assumed overall irrigation efficiency of 75% (20% deep percolation and 5% evaporation).
Source: Adapted from Refs. 97 and 98.

An optional semiempirical method can also be employed to help in sizing the drip system to meet the maximum evapotranspiration (ET) rate for the vine, landscape shrub, or tree. ET rates can be obtained from published summaries of climatological data. Thus, peak water use per day can be calculated by applying the following formula (99):

$$R = 0.623 \times C \times A$$

where R is the maximum water requirement per plant unit in gallons (liters) per unit per day, C is the peak evapotranspiration (ET) rate in inches (millimeters) per day, and A = plant unit spacing. To illustrate: For a vineyard on a 6-ft × 12-ft (2-m × 4-m) spacing and an assumed maximum ET rate of 0.24 in. (6.25 mm) per day,

$$R = 0.623 \times 0.24 \times 6 \times 12 = 10.8 \text{ gal/day (41 L/day)}$$

Thus, for an emitter rated at 1 gal (3.8 L) per hour an irrigation setting of about 11 h would meet the probable peak plant requirement. Again, as the application of irrigation water by drip irrigation is not 100% efficient, the portion that is not plant-available moisture must be reflected in an increase in water applied. Soil texture will play a significant role in the efficiency of applied water delivered to the vine; that is, sandy soil will transmit applied water to lower regions in the soil profile very efficiently. Thus, the deep percolation component may represent nearly all of the water that is unavailable for vine use. In contrast, clay soils, which have a low porosity and percolation rate, may cause surface ponding of applied water, allowing a greater proportion of water loss to solar evaporation than would prevail under the sandy soil example. Using a 75% efficiency under all soil conditions is conservative enough, so that serious errors in estimating applied irrigation water volumes will not occur. Therefore, the amount of applied water per vine to accommodate the maximum ET day example would be

$$R_{actual} = 10.8 \times 1.25 = 13.5 \text{ gal (51 L)}$$

SYSTEMS AND EQUIPMENT

The biggest operational problem to overcome in the reuse of winery process wastewater for vineyard or landscape irrigation is the plugging of the very small orifices 0.004 in. (0.1 mm) (orifice emitters) to 0.02–0.04 in. (0.5–1.02 mm) (microtubes). [Plugging can occur from mineral deposits at the outlet end of the emitter, fertilizer particles (if fertigation is practiced) and organic slime and algae growth from the bionutrient-rich reclaimed wastewater.] The orifice emitter and long-path, screw-type emitter are designed to be semi-self-cleaning. Figure 7.1 illustrates the two types of drip emitters that would provide the lowest clogging potential of any emitters, that to the author's knowledge, are currently available on the commercial market.

As suggested in the introduction to this chapter, the direct reuse of winery process wastewater for vineyard irrigation is made difficult because the large volumes of process wastewater associated with harvest cannot be equitably distributed to vineyards, as the vines are only 30–60 days from complete dormancy, when the largest component of reclaimed process wastewater is produced. Some viticulturists recommend a postharvest application of water to the vines, whereas others caution that waterlogging of heavy vineyard soils in young vines (less than 6 years old) can produce collar rot. If the total salinity of the seasonal irrigation water has been high, there are potential benefits from the postharvest application of water to leach accumulated salts from the root zone (97). The ultimate decision on postharvest irrigation should rest with the vineyard manager. A better scheme with the least risk to vineyard health involved is the blending of reclaimed process wastewater and the primary irrigation water supply. Figure 7.2 is a schematic plan of an irrigation blending system that has been employed successfully for several western U.S. wineries.

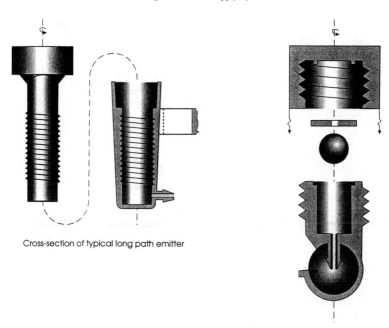

Cross-section of typical long path emitter

Cross-section of orifice emitter device with
ball and diaphragm to regulate flow

Fig. 7.1.   Semi-non-clog emitters best suited for use with reclaimed winery wastewater.

The term "water balance" will be used in this chapter and subsequently in Chapter 8 to describe the hydraulic and hydrologic conditions that must be met to achieve a zero or near-zero discharge, reclaimed wastewater storage system. Water balance is an accounting process and a simulation of a reclaimed wastewater storage reservoir operation over a specified period, usually a year. Most state and federal regulatory agencies require that the reservoir *not* spill or produce an unregulated discharge for a winter rainfall pattern that has a 1 in 25-year return frequency. Table 7.2 is the accounting format for a typical water balance and indicates how the values of each element of the balance equation are applied.

Table 7.2.   Water Balance Simulation for Reclaimed Wastewater Reservoir Sizing.

Month	V_{BM} (beginning month storage)	W_{IN} (wastewater r inflow)	E_L (evaporation)	R_{IN} (rainfall)	I_{OUT} (irrigation draught)	S_{OUT} (reservoir spill)	V_{EM} (end of month storage)

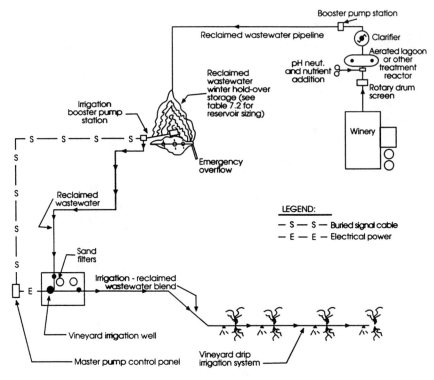

Fig. 7.2.   Schematic plan of reclaimed wastewater–primary irrigation water blending system.

Water Balance Equation:

$$V_{BM} + W_{IN} - E_L + R_{IN} - I_{OUT} - S_{OUT} = V_{EM}$$

Thus, if the hypothetical reservoir with an assumed storage volume is operated for 2 or 3 calendar years through a historical wet period (1 in 25-year return frequency) without the reservoir spilling any of its contents, then it would be prudent to make a subsequent second trial run with a reduced storage volume until some spill is calculated. The approach for this series of successive approximations to determine reclaimed wastewater storage volume can be streamlined and programmed for a PC to accomplish by setting "spill" equal to zero in the water balance equation and solving for end of month storage.

The system shown in Fig. 7.2 will operate successfully if:

1. Each year before the end of the vineyard and/or landscape irrigation season, the total storage volume of reclaimed wastewater is withdrawn from the reservoir (see Table 7.2 for storage reservoir volume calculations).

2. An active program of plankton and emergent aquatic vegetation control at the storage reservoir is undertaken to reduce the production of organic detritus in the storage pond and its carryover to the drip irrigation filter system;

3. Adopt a water quality monitoring program that will provide a long-term database on the reclaimed wastewater (alone) and irrigation water blend for required reporting. Permit the vineyard manager the greatest degree of flexibility in controlling the irrigation water blend for the ultimate benefit of the vineyard. [If fertigation is practiced (i.e., the introduction of vine nutrients and soil amendments, like gypsum, through the drip system), the nutrient combination of the reclaimed wastewater can be credited in the nutrient balance requirements for the vineyard.]

The problem of weed proliferation around drip emitters should be mentioned before the discussion of drip irrigation system performance is closed. Herbicides have been introduced into drip systems by means of the same drip system control head complex that is used to apply vine nutrients and chelating agents for improvement in soil permeability. Figure 7.3 shows a typical control head arrangement and the essential piping, valving, and filters that are common to all drip systems. Lange et al. (100) found that in heavier clay soils preemergent herbicides become partly adsorbed on soil particles and did not move readily with the irrigation water. The macronutrient and micronutrient supply that may be created with the

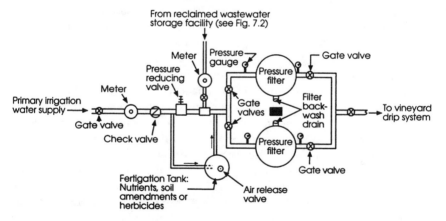

NOTE:
Depending upon the degree of instrumentation and control sophistication desired, drip systems can be fully automated. For illustrative purposes and simplicity the manually operated control headworks scheme is shown.

Fig. 7.3. Typical control headworks for a drip irrigation system arrangement (reclaimed wastewater blending).

reclaimed wastewater and primary irrigation water blend may be just the stimulus that the native weed species need to produce exceptional and troublesome growth. The standard U.S. practice of strip spraying with herbicides may be the best control measure for extraordinary weed propagation.

OPERATION AND MAINTENANCE OF RECLAIMED WASTEWATER DRIP IRRIGATION SYSTEMS

The most important maintenance task for the reclaimed wastewater irrigation system is the coordinated management of the winery's waste treatment system (see Chapter 8) and the winter hold-over storage facility. Inadequate treatment of the winery's process wastewater will add unwanted suspended solids to the storage pond, and further stabilization of those waste products in the pond could produce odors and suspended solid residues that would impair the normal operation of the drip system control headwork filters. The filter media would clog rapidly and the filter backwash cycle would be almost continuous. Thus, just prior to the beginning of the vineyard irrigation system, the reclaimed wastewater storage pond should be inspected and samples taken for qualitative analysis (pH, plant nutrients, and suspended solids). If suspended solids are greater than 30 mg/L, the use of settling aids should be considered to flocculate and agglomerate the suspended material and accelerate the sedimentation process in the pond. Simple "jar tests" using several concentrations of some of the more commonly used flocculents will quickly reveal the amount of flocculent to be added. (Settling aids will be discussed in detail in Chapter 8.) Although this posttreatment clarification step may seem troublesome, it is necessary to have a happy and continuous marriage of wastewater recycling and viticulture. Plankton growth and emergent shoreline vegetation should be controlled with a year-round program with biological controls preferred over chemical ones (101). Fine-tuning of any nutrient additions to obtain treatability in the winery wastestream (see Fig. 7.2 detail on treatment process) will prevent carryover of excess nutrient loads to the reclaimed wastewater storage pond and assist in the control of unwanted blooms of algae. Periodic (every 4–5 years) dredging of the pond's bottom sediments will permanently remove a large portion of the phosphorous–nitrogen compounds that can continue to make management of the storage pond very difficult.

Each year, the vineyard manager should be made aware of the "water balance" objectives and a month-by-month program should be established between the wastewater treatment system supervisor and the vineyard manager to ensure that the storage reservoir will be virtually empty at the end of the irrigation season. The operation and management document (so-called O&M manual) to be prepared by the winery's utilities design engineer should include a subsection that details the management of the storage pond and reclaimed wastewater for irrigation and ancillary subsystems. The respective responsibilities of the cognizant vineyard and utility operations personnel should be clearly stated.

Record keeping is important not only to provide an historical database for managing the systems but also to facilitate the reporting of quantities and qualities of reclaimed water to local and regional water pollution control agencies. The software programs for operation and maintenance, that have been referenced in previous chapters are readily capable of managing the data for the reclaimed wastewater subsystem.

GLOSSARY

Activated sludge process A biological wastewater treatment process in which a mixture of wastewater and activated sludge is blended and aerated.

Advanced wastewater treatment Any physical, chemical, or biological treatment process used to accomplish a degree of treatment greater than that achieved by secondary treatment. Usually implies removal of nutrients and a high percentage of suspended solids.

Alkalinity The capacity of water to neutralize acids; a property imparted by carbonates, bicarbonates, hydroxides, and occasionally borates, silicates, and phosphates. It is expressed in milligrams of equivalent calcium carbonate per liter.

Available water The portion of water in a soil that can be readily absorbed by plant roots or that water held in the soil against soil moisture tension.

Applied water The total amount of water applied to a vineyard, pasture, or other crop to meet plant requirements, including irrigation losses such as spray loss and deep percolation.

Blending schemes Methods for blending and diluting less desirable reclaimed wastewater with a primary irrigation supply of good chemical quality; i.e., reservoir blending, pipeline injection blending, etc.

BOD (1) Biochemical oxygen demand. The quantity of oxygen used in the biochemical oxidation of organic matter in a specified time, at a specified temperature, and under specified conditions. (2) A standard test used in assessing wastewater strength, usually 5 days at 20°C.

Cation-exchange capacity (CEC) The sum of exchangeable cations that a soil can adsorb, expressed in millequivalents per 100 grams of soil or in millimoles of positive charge per kilogram of soil. CEC is directly related to a soil's ability to retain cation against leaching. CEC is also used in calculating exchangeable sodium percentage (ESP), a measure of excessive sodium hazard in the soil.

COD Chemical oxygen demand. A quantitative measure of the amount of oxygen required for the chemical oxidation of carbonaceous (organic) material in wastewater using dichromate or permanganate salts as oxidants in a 2-h test. (BOD ≈ 0.5 COD.)

Dentrification   The biological conversion of nitrate or nitrite to gaseous N_2 or N_2O.

Effluent   Partially or completely treated wastewater flowing out of a treatment plant, reservoir, or basin.

Electrical conductivity (EC_w for water, EC_e for the soil saturation extract)   A measure of salinity expressed in millimhos per centimeter (mmho/cm) or decisiemens per meter (dS/m) at 25°C. Empirically related to total dissolved solids (TDS) divided by 640.

Emitter   A calibrated orifice device inserted in a drip irrigation tube to discharge a given rate of water (0.5 gal/h or more).

Evapotranspiration (ET)   The combined loss of water from a given area and during a specified period of time by evaporation from the soil surface and by transpiration from plants. ET_0 is reference ET, defined as the ET from an extended surface of 3–6-in.-tall green grass cover of uniform height, actively growing, completely shading the ground and not short of water. E_p or E_{pan} is evaporation from a standard evaporation pan.

Exchangeable sodium percentage (ESP)   The ratio (as percent) of exchangeable sodium to the remaining exchangeable cations in the soil. (See *Sodium adsorption ratio*.)

Fertigation   Application of plant nutrients or soil amendments using the irrigation system as the delivery mechanism.

Field capacity (FC)   The percentage of water (either weight or volume) remaining in a soil 2 or 3 days after having been saturated and after free drainage has practically ceased.

Gypsum requirement   The quantity of gypsum or its equivalent required to reduce the exchangeable sodium fraction of a given increment of soil to an acceptable level.

Horizon   A layer of soil differing from adjacent genetically related layers in properties such as color, structure, texture, consistency, and pH.

Hydraulic conductivity   The rate of water flow in soil per unit gradient of hydraulic head or potential.

Impact sprinkler   Sprinklers which apply irrigation water overhead using water pressure against a spring-loaded plate to achieve 360° rotation and uniform water distribution.

Infiltration   The downward entry of water into soil.

Infiltration rate   A soil characteristic describing the rate at which water can enter the soil under specific conditions, including the presence of excess water.

Infiltrometer   Any of several devices for measuring the rate of entry of water into a soil.

Influent   Wastewater flowing into a treatment plant, or treatment process.

Land application   The recycling, treatment, or disposal of wastewater or wastewater solids to the land under controlled conditions.

Land treatment   Irrigation with partially treated wastewater on land where additional treatment is provided by soil, microorganisms, and crops which metabolize plant available nutrients.

Leaching fraction (LF)   The fraction of water applied to soil that leaches below a depth of interest such as the rooting depth.

Leaching requirement (LR)   The leaching fraction required to maintain average root zone salinity below a plant toxicity value.

Mineralization   The conversion of an element form an organic to an inorganic form (e.g., the conversion of organic nitrogen in wastewater to ammonium nitrogen by microbial decomposition).

Oxidation pond   A relatively shallow pond or basin in which biological oxidation is effected by natural or artificially accelerated transfer of oxygen (e.g., algae pond, lagoon).

Permanent wilting point (PWP)   A plant physiology term referring to the soil water content at which plants wilt and do not recover.

Permeability   The ease with which gas, liquids, or plant roots penetrate or pass through a soil horizon.

pH   The degree of acidity or alkalinity, defined as the negative logarithm of hydrogen ion activity of water.

Primary treatment   The first major treatment in a wastewater treatment facility, usually sedimentation, but not biological oxidation. Example: The effluent from a conventional septic tank has received primary treatment.

Rapid infiltration   A type of land treatment in which treated or untreated wastewater is applied to relatively porous soil at rates far in excess of normal crop irrigation.

Reclaimed wastewater   Wastewater that, as a result of treatment, is suitable for a beneficial use.

Secondary treatment   Generally, a level of treatment which produces removal efficiencies for BOD and suspended solids of 85%.

Slow rate land treatment process   A type of land treatment of wastewater, where wastewater is applied to a vegetated surface with the applied wastewater being treated as it flows through the plant–soil matrix.

Sodium adsorption ratio (R_{Na} or SAR)   A measure of the amount of sodium relative to the amount of calcium and magnesium in the water or in a saturated soil extract.

$$R_{Na} \text{ or } SAR = \frac{Na}{\sqrt{(ca + Mg)/2}}$$

where the quantities Na, Ca, and Mg are expressed as milliequivalent per liter.

Soil moisture tension (or pressure)   The soil water content expressed as an equivalent negative pressure. It is equal to equivalent pressure that must be applied to soil water to bring it to equilibrium, through a permeable membrane—a pool of water of the same composition.

Soil structure The combination or arrangement of primary particles into secondary particles, aggregates.

Soil texture The relative proportions in a soil of sand, silt, and clay-sized mineral particles (also known as grain-size distribution).

Soil water content The amount of water lost from the soil upon drying to constant weight at 105°C, expressed as grams of water per gram of dry soil or cubic centimeter of water per cubic centimeter of bulk soil.

Tailwater Excess irrigation water that does not enter the soil profile, but emerges as surface runoff to man-made or natural drainage courses. More likely to occur in soils in steeply sloping terrain with high clay content. In some jurisdictions, reclaimed wastewater tailwater may require special handling, regulation, and capture for secondary reuse.

Total dissolved solids (TDS) The sum of all dissolved solids in a water or wastewater and an expression of water salinity in mg/L. Empirically related to electrical conductivity (EC in dS/m) multiplied by 640.

Veraison That time in the growth and maturity cycle for red grape varieties when the berries change color from green to red.

Wastewater irrigation Land application of wastewater for the purpose of crop production. Often used in a broader sense to mean land treatment and disposal of wastewater where maximum crop production is a secondary objective.

Wastewater reclamation The process of treating wastewater to produce water for beneficial uses, its transportation to the place of use, and its actual use.

Wastewater reuse The additional use of once-used water.

REFERENCES

97. WINKLER, A.J., COOK, J.A., KLIEWER, W.M., and LIDER, L.A. 1974. *General Viticulture.* Berkeley: University of California Press.

98. WEAVER, R.J. 1976. *Grape Growing.* New York: John Wiley and Sons.

99. _____ 1984. *Drip Irrigation Management.* U.C. Davis Cooperative Extension, Division of Agriculture and Natural Resources.

100. LANGE. 1974. Weed control under drip irrigation in orchard and vineyard crops. *2nd International Drip Irrigation Congress Proceedings*, pp. 422–424.

101. STORM, D.R. 1989. Plankton and aquatic weed control. *Pract. Winery Vineyard* 10(4).

CHAPTER 8

WASTEWATER SYSTEMS

OVERVIEW AND SYSTEM SEPARATION PHILOSOPHY

Chapter 7 bridged the gap between purely water supply engineering and wastewater engineering by presenting a framework for reclaimed wastewater recycling as a separate but related subcategory of the total water utility system model. Although it is well within the realm of technical possibilities to recycle all wastewater from wineries, from a practical, microbiological safety, and economic standpoint only process wastewater reclamation will be dealt with in this chapter.

Early prototype systems (ca. 1970) for sanitary wastewater recycling (i.e., the Santee Project near San Diego, California and the Windhoeck Project in South Africa) demonstrated that direct reuse of domestic wastewater for potable purposes could occur without endangering public health (102). A resurgence of dangerous blood-borne diseases worldwide in the 1980s, however, has required a reexamination of the prudence of sanitary wastewater recycling and direct reuse for culinary purposes. Further, constructing a true "fail-safe" treatment system, which would contain the necessary redundant critical components and multiple unit, backup equipment arrangements still would not necessarily guarantee that pathogen-free water would be produced 100% of the time. The bottled-water industry has, for part of its success (see "Bottled Water Service" in Chapter 6), capitalized on current public reluctance to trust what municipal waterworks systems produce at the tap (103).

Well-traveled individuals with a worldwide view find in most countries outside of the United States that bottled water is, in fact, the safest source of drinking water, a habit that returning U.S. travelers readily embrace. This large increase in bottled drinking water consumption reinforces the perceived movement away from public water supply systems. The water works industry has a large public relations task ahead in order to restore confidence in municipally owned and operated water supply systems.

From the first meeting with the winery's architect until the last utility system drawing is completed, the complete separation of sanitary and process sewage should direct the master planning of the sewer network. The only exception to the foregoing principle would be a winery that can discharge both of its wastestreams to a municipal or special district, publicly owned treatment works (POTW). It is also strongly recommended, even for small wineries (less than 40,000-case annual production per year) which might be able to employ a combined soil absorption system for disposal of both sanitary and process wastewater, that the physical separation of both the treatment and disposal works (separate septic tanks and leachfields) be maintained. The logic for system separation can be readily understood if the consequences of a combined system failure are analyzed from both a winery operations and public health hazard standpoint. The failure of a combined system leachfield, with, for example, surfacing effluent, would present an immediate health hazard situation at the leachfield, where direct contact would pose a disease risk. Runoff of this leachate to a natural drainage course would represent a potential contamination of related tributary streams. Local environmental health officials would have no choice but to require a cease and desist order for the winery, and operations would have to be terminated until the system was repaired or replaced. If, on the other hand, the systems were separate and the sanitary system failed, there would not be a statutory public health requirement to have the winery cease production operations. Winery bathrooms could be closed temporarily, and portable chemical toilets brought to the site for use by visitors and employees until the failed system was repaired or replaced.

TREATMENT AND DISPOSAL OF SANITARY WASTEWATER

Chapter 5 dealt with the subject of water supply and demand for sanitary purposes, as produced by winery visitors and full-time employees. Underestimating sanitary wastewater flows can produce a hydraulic and solids overload condition to an underdesigned system that would lead to eventual failure. Almost all of the design criteria for on-site wastewater systems are empirical, tempered with a healthy, built-in factor-of-safety and the conservative judgment and experience of the design engineer. For example, the standard soil percolation test uses "clean" water to measure the permeability of a soil. The solids content of the percolate (septic tank effluent) will not be entirely free of solids. There are permitted variations to the percolation test from one governmental jurisdiction to another, some of which allow a high degree of error to enter the testing protocol. One of the most glaring procedural errors is to allow the use of mechanically bored auger holes to serve not only as indicators of the soil profile characteristics but also to allow the same auger hole to serve as the vessel for the percolation test. As the sidewalls of the standard 4-in.-diameter auger hole cannot be carefully inspected from the surface, the resultant test may err on either side of the actual porosity of the soil. An unseen lens of pervious sand or gravel can overpredict permeability, and the "slickened" sidewalls of the auger hole in soils containing clay can effectively seal the soil, giving results that reflect an unusually low and artificial permeability. The State of Oregon has developed a unique system of determining soil suitability for septic systems. In lieu of the standard field percolation test, the soil is classified (grain-size analysis and other determinations of the soil's properties). Once classified, the soil is compared to a set of laboratory standard permeabilities for soil type. The Oregon State standard permeability is then used for the design of the soil absorption system. (The published standard permeabilities by soil type are based on both laboratory controlled permeability tests and record experience with that soil for long-term septic tank system performance.) Percolation test results will be discussed in more detail in subsequent sections on soil absorption systems design for both sanitary and process wastewater systems.

Design Flows

Reference is made to Table 6.4, Chapter 6, which illustrates the derivation of a typical winery potable water demand and equivalent sanitary wastewater discharge value. As suggested previously in the Chapter 6 section entitled "Water Conservation," the unit flows assigned to each winery visitor and employee should not reflect the net water saving afforded by low-water-use or mini-water-use fixtures. This justifiable "design cushion" in flow and solids loading for the system will produce an additional factor of safety of 1.25 or more. Table 8.1 provides a sample format and a step-by-step process for developing maximum sanitary wastewater discharges. The winery's de-

Table 8.1. Derivation of Maximum Sanitary System Wastewater Flows for Wineries with Tours and Tastings

	Maximum flow in gallons (liters) per day	
A. Number of employees (full time)	_____ × 25^a (100) =	_____
B. Number of seasonal employees (crush or bottling)	_____ × 25^a (100) =	_____
Total		_____
C. Number of visitors per maximum day (Labor Day, Fourth of July, Memorial Day, etc.)	_____ × 6.0^b (24) =	_____
or		
D. Special winery events (new wine release, concert, or winery-media day)	× 25^a (100) =	_____ gpd(L/day) (L/day)
E. Misc. (winery kitchen, tasting room^c or laboratory)	_____	_____ gpd(L/day)
Grand Total A+B+(C or D whichever is greater)	+ E =	_____ gpd(L/day)

Note: For new winery planning, visitor values will have to be estimated based on a percentage of traffic flow on a major thoroughfare near the winery or on visitor records for other commercial or recreational facilities in the area.

^{a}Four flushes at 5 gal (20 L) plus five handwashes at 1.0 gal (4 L).

^{b}One flush at 5 gal (20 L) plus one handwash at 1 gal (4 L).

^{c}Additional sanitary wastewater flows from tasting room glass washer or from winery kitchen or laboratory should be taken into account. [Flows from the kitchen or laboratory can be estimated using the number of installed "fixture units" and their respective flow rates as given in Tables 7.3 and 7.9 of Section 703.2 of the Uniform Plumbing Code (106).]

sign environmental engineer may have a different method for estimating sanitary wastewater flows other than that suggested in Table 8.1; however, the magnitude of the numbers developed by both methods should not vary by more than plus or minus 5%. Also, the derived flow values must stand the scrutiny of the local governmental agency that is responsible for review and approval of the winery.

Sanitary Wastewater Characteristics

Other characteristics of winery sanitary wastewater which influence the system design should be noted. The winery's sanitary discharge will be very similar to the physical and chemical properties published in most textbooks and references which detail the nature of the solids content and chemistry of the average, unadulterated domestic wastewater (104, 105). One sanitary wastewater constituent that should be given careful consideration by the design engineer is grease (fats, oil, and waxes) that result from restaurant or food-service activities. An employee lunchroom would not qualify for the latter category unless food preparation and service was performed. The grease load from a 150–200-meal/day restaurant is impressive even in this era of the desirable low-fat diet. At the required commercial dishwasher wash cycle and rinsing temperature of 150°F–180°F, grease is solubilized, but when discharged to the colder sewer system, it has a tendency to congeal, producing fatty, soapy deposits on the walls of the sewer. The grease that does remain in the waste stream and which reaches the septic tank (in the case of an on-site winery system) becomes a long-term problem in the digestion compartment, as decomposition of fats and oils occur at a much slower rate than do the other forms of carbonaceous waste. Should hydraulic overloading of the septic tank system occur, there is a relatively high probability that any undigested grease (sometimes in the form of an oily scum on the septic tank's free-water surface) might be transported to the soil absorption portion of the treatment/disposal works with almost certain reduction of soil permeability on the leaching trench bottom and sidewalls. Thus, for the grease-vulnerable septic system, entrapment of grease upstream from the septic tank must be a feature of the design. This is most often accomplished by the installation of a so-called grease trap, which by means of deliberate hydraulic detention and baffling, a properly sized unit can effectively remove 80–90% of the grease content. Both the Uniform Plumbing Code (106) and the Design Manual for Onsite Wastewater Treatment and Disposal (104) suggest that the grease trap be designed to store a grease mass in pounds equal to two times the maximum kitchen fixture discharge rate in gallons per minute (grease mass to 13.6 times the maximum rate in liters per second). Whereas this apparent mismatch of dissimilar units of measure seems less than a rigorous scientific relationship of variables, it does provide an empirical guide to selecting a grease trap that will serve the winery's intended restaurant or food service discharges.

A more mathematically correct approach is outlined in Ref. 104 and is identified as being appropriate for use in estimating the capacity of grease traps for restaurants, exclusively. The relationship is outlined as follows:

Grease trap capacity in gallons (liters) = D GL (L) × ST × HR/2 × LF

where D is the number of dining room seats, GL (L) is the gallons (liters) of wastewater produced per meal served (usually 5 gal/20 L for food preparation and dishwashing), ST is the grease storage capacity factor (2.5 for disposal to septic tank and leachfield system), HR is the number of hours of restaurant operation, and LF is the loading factor (from 1.25 for the busiest freeway location to 0.5 for rural roads and highways)

The minimum acceptable grease trap size suggested is 750 gal (2840 L); however, the local statutes and ordinances which govern plumbing practice should be compared by the winery's architect and utility systems designer to ensure that the grease interceptor will comply in all respects. Figure 8.1 shows the construction features of a typical grease trap (which will be later seen to closely resemble a conventional two-compartment septic tank). As noted in Fig. 8.1, for wineries with food service facilities and with grease traps installed on the kitchen sewer, the installation of garbage grinders is not recommended, as the high organic solids loading interferes with the performance of the grease trap. Grease accumulation in the trap should be

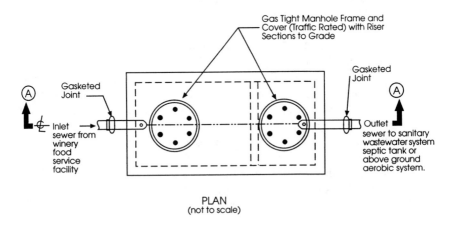

PLAN
(not to scale)

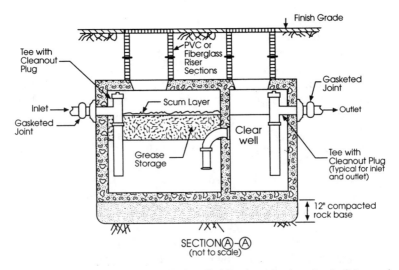

SECTION (A)-(A)
(not to scale)

NOTE: Garbage grinders should not be installed in winery food service facilities, as the high solids loading interferes with the operation of the grease trap.

Fig. 8.1.   Typical precast concrete two-compartment grease trap.

checked periodically and removed by a commercial septic tank pumping service when the floating grease mass exceeds 8 in. (200 mm).

In addition to grease, there are other substances that can find their way into the sanitary wastewater treatment disposal system that could seriously hinder the orderly digestion of wastes in the solids chamber of the septic tank. It is probably more appropriate to discuss *what not* to pour down the drain in the section entitled "Operation and Maintenance of Wastewater Systems," but it is important enough to repeat here for emphasis. Spent brine discharges from water softeners are sometimes plumbed to the sanitary wastewater system for construction convenience. There is no unanimous agreement among sanitary/environmental engineers and soil scientists on whether there are, in fact, adverse effects on the performance of septic tanks and on the soils in the leachfield from a periodic assault of water-softener-originated sodium chloride brine (107). Of lesser importance will be the small quantities of chemical reagents from the winery's laboratory and the similarly small volumes of wine from the tasting room bottles that are routinely dumped at the end of each tasting day. Finally, the choice of toilet paper, strictly a system operational choice, can minimize the problems of paper carryover to the leachfield, which can, on occasion, occur with non-septic-tank approved papers (108). Although the foregoing discussion may imply that conventional septic systems are fragile—they are—and every design and operational opportunity should be taken to sustain their optimum performance. Recovery from abuses are troublesome and costly. Well-designed and prudently operated on-site systems should last 20–25 years, if given reasonable care.

Conventional Septic Tank and Leachfield System Components

The conventional septic tank and leachfield system will be an acceptable design option only when rather specific design criteria are met. The design alternatives to a conventional system are discussed in detail in the section entitled "Alternative Wastewater Systems." A listing of the most generally accepted design criteria are included in Table 8.2. Not all codes, ordinances, and design guidelines can be covered within the scope of this book. As suggested previously, the specific regulations, which apply in the particular governmental jurisdiction, should be sought by the winery's utility design team, so that the designs submitted for review and approval will reflect the local statutory requirements. If they vary significantly from the values reported in Table 8.2, the designer should make a formal challenge and determine the source and logic for the statutory requirement.

Two of the most successful design features to sustain the useful life for conventional septic tank–leachfield systems are the construction of dupli-

Table 8.2. Generally Accepted Design Criteria for Conventional Septic Tank and Leachfield Systems

Minimum horizontal distance in clear required from:	Building sewer	Septic tank	Disposal field	Seepage pit or cesspool
Buildings or structures[a]	2 ft (0.6 m)	5 ft (1.5 m)	8 ft (2.4 m)	8 ft (2.4 m)
Property line adjoining private property	Clear[b] (15.2 m)	5 ft (1.5 m)	5 ft (1.5 m)	8 ft (2.4 m)
Water supply wells	50 ft (15.2 m)	50 ft (15.2 m)	100 ft[g] (30.5 m)[g]	150 ft (45.7 m)
Streams	50 ft[c] (15.2 m)	50 ft (15.2 m)	50 ft[g] (15.2 m)[g]	100 ft[g] (30.5 m)[g]
Trees	—	10 ft (3.0 m)	—	10 ft (3.0 m)
Seepage pits or cesspools	50 ft	5 ft (1.5 m)	5 ft (1.5 m)	12 ft (3.7 m)
Disposal field	—	5 ft (1.5 m)	4 ft[d] (1.2 m)	5 ft (1.5 m)
On-site domestic water service line	1 ft[d] (0.3 m)	5 ft (1.5 m)	5 ft (1.5 m)	5 ft (1.5 m)
Distribution box	—	5 ft (1.5 m)	5 ft (1.5 m)	5 ft (1.5 m)
Pressure public water main	10 ft[e] (3.0 m)	10 ft (3.0 m)	10 ft (3.0 m)	10 ft (3.0 m)

(see footnotes on pg. 203)

Septic tank capacity	1.5 times maximum daily flow (up to 1500 gpd); 0.75 times maximum daily flow + 1125 (>1500 gpd)
Leachfield details	Maximum length of line—50 ft.
	Slope of leachfield trenches and leachlines—0%
	Leachfield replacement area reserve—100%
	Leachline pipe perforations—⅜-in. diameter to ¾-in. holes at 6–12 in. spacing
	Leachline intervals—• PVC Schedule 40 (ASTM D2665) • ABS Schedule 40 (ASTM 2661)
	Leachline spacing—3–6 ft outer to center
	Installation—laid with perforations down
Slope	In deep soils—>8 ft soil depth: 0–15%; in shallow soils—<8 ft soil depth: 0–10%
Soil depth	Slope flat to 8%—5 ft minimum depth of soil meeting permeability requirements below the trench bottom
	Slope 9–15%—8 ft minimum depth of soil meeting permeability requirements below the trench bottom
Permeability	5–120 min/in. (lower permeabilities are permitted in some jurisdictions if extraordinary low applications rates are used and duplicate leachfields are constructed to permit alternate use)
Depth to groundwater	Depth to seasonal (wet season) groundwater must be 5 ft or greater below the bottom of the trench
Permitted artificial drainage	On sloping sites with seasonal or perennial subsurface drainage problems, so-called curtain drains or interceptor drains may be constructed to control soil moisture within the proposed leachfield area. (See Ref. 104, Section 7, p. 261.)

NOTES: [These notes refer to Table 8.2 which is a direct citation from the Uniform Plumbing Code.] When disposal fields and/or seepage pits are installed in sloping ground, the minimum horizontal distance between any part of the leaching system and ground surface shall be 15 ft (4.6 m).

*Including porches and steps, whether covered or uncovered, breezeways, roofed porte-cocheres, roofed patios, carports, covered walks, covered driveways, and similar structures or appurtenances.

*See also Section 313.3 of the Uniform Plumbing Code.

*All drainage piping shall clear domestic water supply wells by at least 50 ft (15.2 m). This distance may be reduced to not less than 25 ft (7.6 m) when the drainage piping is constructed of materials approved for use within a building.

*Plus 2 ft (0.6 m) for each additional foot (0.3 m) of depth in excess of 1 ft (0.3 m) below the bottom of the drain line. (See also Section 16.)

*See Section 720.0 of the Uniform Plumbing Code.

*For parallel construction. For crossings, approval by the Health Department shall be required.

*These minimum clear horizontal distances shall also apply between disposal field, seepage pits, and the ocean mean higher hightide line.

Source: Table 8.2 is reprinted in its entirety from the Uniform Plumbing Code with permission of the International Association of Plumbing and Mechanical Officials©, copyright 1994.

cate leachfields to provide both a dosing and a resting cycle, and a "dosing tank," a wastewater accumulator which fully loads the leachfield periodically. The key feature of the dual leachfield system is a hand-operated diversion valve that can shunt the effluent from one field to another. This feature becomes particularly important for sites which have conditions (soil, slope, or soil permeability) that are near the design limit envelope. For example, for a winery whose visitor traffic peaks can be anticipated, the alternate use of duplicate leachfields during high-attendance days, can often prevent hydraulic overloading and permit the soil absorption system to continue to operate at maximum efficiency. Alternating leachfields weekly or monthly might be the normal cycle for non-peak periods.

The dosing siphon is an adjustable hydraulic accumulator, which when properly sized can periodically fully charge the perforated leachfield piping. Ideally, two charge cycles per day have been found to produce the maximum performance in leachfield systems (104). Without the dosing siphon feature, the leachfield is subject to frequent, but low-volume flows during the winery's working day. The net negative impact of the small volumes is to maintain the first 10–15 ft of the leachline network in a nearly constant hydraulic loading state, leaving the balance of the leaching trenches relatively untouched. This hydraulic and solids overloading of a small sector of the leachfield can lead to eventual failure of that sector, with the failure eventually extending the entire length, as the short perforated pipe segments shunt the flow to the trenches immediately downstream. For the conventional leachfield systems that require a charge pump (i.e., leachfield situated topographically above the septic tank), the pump performs the measured dose function, if the pump wetwell has the requisite storage volume. Pump stations are described in the following subsection. The leachfield charge volume is generally calculated at about 75% of the entire leachfield perforated pipe volume. Other nonconventional or so-called engineered alternative on-site systems have a dosing volume criterion that differs from the 75% nominal pipe volume value cited in this subsection.

A typical conventional septic tank and leachfield system is shown in Fig. 8.2 and the principal components of a dosing siphon are shown in Fig. 8.3. Figure 8.4 depicts the details of a typical concrete, two-compartment septic tank.

SEWAGE PUMP STATIONS

Prudent siting of the winery on the gross parcel area by winery architect and owners can often give the utility design engineer the luxury of creating a wastewater system which functions entirely by gravity. Having to resort to

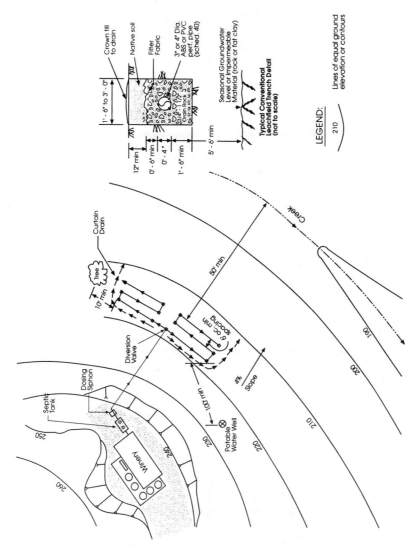

Fig. 8.2. Typical sanitary wastewater conventional septic tank and leachfield system showing required setbacks and duplicate leachfield arrangement (not to scale).

205

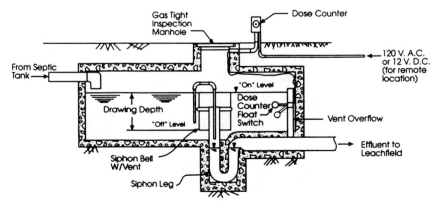

NOTE:
• The volume cycled with each drawdown can be varied by adjusting the height of the siphon bell.
• Float activated dose counters are a simple method of monitoring flow to the leachfield. This inexpensive accessory is a must item for a winery with a short-handed utility maintenance staff.

Fig. 8.3. Principal features of a leachfield dosing siphon.

pumping is often the only solution, albeit a future long-term commitment to energy use and a maintenance and operation worry for the life of the winery. Involving the utility design engineer very early in the site planning process can often result in a winery development that reduces construction and future wine production and utility system operational conflicts to a near absolute minimum. The nature of the potential site planning compromises and benefits has been described, wherein the fundamental philosophy of shared site planning by all the design team professionals is expressed (109, 110).

Pump stations can be of several types and subtypes. The most advantageous pump arrangement is one in which the suction end of the pumping equipment is "flooded" or where a positive suction pressure (or head, as it would be expressed in hydraulic nomenclature) permits the pump to operate without the need to expend energy to literally pull the fluid into the volute or impeller region of the pump. The pump is, after all, adding kinetic energy to the fluid by converting rotational, mechanical energy to hydraulic energy (less the ever-present friction losses, of course). Thus, for a given horsepower, the difference in energy levels between the suction (inlet) side and the discharge (outlet) side as indicated by pressure levels at those locations favors optimum energy efficiency if the suction pressure is a positive value and not a negative or vacuum. This ideal situation can be achieved in one of three ways:

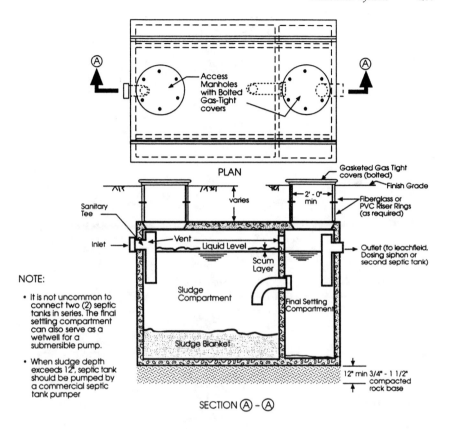

PLAN

NOTE:

- It is not uncommon to connect two (2) septic tanks in series. The final settling compartment can also serve as a wetwell for a submersible pump.

- When sludge depth exceeds 12", septic tank should be pumped by a commercial septic tank pumper

SECTION Ⓐ - Ⓐ

(1) Ready access to the septic tank is necessary, preferably by means of an all-weather paved access. It is an important siting issue to be jointly decided by the winery architect and utility design engineer.
(2) Some governmental jurisdictions require the use of chemical sealants on the inside of septic tanks and dosing siphons to prevent any exfiltration of contents.
(3) Septic tanks constructed from special plastics and fiberglass are also commercially available in a variety of capacities. Price and governmental acceptability of the alternative to concrete will dictate the winery's choice of one product over the other.

Fig. 8.4. Configuration for typical concrete septic tank. (not to scale)

1. By use of submersible pump in a wet pit or vault that is suitable for sewage service
2. By use of a dry pit for close-coupled sewage pumps in a partitioned wet and dry pit vault arrangement
3. By use of a turbine-type pump installed in the wet well with the drive motor at the surface connected to the pump by a shaft

There are several other combinations of pump and storage pit arrangement, but those shown schematically in Fig. 8.5 are the most frequently

A. Wetwell using precast concrete manhole, modified septic tank, precast concrete vault or poured-in-place concrete vault with a submersible sewage pump.

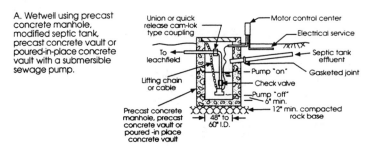

B. Poured-in-place wetwell and dry pit with close-coupled centrifugal sewage pump.

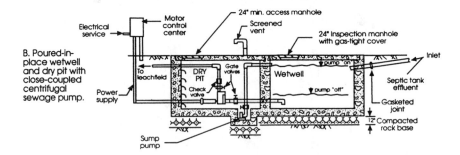

C. Wetwell using precast concrete manhole, vault, septic tank or poured-in-place vault with turbine-type sewage pump.

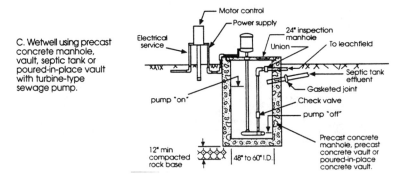

NOTE: Pump machinery for raw sewage pump stations differs from that shown in this figure. (See text for discussion of raw sewage pump stations.)

Fig. 8.5.   Typical septic tank effluent pump station configurations showing possible wet well types and pump machinery options.

used by design engineers. As suggested previously and shown in Fig. 8.4, the second compartment in a series arrangement of two septic tanks can also serve as a wet well for a submersible pump. The transverse concrete wall separating the tank into two compartments is breached, so that the total volume of the septic tank can be utilized in the pump drawdown cycle.

In this situation, it is also prudent to set the submersible pump 6–12 in. (150–300 mm) above the bottom to prevent the discharge of any solids carryover that might occur from the first septic tank in the series. Economy of construction is often the criterion which favors the use of a septic tank for a pump wet well. The precast, concrete construction can often be considerably less expensive than a poured-in-place vault of the same size. Local availability of good quality, concrete septic tanks and the designer's own preference will ultimately influence the final choice. Fiberglass and other plastic resins have also been used for septic tank construction. Because of their lightweight construction, they can, if necessary, be lifted by four laborers and placed in the finished excavation, whereas a crane is required to place the 10–12-ton concrete equivalents.

An important issue that must be addressed by the design engineer is the control of odors produced by effluent that has become septic due to effluent pump stations that are infrequently cycled (111). The effluent leaving the septic tank is anaerobic and with warm summer temperatures and as little as a 10–12-h quiescent period in the wet well, methane and hydrogen sulfide can be generated. Wineries that have this as a chronic problem resulting from poor siting of pump stations upwind from public areas (i.e., parking lots, tasting rooms, and picnic areas) have several choices for abating the nuisance odors:

1. Move the pump station to a less vulnerable location (probably the most expensive and disruptive solution)
2. Install a hydrogen peroxide (H_2O_2) injection system to prevent the formation of hydrogen sulfide gas (111, 112).
3. Install a landscape or barrier wall to promote vertical dilution of nuisance odors (112–114).
4. Install a programmable pump controller that can override the preset wet-well level controls when the pump cycle interval exceeds a programmed value. A second-stage control parameter could also be included that would energize the pump at the end of each winery working day, so that the partial leachfield dose volume would not remain overnight in the wet well.
5. Do not resort to the use of industrial perfumes to mask the offensive odors, as the resulting aromas are likely to be worse than the hydrogen sulfide alone.

Of all the odor suppression solutions for pump stations, the hydrogen peroxide injection scheme is the most effective. The limitation on its use is that H_2O_2 is extremely corrosive, even at the 40% dilution normally used for small odor suppression problems. Safe handling of the material by the winery staff and the implementation of special security measures to exclude winery visitors from the vicinity of the H_2O_2 injection system are essential items with use of this odor control alternative. Check with the

chemical supplier for safe handling instructions and the availability of
H_2O_2 in your locale. Finally, enlist professional guidance from a competent
utility engineer on the design of the system.

The pump station for the sanitary wastewater effluent is classified as a
critical system component. Its failure from a loss of electrical power or a
mechanical/electrical malfunction could result in a sewage overflow and
the health risks associated with such an event. The design engineer can
increase the reliability of the system in a number of ways:

1. Provide for additional storage volume in the wet well for a 1- or 2-day
   emergency at maximum flow.
2. Provide a separate emergency overflow system from the pump wet well to an
   adjacent storage tank that is earmarked only for temporary storage (the
   overflow tank would have to be equipped with a small submersible pump
   and piping that would reintroduce the stored effluent into the pump
   wetwell after the malfunction is repaired).
3. Install duplicate pumps that would be identical in service characteristics that
   would alternate use on each required pump cycle. This system requires the
   installation of a duplex controller that performs the alternating use function. Over the useful life of the project (20–25 years), the wear on each
   pump is only one-half of the wear that would prevail under a simplex system.
4. Finally, for wineries in areas that are subject to chronic power outages, the
   installation of a standby motor–generator set may be a necessity. (This aspect of a nearly fail-safe supply of electrical energy was previously covered in
   Chapters 2 and 6.)

There are distinct differences in the construction and capabilities of
sewage pumps used for pumping septic tank effluent (primary effluent)
and raw sewage. Even for wineries on small, steeply sloping parcels (and
there are many in this category in the coastal regions of California), there
is generally enough space to allow gravity flow of raw sewage from the
building sewer to the septic tank. Almost all of the previous discussion of
effluent pump stations applies to raw sewage pump stations as well. The
major differences are the following:

1. Because of the solids content of the raw sewage, a pump capable of passing
   a specified size of solids (manufacturers specify pumps solid handling capability as being able to pass a 3-in. ball or 1½-in. ball, etc.) must be selected.
   Grinder pumps, which are a subcategory of sewage pumps, have a grinding
   stage installed ahead of the pump impeller, which reduces the solids to a size
   which can be more easily pumped. Also, sewage pumps are often equipped
   with a face plate on the pump housing that is easily removed in case the
   impeller is jammed or the volute clogged with rags or other stringy debris.
2. Raw sewage wet wells will have a greater tendency to produce odors than

septic tank effluent wet wells. The solutions offered for odor control in this subsection previously apply equally well to raw sewage pump stations.

3. Grinder pumps have a secondary benefit in that they augment the anaerobic digestion of solids to take place in the septic tank at a slightly higher rate, because bacteria will metabolize the smaller-sized waste solids more quickly, all other things being equal.

Instrumentation and controls for effluent and raw sewage pump stations consist of a basic package of liquid level controls ("off," "on," and "high water alarm"), the master motor electrical control panel necessities of a starter (either across-the-line or reduced voltage type, depending on pump power), and a circuit breaker to disconnect the pump motor electrically in the event of a system malfunction (115). The basic control and instrumentation package can be augmented to do the following:

1. Send an alarm signal from the pump station to a master display panel at the winery, if the two locations are remote.

   (a) The signal can be an audio or visual or both.

   (b) The alarm signal can also be transmitted via the winery's commercial security and fire alarm system to the 24-h dispatcher.

   (c) The alarm signal can energize an automatic telephone dialer (described previously in Chapter 6) that will dial a series of preprogrammed numbers until it reaches one of the key staff in the dialing index.

2. If the pump station is not remote from the winery and the wet-well system has been designed with an emergency overflow tank, a visual alarm [red warning light on a mast, clearly visible to the responsible employee(s) entering the winery each workday] is probably sufficient.

The control and instrumentation elements for the pump station can be expensive if all the possible human alert and alarm options are incorporated into the design. The approving agency will be looking for a well-designed system with a high degree of reliability, such as the duplex pump installations previously described. Thus, the system designer should try to balance total system cost by maximizing pump station reliability and including possibly only the basic control and alarm package. If the approving agency is insistent on all the bells and whistles for controls and alarms, then the designer might consider deleting standby emergency power and pump duplexing to prevent the ballooning of construction costs beyond all reason. A good designer will understand the technical issues and their cost implications and provide a system that will perform well by any set of acceptable engineering quality standards. An important aspect of the designer's role in plan review are the sometimes protracted negotiations with the responsible agency staff, which often take the form of trade-offs of "preventative comfort" for higher quality and more reliable pump system features and components. (When a winery developer is in the very early process of screening candidates for the design team, a reasonable question that should be asked is "What is your philosophy and methodology in coping with the oftentimes lengthy process of design review and the procurement of permits and approvals?)

To avoid the cost of long runs of buried signal cable to transmit pump station status and alarm signals from a remote location, the author has, on occasion, elected to use local telephone company lines (where capacity exists) to transmit signals from one location to one or more supervisory control centers. The annualized costs of leasing the telephone circuits and the use of single-purpose buried signal cable are easily compared to find the least expensive alternative (construction and future operation and maintenance cost over the life of the project). This scheme is much like the commercial security system method of alarm signal transmission either through the use of dedicated telephone lines or alternatively, microwave transmitters. The potential use of leased telephone lines by wineries was discussed briefly in Chapter 3.

Before leaving the subject of sanitary wastewater septic tank systems, mention will be made of the role of bioactivators (freeze-dried bacteria/ yeasts and enzyme blends) in the performance and operation of septic tanks. Commercially prepared bioactivators are in the marketplace under a number of brand names. A variety of claims are made regarding the benefits of using these products, which range from dissolving grease to unplugging leachfields. The sanitary waste stream and healthy septic tanks are already abundantly endowed with a superflora of microorganisms, so that providing an additional freeze-dried megapopulation does little to improve septic tank digestion performance unless biotoxic materials have produced a massive die-off of all the system's flora. The companion enzyme additives in some bioactivators may, in fact, assist in the breakdown of complex fats and oils, but measuring the benefit against the cost of the additive is a difficult if not impossible task. The use of bioactivators will be discussed in detail, as their role in the start-up and maintenance of a standing crop of organisms in above-ground, aerobic treatment reactors for winery process wastewater is of special importance (116).

Package Treatment Plants and Other Above-Ground Alternatives (Sanitary Wastewater)

Above-ground treatment systems including small package treatment plants (systems that consist of prefabricated steel tanks that are supplied by the manufacturer and factory assembled and that can be installed and made operational in a very short time period) have not been used extensively for either sanitary or process wastewater systems in wine regions of the western United States. They are, in fact, prohibited in some California jurisdictions, unless the intended user is a public entity (i.e., a special district or municipality) (117). The package sewage treatment plants also fail to meet the design criteria for specially engineered wastewater systems (alternative systems) in the guidelines of both the Napa and Sonoma County wastewater system development code. There are states which do permit the use of package plants. Your designer or the local water pollution control agency can explain the policy regarding their possible application for wineries. The exception to the latter statement is for large wineries; because of sheer size (number of employees, etc.), remote location, and volume of sanitary wastewater, they may be forced to install what amounts to a small municipal-type waste treatment system (aerated lagoon, activated sludge, or trickling filters) to achieve the proper level of treatment.

Although the septic tank–soil absorption disposal systems require only modest care by winery maintenance staff, aboveground systems require a higher level of monitoring and management attention. Specially trained operators for a wastewater treatment plant, often with state-sanctioned certifications, are necessary to ensure that the wastewater receives adequate treatment and disinfection 100% of the time. In the Napa–Sonoma County area of northern California, an enterprising temporary employment firm established a wastewater treatment system operator division, which could tailor the part-time technical help and assistance to the needs of the system, local staff maintenance personnel capabilities, and workload (118). As most winery utility maintenance personnel also have duties for mechanical and electrical equipment, the time that can be devoted to wastewater system management and operations is highly restricted, except for large to very large wineries with well-trained and staffed maintenance specialists. Above-ground systems for sanitary wastewater can be made to work for small wineries, but the following problems need to be recognized and overcome by the designer:

- Capital and annual operating costs will be considerably greater than for a comparable septic tank and leachfield system.
- If local ordinances and water pollution control regulations do not prohibit package treatment plants, ensure that the company supplying the equipment has a reasonable chance of surviving in the future, so that service and replacement parts can be obtained when required.
- With the proper combination of site attributes of soil, slope, and drainage, some rather creative methods of disposal of disinfected package treatment plant effluent may be possible:

(a) "Overland flow" for seasonal surface filtration and nutrient removal in a growing, managed grass cover crop

(b) Creation of seasonal artificial wetlands for plant propagation nutrient removal and habitat creation

- Certainly, the treated and disinfected effluent from a package treatment plant could be diverted to a conventional or specially engineered soil absorption system (specially engineered or so-called alternative systems are described in a subsequent section).

The "overland flow" and artificially created wetland disposal options are limited to seasonal use in Mediterranean-type climates, where winter rainfall would at times render the two disposal sites rainfall-saturated and thus unsuitable for disposal purposes. In many wine producing areas with subnormal rainfall and where solar evaporation exceeds rainfall by a factor of 10–12, disposal of effluent by evaporation is genuinely possible. The wastewater production–solar evaporation hydrologic balance is possible in the Temecula grape growing region of California's San Diego and Riverside Counties and elsewhere in selected regions worldwide.

The use and limitations of conventional oxidation ponds and facultative ponds for treatment of winery sanitary wastewater will be discussed in detail as a treatment option in the section entitled "Process Wastewater Characterization." Ponds will be discussed briefly and their limitations for treatment of sanitary wastewater expressed particularly as to the potential for nuisance odor generation during seasonal "turnovers" (limnologic reversal of top and bottom pond layers in response to ambient temperature changes) and their rather large spatial requirements for proper operation. Table 8.3 illustrates the space/area requirements for a conventional oxidation pond and compares it to a conventional septic tank and leachfield system to accomplish approximately the same level of primary treatment.

As shown in Table 8.3, nearly 200% more land area is required for pond treatment (treatment only and no disposal) by a conventional oxidation pond than for a complete septic tank and leachfield system for comparable flow rate and solids loading. Thus, it can be concluded that if abundant land area exists on the winery site, an oxidation pond could be accommodated. The designer, if given this option for study and evaluation, must balance all the other design criteria previously mentioned (preferred gravity flow, visual impact, length of raw and treated wastewater sewer, and nuisance odor impacts on winery visitor and neighboring property). Mitigating features, such as landscape odor screens and barriers, must be included in the overall economic analysis of the pond alternative.

The facultative pond is a subgroup of the oxidation pond classification. Whereas the shallow, 3–5 ft (0.9–1.5 m) depth of the true oxidation pond

Table 8.3. Comparison of Spatial Requirements for a Conventional Oxidation Pond and Conventional Septic Tank and Leachfield System for Winery Sanitary Wastewater

	Oxidation pond	Septic tank and leachfield	Remarks
Daily flow (gal)	1000	1000	
Estimated 5-day BOD[a] (lbs/day)	7	7	Based on daily per capita BOD of 70 g
No. of employees	40	40	
Pond acreage required for surface loading rate at 150 lb/day/acre	2033 ft² ± *or* 0.05 ac ±	—	Includes pond surface area only and not perimeter levees (add 30%)
Septic tank and leachfield area requirements (ac.) • 1500-gal septic tank • Load rate: 1 gal/sf/day • trench required: 250 ft	—	900 ft² ± *or* 0.02 ac	Use a trench sidewall area only at 4 ft²/ft of trench

[a]BOD = biochemical oxygen demand.

generally allows aerobic conditions to prevail throughout the full pond depth (reaeration by wind and wave action and algae respiration), the facultative pond at 10–12 ft (3–3.7 m) in depth has two zones [i.e., a shallow 3–5-ft (0.9–1.5 m) aerobic zone and a 7-ft (2 m) anaerobic zone]. The dissolved oxygen level in the aerobic zone is often augmented by use of mechanical aerators or diffused air, injected by submerged laser-out tubing in a grid pattern throughout the pond at or near the aerobic–anaerobic zone interface. Again, the disposal scheme for this type of pond treatment system must employ one of the acceptable options previously described for above-ground alternatives.

For larger wineries with a good grasp of the complicated logistics that sometimes accompany the dispersal of operations from one centralized location to several satellite locations, the quartering (or greater subdivision) of the wastewater production load can often open up a new series of wastewater treatment and disposal possibilities, just because of the reduction in overall scale of each subunit operation. This is often done to bring a particular wine grape processing function closer to the vineyard source. J. Lohr and Glen Ellen wineries, for example, have advanced this production technology innovation to its highest level. For J. Lohr, white grape variety pressing and red grape crushing/destemming/fermenting are performed at two distant vineyard locations with wine finishing performed at the home winery in San Jose, California. Glen Ellen's bottling operations have been relocated from the central processing facility to a location some 2 miles distant from the home winery. The reader is left to place their particular mix of production and wastewater treatment and disposal problems in the framework of potential utility system simplification from process operations dispersal. Certainly, there are more variables to be entered into the dispersal concept equation such as transportation costs, local versus dispersed site real estate costs and taxes, and differences in local governmental interest and oversight in winery operations from one locale to another.

Alternative Wastewater Systems (Specially Engineered Systems; Sanitary Wastewater)

Alternative or specially engineered systems are often the only recourse for sanitary wastewater systems when the design criteria previously tabulated in Table 8.2 cannot be met. Table 8.4 summarizes the principal features of the most commonly used alternative wastewater systems. It should be noted that a requirement that is common to all systems, except the ET Bed, is the need for a dosing unit (pump or dosing siphon) to periodically load the system to a specified volume. This hydraulic requirement was discussed previously under the section entitled "Conventional Septic Tank and Leachfield System Components," and a typical dosing siphon was depicted in Fig. 8.3.

One of the safety factors that should be incorporated in any alternative system design is the duplicate system concept suggested previously as a desirable feature under the section entitled "Conventional Septic Tank and Leachfield Systems." Even though the accepted design criteria (104, 119–121) for specially engineered systems are very conservative, the alternate dosing and drying of mounds, beds, and pressure distribution trenches in low-permeability soils, offers the best chance for long-term, optimum performance of the systems. If duplicate systems are not installed at the time of initial construction, then, as a minimum, reserve areas equal to at least 100–200% of the land area required for the single system should be set aside for future system replication if failure occurs or an increase in capacity becomes necessary. Although this provision may appear to predict a future system failure, it is more prudent to reserve open areas as future construction easements rather than having to choose the alternative of destroying producing vines or undertaking the expensive moving of infrastructure or buildings to accommodate a replicate system.

Monitoring wells placed at strategic locations in and around a specially engineered system allow the hydraulic performance of the system to be checked and allow early detection of problems before they become chronic and irreparable. Figure 8.6 shows a typical monitoring well and construction details.

One of the most important post-construction tests that is made on the dosed, alternative systems is the so-called "squirt test." This test verifies to the design engineer, contractor, and the responsible local governmental agency that, when operated, the measured dose will be distributed evenly throughout the system with approximately a 2-ft (0.61-m)-high column (squirt) above the pipe. There are two methods of accomplishing the squirt test:

(a) The laterals are rotated so that the orifices are pointing vertically upward. The lateral joints of the main line connection are not solvent-welded until the successful squirt test is completed.

(b) A threaded hole of the proper diameter is drilled in the crown of the lateral pipe just downstream from the blow-off valve. The system is charged and the squirt height noted and adjusted to the 2-ft (0.61-m) requirement. This allows the contractor to connect the laterals permanently before testing, which prevents unwanted backfill soil from entering the trench, as frequently happens if squirt test method (a) is used. The plugged orifice arrangement is shown in Fig. 8.7.

The 2-ft (0.61-m) head or pressure is the equivalent of about 0.87 pounds per square inch (psi) at the discharging orifice. In the design and layout of the systems, it is not always possible to get the perforated pipe laterals all the same length and to receive equal pressure throughout the pipe network. Balancing the orifice flows during the squirt test is accomplished by installing PVC ball or globe valves at the inlet to each lateral from the main distribution pipe. The valve is then adjusted to the proper flow for the squirt height to be achieved. Also, PVC ball valves are installed at the end of each lateral in a fiberglass or PVC valve box with a cover, so that, periodically, lines may be flushed of any accumulated biological slimes or other solids. The design guidelines for alternative systems and the references cited do not incorporate the blow-off valve detail suggested, but it has been found to be an essential design feature not only for future maintenance but also to flush the lines of any soil, sand, or drain rock that may have inadvertently found its way into the pipe system during construction. A detail of the lateral blow-off valves and the test orifice arrangement for Wisconsin Mound and Shallow Trench Pressure Distribution Systems as used for squirt test method (b) are shown in Figs. 8.7a and 8.7b.

Preferences for the size (diameter) of orifices (perforations) in the distribution laterals for mounds, Evapotranspiration-Infiltration/Evapotranspiration Beds (ETI/ET beds) or pressure trenches are most often based on the designer's experience and understanding of the operational rigors to which the system will be subjected. The designer must balance the

Table 8.4. Characteristics of Typical Alternative (Specially Engineered) Winery Sanitary Wastewater System Disposal Options

Common Name(s)	Soil Permeability Limits (min/inch)	Design Detail	Other	Remarks
1. Wisconsin Mound	Used in areas with low soil permeability and/or slopes up to 6% (12% on permeable solids ≤120 min/inch). Also used in areas with shallow soils and high groundwater		a) Supplemental irrigation of perennial cover may be required b) Requires a charge pump system c) Mounds are longer than wide in tight soils	• Must be designed by registered civil engineer or sanitarian • A 200% reserve area set aside • Classified as an "innovative" alternative system
2. Shallow Trench Pressure Distribution System	Can be thought of as the trench version of a mound system. Used in tight soils on slopes up to 30% and percolation rates of 120 min/inch		a) 2' min. depth to groundwater, bedrock or imperm. soil b) Requires a charge pump system c) Require curtain drains upslope from drainfield to intercept groundwater interflow d) A minimum of 6 monitor wells shall be installed within and downgrade from the trenches	• Must be designed by registered civil engineer or sanitarian • A 200% reserve area set aside • Not a recommended option when slope requirements can be met for a mound system • Classified as an "innovative" alternative system

Common Name(s)	Soil Permeability Limits (min/inch)	Design Detail	Other	Remarks
3. Trickle or Drip Tape	Used in areas with low soil permeability and shallow depth		a) Tape is constructed of polyethylene tubing with small diameter emitters on 2 ft. spacing b) Tape system is of the close-loop type to prevent orifice plugging c) Charge pump and filter or strainer required d) Still undergoing performance testing and evaluation in New Jersey and Texas	• Must be designed by registered civil engineer or sanitarian • A 200% reserve area set aside • See Ref. 129 for further reading on drip tape systems
4. ETI Beds (Evaporation Transpiration, Infiltration Bed)	Used in areas with low soil permeability (> 120 min/in)		a) Supplemental irrigation of perennial cover may be required b) Slopes to 6% if curtain drain installed on upgrade side of bed c) Depth to seasonal groundwater ≥ 5 ft. below bed bottom	• Must be designed by registered civil engineer or sanitarian • Dosing siphon or charge pump required to fully load bed hydraulically for each cycle • Double bed size and alternative cycling between 1/2 beds during maximum loading • See Ref. 121 for approved design details for California practice
5. ET Beds (Evapotranspiration Beds)	Used in areas with very low or no soil permeability		a) ET rate for project area must equal or exceed the rainfall plus wastewater volume for the year b) Good design sense would have a second ET bed in series with overflow to accommodate very wet rainfall cycles c) Monitoring wells are required (see Figure 8.6)	• Must be designed by registered civil engineer or sanitarian • Dosing siphon or charge pump nut required • Plant cover must be salt tolerant, as salt build-up will occur in ET beds • See Ref. 121 for additional design details

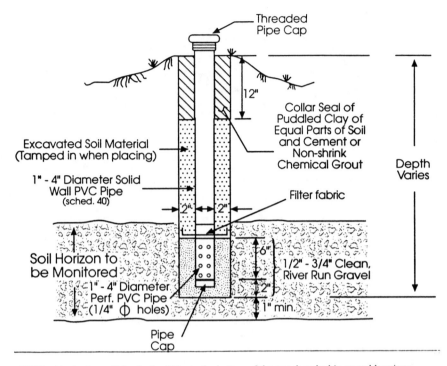

NOTE: Monitoring well depth should be to the bottom of the trench or bed in several locations within the bed or mound system and penetrate to a depth 6 inches to 12 inches below trench or bed depth outside the bed or mound system (upgrade or above and downgrade or below the trench or bed construction area).

Fig. 8.6. Typical water level monitoring wells for installation within and outside of specially engineered soil absorption systems.

desirability of having a large, non-plugging orifice of ¼ in. (6 mm) to a near microhole of ⅛ in. (3 mm). The larger orifice [¼ in. (6 mm)] will, all other things being equal, for the same orifice spacing [30–60 in./(0.75–1.5 m) nominal spacing] require nearly 50% more flow per orifice than say a ³⁄₁₆-in. (4.5-mm)-diameter orifice. If there is a substantial difference in elevation between the pump vault and the disposal field, the power of the pump motor (installed horsepower) has to increase dramatically to accommodate both the head (lift plus pipe friction loss) and flow requirements associated with the larger ¼-in. (6-mm) perforations. The ³⁄₁₆-in. (4.5-mm) orifices are a reasonable compromise between the nearly stoppage-free ¼-in. (6-mm) holes and the prone-to-eventual-blockage ⅛-in. (3-mm) perforations.

There are two additional alternative systems that will be briefly mentioned; however, their use in commercial/industrial applications would be extremely rare. Vault systems have been used in alpine regions, where subsurface conditions consist of uniformly impervious rock. The vault serves

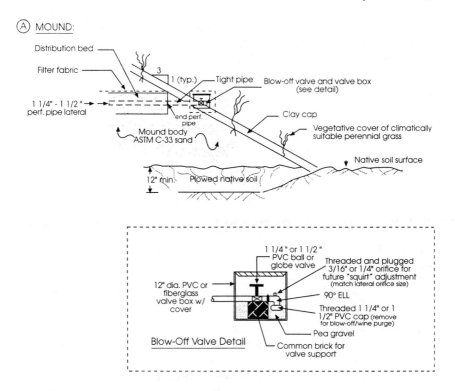

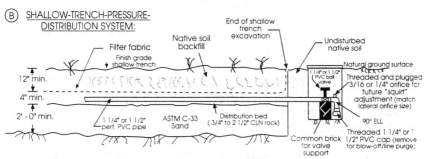

Fig. 8.7. Blow-off valve installation for pressure laterals in mounds and shallow trench–pressure distribution systems.

merely as a storage facility or holding tank, usually consisting of a two-compartment septic tank with no outlet and gas-tight manhole covers or a single-compartment utility vault with gas-tight manhole covers, and all knock-outs where pipe penetrations would usually be made kept intact except for the inlet pipe. Remote alarms and status on contents in the vault would be an essential control/instrumentation feature for a vault system.

The annualized costs for pumping the vault and disposal by a commercial septic tank pumper would be extremely high at $200–$300 per event (1996 U.S. price levels) for 1500–2000 gal (5679 to 7572 L). The vault system should be considered only after all other possible options have been exhausted.

Black-water/gray-water (B-W/G-W) systems are used in many regions worldwide. Because the component of domestic (sanitary) wastewater flow associated with bodily waste is only about 8%, the pure hydraulics of treatment and disposal of black-water is reduced considerably. As with vault systems, their application to commercial/industrial activities is virtually unknown, at least in the United States, although recent federal legislation may create the necessary incentives for serious consideration of the B-W/G-W concept by design engineers in the next decade (124). When used as a solution to a purely domestic wastewater problem, space is generally the limiting factor that allows successful consideration of the separation of sanitary sewage into two waste-stream components. Black-water or toilet waste is given primary treatment in a septic tank and the hydraulic volume, reduced approximately 90% by the elimination of shower, dishwashing, clothes washing, and personal sanitation sink waste, the smaller sized leach-field can often be accommodated on a portion of a small, substandard residential lot (123). The gray-water with a very modest amount of settleable solids (mostly dissolved and suspended solids) and a much smaller titer of pathogenic organisms, can be more safely disposed of on the balance of the parcel by the use of trickle systems (122) (drip tape; see Table 8.4) or other shallow-trench disposal alternatives. The incremental expense of modifying interior plumbing for the residence or winery (if local environmental health regulations would allow consideration of the method) for separate flows is expensive and often impossible once the winery structure has been completed. Nonetheless, a gray-water/black-water system may have its uses in wine producing regions of the world where statutes permit a wider application of alternative technologies than are currently acceptable in the United States.

Operational and Maintenance of Sanitary Wastewater Systems

One feature of septic tank and soil absorption systems that works against consistent operation and maintenance (aka: system monitoring) is the fact that the entire system is underground and out of sight. Therefore, even with small wineries, a formal operation and maintenance (O&M) plan must be prepared and a staff member assigned for the periodic tasks that must be performed (125).

The minimum number tasks that must be performed and recorded are as detailed in Table 8.5.

Table 8.5. Septic tank and soil absorption system O&M tasks

O&M task	Gravity sewers with dosing siphon	Frequency	Pressure sewers with booster pumps(s)	Frequency
(1) Check monitoring wells	Prior to and immediately after peak use (special events, holidays, etc.)	Monthly	Prior to and immediately after peak use (special events, holidays, etc.)	Monthly
(2) Septic tank sludge level	Check dosing siphon for any solids accumulation and record sludge depth in siphon and septic tank	Semiannually	Check pump station wet well for any solids accumulated and record sludge depth in wet well and septic tank	Semiannually
(3) Check and record dose counter reading	Prior to and immediately after peak use (special events, holidays, etc.)	Monthly and as required for peak use days	Prior to and immediately after peak use (special events, holidays, etc.)	Monthly and as required for peak use days
(4) Run electrical check on high water audio/visual alarm	N.A.[a]	N.A.	Override level controls and activate alarm system	Monthly
(5) Check for surface wet spots at base of mound or in leachfield area [correlate with item (1)]	Best performed during nonrain season; also look for proliferation of wetland plant species	Monthly during dry season	Best performed during nonrain season; also look for proliferation of wetland plant species	Monthly during dry season
(6) Sample second compartment of septic tank after peak use	Pull manhole cover and sample second compartment using long-handled dipper; analyze for settleable and suspended solids	After peak use, twice yearly	Pull manhole cover and sample second compartment using long-handled dipper; analyze for settleable and suspended solids	After peak use, twice yearly
(7) Verify operation of curtain drain systems (if any)	Check volume of drainage from outlet of curtain drain after a major storm event with saturated soil conditions	Once yearly when soil is saturated (min)	Check volume of drainage from outlet of curtain drain after a major storm event, with saturated soil conditions	Once yearly when soil is saturated
(8) Check contents of grease trap (for wineries with food service)	—	Monthly (min.)	—	Monthly (min.)
(9) Check hydrogen peroxide (H_2O_2) injection system	N.A.	N.A.	Verify contents of H_2O_2 drum and cycle injection pump (record volume of H_2O_2 used since last reading)	Monthly

[a]N.A. = not applicable

A master utility system O&M status board should be posted in the office of the appropriate manager, who can, by a rapid visual scan, observe when and by whom the listed O&M functions were performed. The recorded data should be logged into the winery's computer with a hard-copy version kept by the staff member performing the field work.

At the conclusion of the construction phase, the design engineer should provide the winery with a copy of an O&M manual, usually a looseleaf binder, which will contain manufacturers' parts lists, lubrication instructions, and circuit diagrams for all mechanical/electrical equipment. Included with the O&M package should be a set of "as-constructed" drawings and equipment specifications for future reference. Any changes or additions to the system performed after the initial construction should be transcribed to these drawings and dated. A special section of the O&M manual should be set aside for maintenance history records for any electrical or mechanical equipment for which repairs or replacements are made. Preventative maintenance software for a PC was mentioned in Chapter 5. The basic program can handle any maintenance function and includes standard formatting for such things as repair history and repair work orders.

Finally, a modest educational program for management and staff on the "do's" and "don'ts" for sanitary wastewater septic tank and leachfield systems should be implemented. For example, posters mounted appropriately in the bathrooms can list the items *not* to be flushed down the toilet— cigarette butts, paper towels, any cardboard or other bulky cellulose products, or toxic chemicals.

Worker safety should be of primary concern to winery management, and staff maintenance personnel should be provided with disposable protective clothing when performing sewer or sewage treatment facility maintenance/repair tasks (goggles, disposable work suit, and rubber gloves; see chapt 12 for details). The OSHA agency in your region will have current guidelines on safety requirements for wastewater system workers. If maintenance personnel are squeamish about sewer cleaning tasks, commercial sewer cleaning firms are available nationwide, whose one-man mobile units are well equipped to handle even the most troublesome of line stoppages.

It is imperative that any disposal of sewage solids (septage from septic tank sludge compartments or grease from grease traps) will be removed and disposed of by a licensed contractor authorized to perform such work in your locale. Septic tank pumping records in some regions of California, at least, have to be submitted to the local environmental health agency for their information and files. The frequency of septic tank pumping is one way by which water pollution enforcement agencies can become alert to a potential septic system failure. Excessive pumping (more than two times per year) can mean either excessive solids loading of the system or that the leachfield has failed and the septic tank is now being used as a vault system (see previous discussion on vault systems).

WINERY PROCESS WASTEWATER

Wastewater Flows

In Chapter 6, a complete discussion of methods for estimating the water supply requirements for the production phase of winery activity was given. These water demand values, as detailed in Tables 6.4 through 6.7, are also the basis for determining the waste discharges that must be handled by the process wastewater treatment and disposal system. For small wineries, the choices of treatment and disposal by soil absorption are similar to the conventional and specially engineered systems described in the section entitled "Sanitary Wastewater." Wineries which exceed the small classification size will generally require a more sophisticated system of above-ground, aerobic reactors, clarifiers, and reclaimed wastewater recycling/reuse subsystems, discussed previously in Chapter 7. A small winery is one for which peak daily process wastewater discharge does not exceed 1500–2500[†] gal (5679–9465 L) per day.

Converting peak discharge to cases of (12–750-mL bottles) can be accomplished using the widely accepted Napa County Formula:

$$\begin{array}{l} \text{Maximum daily} \\ \text{winery wastewater} \\ \text{discharge in} \\ \text{gallons per day} \\ \text{(gpd) (L/day)} \end{array} = \frac{1.5\ [\text{winery annual production in wine gallons}]}{60\ \text{days}}$$

$$\therefore \quad \text{Winery production} = \frac{60\ (\text{winery wastewater discharge})}{1.5}$$

$$\begin{array}{l} \text{For 1500 gpd} \\ \text{(5679 L/day):} \end{array} = \frac{60(1500)}{1.5} = 60{,}000 \text{ wine or } 227{,}160 \text{ L}$$

$$= 25{,}000 \text{ cases}$$

$$\begin{array}{l} \text{For 2500 gpd} \\ \text{(9465 L/day):} \end{array} = \frac{60(2500)}{1.5} = 100{,}000 \text{ wine or } 378{,}600 \text{ L}$$

$$= 41{,}667 \text{ cases}$$

Note: Wine gallon to case conversion = 2.38 U.S. gal per 9L case.

[†](This range in allowable maximum daily flow for subsurface/soil absorption systems is as prescribed by typical Water Pollution Basin Plans for California Counties that have significant winery development.)

The statutory limit for conventional soil absorption systems in Napa County, California, is approximately 1500 gal (5679 L.) per day. For specially engineered systems, the upper limit might be 2500 gal (9465 L.) per day, if site conditions (soil permeability, soil depth, slope, and depth to groundwater all exceed the minimum acceptable design values by 10–15%) were close to ideal. The other factor that must be inserted into the equation is the cost for constructing the very extensive, below-ground components for a system that is close to the statutory upper limit for waste discharge. At some point, the spacial requirements for a conventional or alternative soil absorption system becomes self-limiting, and an above-ground system is likely to have a lower construction cost, but a higher commitment to annual operating costs. For wineries approaching the size boundary between small and medium, the skill and imagination of the designer is very important to give the winery developer the optimum system for both initial construction and future operating costs.

For wineries with uncovered crush and press pads, the disposition of storm water from the pad during the non-grape harvest portion of the year is critical to the balance of the hydraulic design of the process wastewater conveyance, treatment, and disposal elements. The process wastewater system network should not be subjected to water from rainfall runoff. The least expensive solution to the problem is the use of a two-way valve that can shunt the post-crush period storm water to a separate stormwater drainage system. The intended diversion of stormwater away from the winery's treatment works post-crush can only be successful if the two-way valve is, in fact, changed. An above-ground visual indicator should provide easy verification of valve status. Prior to 1991, stormwater discharges from industrial facilities were not regulated. A second stage of federal water pollution control legislation provides for monitoring and, in some cases, treatment of stormwater discharges. All wineries with stormwater discharges are required under state models of the federal statutes (Clean Water Act 1987, Section 405) and 40CFR 122.26 to file for and to receive a permit for stormwater discharges (126, 127). It appears that the United States is the only country with industrial stormwater waste discharge regulations. If the level of anticipated improvement in receiving water quality (reduction in toxic heavy metals and petroleum residuals) is as dramatic as is predicted by the U.S. EPA, then other countries with large industrial bases may follow suit, including wineries, as the United States did in their assessment of a particular industry's potential for producing contaminated stormwater runoff.

Other wastewater discharge hydraulic peaks, which are often not anticipated by a designer who might be unfamiliar with the cooperage aspects of wine production, may find that new oak barrels which are filled with water (sometimes a small amount of sulfur dioxide and soda ash, depending on the winemaker's preferences) to verify water tightness, if emptied simultaneously, can quickly exceed the design capacity of raw process wastewater pump stations and connecting gravity sewers. If the winemaker is made aware of the potential for overtaxing the system hydraulically, metering-in the flow by gradual barrel emptying can offer a ready solution to the unwanted hydraulic peaking. The large volume of water without any BOD (biochemical oxygen demand) or food (carbonaceous waste) for the biomass in the winery's treatment reactor will upset the food to microorganism (F/M) balance and may require a restart of the reactor. The addition of wine lees or other solids and possibly a measured amount of a dried bacteria culture (see discussion of bioactivators earlier in this chapter) can restore bioreduction activity and stabilize the system. Water pollution control agencies strongly dislike any mention of a "bypass" as a feature of a winery process wastewater (or any other type of wastewater system). With full knowledge and concurrence of the regulatory agency, bypassing cooperage "swell water" has been incorporated into several designs. This was because the bypass was directly connected to a reservoir for reclaimed process wastewater. Also, if the winery offers an analysis of a sample of the barrel water prior to discharge, it can help to alleviate any fears that the agency enforcement officer might have over the improper handling of the benign water. The usual swell water contains molecular SO_2 and oak tannin, both in very low concentrations. Finally, as this operation applies only to new cooperage, it usually occurs once or twice per winery production year, a strong argument for bypassing this large volume of solids-free water can be made.

Process Wastewater Characterization

It is rather senseless for the designer to look at *average* conditions for the winery process wastestream. One of the principal problems faced by the utility engineer is the highly variable flow rate and solids content, varying even from winery to winery of the same size because of differences in vinification philosophy, equipment, and product line. There is no typical winery wastewater hydrograph. In large wineries, the temporal variations in flow and solids content are less dramatic because many process functions proceed continuously throughout the year. For example, bottling lines in large wineries are often running daily, and in very large wineries, three 8-h work shifts for bottling operations are not uncommon. Table 8.6 is an array of typical winery process functions and the associated relative concentrations of the major constituents likely to be found in the wastestream. Sparking wine facilities and wine cellars will not necessarily fit this process profile. Whereas sanitary wastewater is 70–80% settleable solids and 20–30%

Table 8.6. Typical Table Winery Process Wastewater Constituency (Qualitative)

Wastestream Constituents

Process Step	Grape Juice Sugar	Seeds	Stems	Grape Leaves	Skins	Closure Residuals*	Alcohol	Yeast Meta-bolites	Potassium Bitartrate	Molecular Sulfur	Fining Agents**	D.E.	Bentonite	Oak Tannins	Iodine	Soda Ash	Lube***	Clng. Agent****
Crush/Destem	75%	75%	10%	10%	75%													
White Grapes or Blush Wine (Press)	75%	75%	10%	10%	75%													
Red Grapes (Press)	50%						25%	25%	25%	10%								
1st Racking							75%	75%	75%									
Other Tank Racking							50%	25%	25%									
Fining							25%				50%				25%			50%
Tank Cleaning																		
General Washdown		25%					10%				25%							10%
Bottling						50%	50%										50%	
Barrel Racking							50%				25%							
Filtration											10%	50%	10%					
Cooperage Swell/Leakage Testing														25%		25%		10%
Process Line/Hose Sterilization															25%			25%
Cold Stabilization									50%									25%
Protein Stabilization													75%					

*(Cork and capsule) **(gelatin, egg whites, etc.) ***Bottling line conveyor lube ****Sodium hydroxide, chlorine & other cleaning agents

Relative Concentration-- ☐ 0%   ☐ 10%   ☐ 25%   ▨ 50%   ■ 75%

Note: Wastestream constituency for sparkling wineries, bonded cellars will vary from this table, because of more or less process steps.

percent suspended or dissolved solids, winery process wastewater has a higher percentage of suspended or dissolved solids at approximately 75–85% with 15–25% settleable solids. Equating this to the treatment process option, unassisted sedimentation (no settling aids or flocculants) will require longer detention times or higher degrees of treatment. Figure 8.8 is a process schematic for a small winery process wastewater treatment and disposal system, that employs available soil absorption technology. The major treatment process differences between the previously described sanitary wastewater systems is the necessity to entrap gross solids during crush (seeds, skins, leaves, stems, and grape pomace), most of which are products of the grape gondola, crusher-destemmer, and screw conveyor/hopper washdown. These solids can either be retained in the winery floor drain system by the use of basket strainers or by the installation of a screening or strainer device immediately upstream from the septic tank array (128). A rotary drum screen requires the least operational attention, whereas the floor strainers and in-line strainers require monitoring and periodic emptying, at least during the crush season. The author has observed crush-fatigued cellar workers removing the basket strainers and turning them upside down into the floor drain, thus defeating the purpose of the device. Rotary drum screens are self-cleaning but usually require a clean water supply to operate the spray-bar nozzles that prevent the slots from being plugged. Drum or inclined gravity screens for winery service are usually constructed with slots no greater than 0.079 in. (2 mm) in width. Ski jump or inclined screens operate on the gravity principle with the flow distribution box at the top of the screen, spreading the wastestream over the full width of the screen. In theory, the solids are retained on the surface of the screen and tumble-down the sloping screen, where they fall off of the screen lip into a container designed for grape solids retention. The liquid wastewater in this case passes through the slots to a collector beneath the screen where it is eventually conveyed to the downstream treatment elements of the system. The screens require hand labor for mechanical (brush or squeegee) or hydraulic (high-pressure water) cleaning to remain operationally effective. Even with partial screen blinding, water exits over the screen tip into the solids container, which requires that the bin or container have a screened bottom drain, which can redirect the wastewater back to the wastestream conveyance system. One other inclined screen shortcoming is that they require 6–7 ft of vertical distance (hydraulic head) for operation. For flat or nearly flat winery parcels, the 6–7-ft head loss may require a raw wastewater pump station to accommodate the open-channel flow hydraulic characteristics of the gravity sewer system. The cost of inclined screens is approximately one-fourth the cost of the self-cleaning rotary drum screen. A self-cleaning strainer of equivalent capacity is about

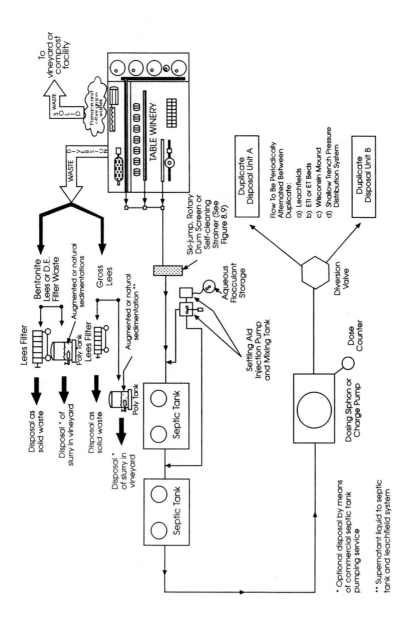

Fig. 8.8. Typical small table winery process wastewater schematic using soil absorption methods of disposal (Small = 25,000 to 42,000 case Annual Production Capacity).

two-thirds the capital cost of the rotary drum screen, and like the drum screen, it requires a clean water supply and motor-driven scraper bar to flush solids from the face of the screen. The self-cleaning strainer requires a secondary storage vault or septic tank to retain the captured solids and flushing water. Each of the foregoing strainer/screening units have their place on the industrial waste treatment arena; however, for the wine industry, the rotary drum screen appears to be the best fit, with well-maintained, floor-drain basket strainers a second choice. Figure 8.9 depicts typical examples of the three major screening/strainer choices.

The settling-aid subsystem shown in Fig. 8.8 together with the series arrangement of septic tanks can provide a relatively clean effluent that will have the least negative impact on soil permeability in the subsurface portion of the disposal system. Generally speaking, the flocculent/settling-aid injection and mixing unit would only be required when certain process steps were being performed in the winery. For crush, press, racking, and protein stabilization with bentonite, in particular, the flocculent addition could increase the efficiency of gravity sedimentation in the septic tanks to a very high degree. The measured improvement in natural sedimentation by flocculation can be determined by jar tests performed in the winery's laboratory with wastestream samples with and without flocculent addition. The amount and type of flocculent is best decided after the winery is completed, in concert with the utility system engineer after wastestream sampling and simple jar tests indicate the proper selection [i.e., (a) size of floc formed and (b) the measured rate of sedimentation] (129, 130).

Special note is made here of the absolute necessity for the careful management of bentonite lees in the winery to prevent their introduction into the wastestream. In the last decade advances in microfiltration and ultrafiltration have provided a soil absorption system safety-net for wineries that have historically used bentonite for protein stabilization and as a clarification aid. Use of ultrafiltration for protein removal might be a very sensible investment if the control of bentonite lees is very difficult for a small winery to achieve. Cost recovery of the filtration equipment could occur rather quickly if the principal benefit is the salvage of a quirky soil absorption system. Bentonite particles, because of their colloidal size (generally less than 0.002 mm in diameter), tend to remain in suspension, unless natural agglomeration occurs or a cationic flocculent is added to produce the coalescence and settling of the negatively charged clay particles. Gross less also fall into the category of winery waste products that are better removed in the winery. By use of a lees filter or by pumping the gross lees to a portable polyethylene tank natural sedimentation and dewatering can occur over a more extended period of time than would be possible in the series-connected septic tanks of the typical small winery wastewater treat-

(A)

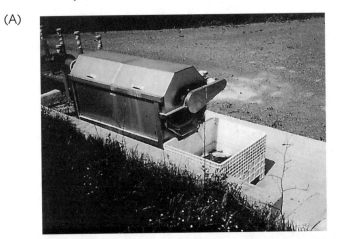

(B)

(C)

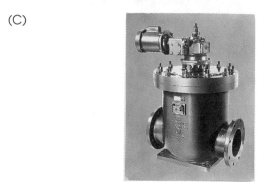

Fig. 8.9.   Strainer/screen system options for winery process wastestream. (a) rotary drum screen. (b) incline screen. (c) self-cleaning strainer. (Courtesy of Hayward Industrial Products, Elizabeth, NJ.)

ment and disposal system, as shown in Fig. 8.8. The filtrate or supernatant liquid with suspended solids load removed can be salvaged as low-grade blending wine or rediverted to the winery's wastestream for disposal.

Of principal interest to the designer of the winery's process wastewater treatment system are four major wastewater parameters (see Ref. 131 for definitions and standard methods for wastewater analysis):

	Units
• BOD or COD* (usually at 20°C)	mg/L
• Settleable solids (SS)	ml/L
• Total suspended solids (TSS)	mg/L
• pH	[reciprocal of the $\log_{10}$ of the hydrogen-ion concentration]

BOD is the biochemical oxygen demand (5-day test at 20°C), COD is the chemical oxygen demand (3-h test), and BOD is approximately (0.4–0.8) COD (comparative laboratory results are required to establish an exact relationship). Values for the above variables need to be determined for the "worst-case" condition, which for most wineries usually occurs during the harvest/crush period and for 60–90 days thereafter, when the process steps that produce the greatest solids mass are performed on the new wine. Once the maximum solids loadings are determined and the peak hydraulic loadings are decided on (see the previous section, in this chapter entitled *Wastewater Flows*), the maximum daily solids load in kilograms or pounds can be determined, and for above-ground systems that utilize aerobic stabilization of waste, the size of tank/reactor components and dissolved oxygen input can be determined. For small systems maximum daily flows equal to or less than 1500–2500 gal (5679–9465 L) per day or 25, 000–42, 000 case annual wine production, the potential solids loading must be analyzed to select the proper size of septic tank (i.e., sufficient detention time at maximum flow to permit 80–90% of settleable solids to settle in the sludge chamber of the recommended two-septic tank series arrangement). Based on the operation modes of typical table wine facilities, the range of maximum wastewater strength values is reported in Table 8.7.

Table 8.7. Typical composition of table winery process wastewater during the crush season (range of maximum daily values)

	Concentration	Units
BOD or COD	3000–7000	mg/L
Settleable solids	5–20	ml/L
Suspended solids	400–500	mg/L
pH	3–5	—

Small, existing wineries are less likely to be able to budget for and afford a complete winery wastestream sampling and analysis program in anticipation of a waste treatment plant expansion or upgrading. Although there is often a free exchange of wine production "tricks" and methods among wine industry members, there is a general reluctance to exchange data on waste discharges and their content, probably because of fear of legal action if the data were improperly used. A search of the literature will show that there is a paucity of wastewater data for wineries, and only one of the wine industry's respected engineering journals or textbooks give space to the subject of wastewater. One California table winery in the 500, 000 case annual production capacity established a baseline process wastewater discharge profile during calendar year 1981 for each of the major wine production steps. The sampling was performed at 1-h intervals during the duration of the process step. For example, the data reported in Table 8.8 is a sample composite for an 8-h bottling-line shift with samples taken in the bottling-room floor-drain outlet. (clean-up and sterilization prior to start-up, washdown between change of wine varieties and cleanup at the end of the shift)

Superimposed over this "best estimate" of hydraulic and solids loading design criteria is the extreme variability in process wastewater flow and content over a full winery production-year cycle. For biological treatment systems, which seek some level of equilibrium for optimum performance (a relatively constant level of food to microorganism ratio or F/M ratio), the feast or famine supply of carbonaceous waste coming from the winery makes it extremely difficult to coax a sustainable level of treatment activity

Table 8.8. Large operating table winery wastestream constituency during bottling operations

Constituent	Unit	Concentration or value
Chemical oxygen demand (COD)	mg/L	9330
Biochemical oxygen demand (BOD_5)	mg/L	5845
Total suspended solids (TSS)	mg/L	1251
Ammonia nitrogen (NH_3)	mg/L	14.6
Volatile suspended solids (VSS)	mg/L	722
Total nitrogen (N)	mg/L	81
Total dissolved solids	mg/L	178.0
Total phosphorus (P)	mg/L	8.8
Settleable solids	mL/L	2.1
pH		5.1

Notes:

[a]An 8-h shift, composite sample with hourly samples.

[b]Laborator methods as per (131).

out of a given system. If the flow/solids variability is coupled with the non-treatable condition of raw winery wastewater, then the utility engineer is presented with a very poor framework for process design.

The septic tank and soil absorption systems are probably best able to cope with flow and solids variability conditions without going into a failure mode. The mere fact that "digesting" and "digested" solids are both resident in the sludge compartment provides the incoming process wastewater with a buffering action of a sludge blanket that has a large capacity for resisting a total biomass annihilation. The more fragile aerobic system and its flora are less robust than the septic system and, thus, are more vulnerable to "upsets" and performance crashes. For example, if the aerobic treatment reactor is deprived of a food substrate for its suspended biomass for 2 or 3 days, the standing crop of organisms could be reduced to critically low levels and a subsequent heavy shock loading of solids would be oxidized at a very low rate until the correct F/M (food to microorganism ratio) was reestablished. F/M ratios for a wide range of aerobic treatment systems of the type normally used for winery wastewater applications vary from 0.2 to 0.6. Because most typical, above-ground aerobic reactor systems designed for medium to large wineries generally receive waste discharges only during the working day (there have been exceptions cited previously, such as the atypical 24-h bottling facility), the treatment system is functioning much like a sequential batch reactor (SBR). The SBR receives a measured volume of waste and solids with no additional inflow during the treatment step, which is then bioreduced until the BOD/COD and suspended solids have been decreased to a specified and acceptable level (usually 80–95% of the raw wastewater concentration). The effluent from the SBR is then discharged to a clarifier for sedimentation and solids removal and the cycle is repeated again. Bergfeld Winery in St. Helena, California has a unique wastewater treatment system based partially on the SBR concept. In all continuous flow-through systems there is a partial loss of the standing crop of treatment organisms. Thus, to maintain a reasonable population of microorganisms, the recirculation of a portion of freshly settled sludge back to the reactor becomes a method to stabilize the biomass. This subsystem feature is called "activated sludge" and can be used effectively in winery waste treatment systems.

There are a number of other factors that influence the design and performance of above-ground aerobic treatment systems, just as there are in the fermentation kinetics for table wine production. The reaction rates and extent of waste stabilization of the several wastewater treatment system choices are driven by very complex microorganism growth kinetics that are similar to the yeast growth cycles in grape juice fermentation that are familiar to all winemakers (i.e., log-growth phase, declining growth phase, and endogenous phase). Reference is made to a comprehensive textbook on microbiology for further reading on growth dynamics of aerobic waste treatment flora.

Above-Ground Aerobic Waste Treatment Systems and Performance Standards

The "zero discharge" concept for winery process wastewater systems is standard practice for most all U.S. wine producing regions. Expressed in water accounting terms, it means that no wastewater discharge is permitted to streams, lakes, bays, or other surface water bodies and that all water that is used for production purposes must be treated and reclaimed for some beneficial use. The use can be either on the winery parcel or through joint use arrangements with other landowners, for which long-term beneficial reuse of the reclaimed wastewater can be demonstrated and ensured by a binding, written agreement between the parties involved.

The system treatment capabilities for the choices of above-ground arrangements of reactor and clarifier generally are in the range of 75–95% efficiency of BOD/COD removal. Thus, for wastewater with a crush period maximum BOD of 5000 mg/L, and if the maximum removal efficiency was achieved, the effluent would still theoretically contain 250 mg/L of 5-day BOD. Because the basic system also includes, at least for the Mediterranean-climate-like western United States (wet winters and dry arid summers), a winter holdover storage facility, additional bioreduction takes place in the storage pond while it performs as a facultative treatment unit. Frequently, there is also an additional 5–10% percent BOD/COD removal in the clarifier subsystem. Therefore, as long as the dissolved oxygen (DO) content of the reactor effluent in the clarifier and in the winter storage pond remains at no less than 3 mg/L, there is not a high probability that nuisance odors would be produced. If the DO concentration did in fact begin to fall, then an injection of a low concentration of hydrogen peroxide (H_2O_2) upstream from the clarifier or in the booster pump station wet well that serves to convey the clarified effluent to the winter holdover storage facility, would be required. Direct injection of the H_2O_2 into the clarifier should be avoided to prevent flow disturbance of the quiescent settling cell and the sedimentation process that takes place in that unit. Likewise, with the storage pond, vortex flow aerators or other devices that would deeply mix the lower anaerobic and upper aerobic pond zones should be avoided. Stability, odor control, and biological equilibrium are best achieved with the two reaction products of the hydrogen peroxide (water and dissolved oxygen). (Safety precautions and other H_2O_2 use guidance was provided in the subsection of this chapter entitled "Sewage Pump Stations.")

Treatability of Winery Process Wastewater

The network of organisms which perform the task of bioreduction of carbonaceous waste from wineries are, as a group, rather fastidious. Except for the acid content of grape juice, which is reflected in a low-pH value that is well below the desired pH for sustained aerobic treatment reactor microorganism growth, the fructose sugar, even after its conversion to ethanol, is a desirable metabolite for the array of bacteria, yeasts (fungi), protozoans, rotifers, and algae found in properly managed reactors. Enzymes literally work hand-in-hand with bacteria to break down the organic substances in the wastestream. On occasion, the waste substance is acted on by an enzyme, outside the microbe cell wall, with the product altered by enzyme action now able to enter the cell wall. The bacterium may also endogenously produce, an enzyme to assist in the metabolism of the adsorbed substance. Thus, enzymes can increase the rate of reaction and hinder it by their absence. Commercially available "bioactivators" have been discussed earlier in this chapter. They often contain proprietary enzymes to ensure a ready supply in complete-mix aerobic reactors. Other essential nutrients for aerobic organisms destined to treat winery wastewater are as shown in Table 8.9.

Winery wastewater may be deficient in both nitrogen and phosphorus because of yeast uptake during fermentation. Although grape juice contains nitrogen and phosphorus, yeast require these essential nutrients in the conversion of glucose and fructose to the major and end products of alcohol, carbon dioxide, and thermal energy. As a winemaking confirmation of this yeast-nutrient-consumption phenomenon, so-called "stuck" fermentations are often re-started with re-inoculation of designer yeast and a yeast nutrient, consisting mainly of di-ammonium phosphate salts and yeast hulls.

Therefore, a wastewater subsystem is required that has the capability to neutralize the acid pH and to provide bio-nutrients in the proper concen-

Table 8.9.   Range of key operating parameters for an aerobic reactor for optimum bioreduction of winery wastewater for a given solids loading

pH	6.5–7.5
BOD_5	5000 mg/L
Temperature	65–75°F
Nitrogen (N)	250 mg/L
Phosphorus (P)	50 mg/L
Dissolved oxygen	3–5 mg/L (never less than 2 mg/L)
BOD:N:P Ratio[a]	100:5:1

[a]BOD:N:P ratio as reported in Ref. 132.

trations to sustain microorganism growth in the reactor, if they are not present in the raw wastewater. Reflecting on the previously described variability in wastewater constituency throughout the production year, the need for pH adjustment and nutrient addition will be directly proportional to the nature of the process events that are occurring in the winery. Coordination between the winery production staff and the utility maintenance group is extremely important to anticipate (through the previous year's operating experience and system performance data) the volume, approximate length of time, and content of the combined wastewater discharge that will likely require treatability adjustment. The arrangement and equipment components for the pH–nutrient addition/adjustment system will be described in more detail in a subsequent section of this chapter.

The dissolved oxygen content in the reactor should be maintained at no less than 2 mg/L under maximum solids loading conditions and preferably at 4–5 mg/L. The success of induced oxygen diffusion in the reactor is a function of temperature, geometry of the reactor to provide complete mixing, and the ratio of mass of oxygen to mass of BOD somewhere in the ratio of 1.5–2.0 to 1.0 (105). In this instance, a safety factor of 1.5–2.0 is strongly suggested to prevent dissolved oxygen concentration declines during periods of low ambient temperature or single-event shock loadings of suspended solids. Because air and not pure oxygen is to be diffused into the reactor or aeration cell, for every 10 pounds (4.6 kg) of air induced only about 2 pounds (0.91 kg) of pure O_2 will be added. Thus, commercially available mechanical aerators are rated on oxygen transfer capability only, with guarantees of a measured number of pounds or kilograms of oxygen transferred per hour per installed aerator horsepower. It is the very important job of the designer to create the dimensions of a reactor that will produce a "complete-mix" condition. A reactor having a length to width ratio of about 2:1 with semi-circular ends (running-track shape) and with an operating depth of 4 ft with a 2-ft freeboard (distance from active water surface to perimeter curb-top of finished aeration cell). The net operating volume of the aeration cell should give wastewater detention times of about 5–8 days [lower value with activated sludge return and longer detention time with no sludge return (sludge return is described later in this subsection)]. Thus, for a maximum daily process wastewater discharge of 4500 gal/day (gpd) (17, 442 L/day), a minimum volume of the reactor should be approximately 22, 500 gal (85, 185 L). Sidewall slopes should not be steeper than 2:1. This allows the concrete lining to be either poured or pneumatically applied (as with gunite or shotcrete systems). A simple skirt, constructed of fiberglass sheets or other corrosion-resistant material, surrounds the overflow weir and serves as both a wave suppressor and as a scum/froth shield. A diagram of the wave suppressor/scum shield is shown in Fig. 8.10 and a photograph of one is shown in Fig. 8.11. Waves from the aerators can be 4–6 in. (100–150 mm) in height, and 6–8 in. (150–200 mm) in height with wind-aided energy. The additional freeboard also provides a surface for foam and scum accumulation, which must be washed back into the reactor daily, or the freeboard zone becomes a breeding area for a host of aquatic insects. The floating, mechanical aerator has been found to be best suited for winery above-ground aerobic, complete-mix reactor systems.

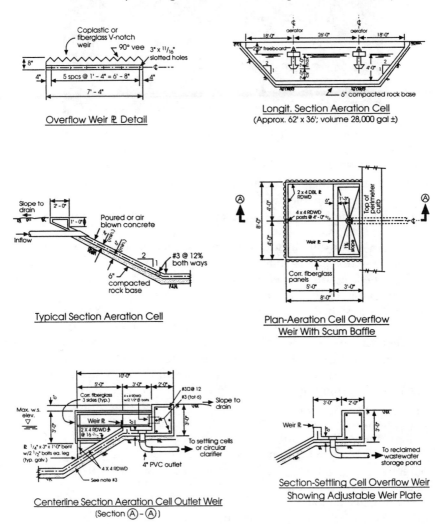

Fig. 8.10. Typical aeration and settling cell construction details (50,000-case annual production capacity).

Fig. 8.11.   Photograph of an existing aeration cell for an 180, 000-case per year annual production capacity winery.

By judicious placement of two aerators, one each in the center of curvature of the rounded reactor ends, no "dead spots" prevail where solids can settle and become anaerobic problems in small, quiescent pockets. The aerators are secured by stainless steel restraining cables; a power-umbilical-cord-strain-buss system provides electrical service. A power disconnect and junction box for each aerator allows the inactivation and aerator movement to the aeration cell perimeter curb for maintenance and lubrication. A photograph of an existing aeration cell for a 180,000-case capacity winery with installed mechanical aerators is shown in Fig. 8.11. Except for aeration ponds excavated in rock, a concrete or gunite lining is recommended. Clay blanket or plastic liners do not offer the service life or low maintenance of concrete liner materials. Unlined aeration cells have been designed by others with a concrete erosion pad installed directly beneath the aerators. Armoring of the pond for several feet at the maximum water surface elevation with rock rip-rap or other wave erosion protection is also required with this design feature to prevent the erosion and introduction of inorganic sediment into the reactor mix.

Figure 8.12 depicts the methods most frequently involved in diffusing oxygen into aeration cells. Aerators can also be installed on fixed platforms of timber or concrete that can be accessed by boat or a permanently fixed walkway. Once operational, removal or replacement of electrical motors or other heavy mechanical components becomes difficult in the extreme by boat or walkway, and places the maintenance staff at some risk. The fixed-

DIFFUSED AERATION SYSTEMS

MECHANICAL AERATION SYSTEMS

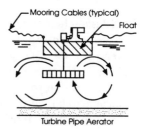

Oxygen Source or Air Compressor

Laser-cut polyethylene tubing

Porous diffuser plate

or

Mooring Cables (typical)

Float

Turbine Pipe Aerator

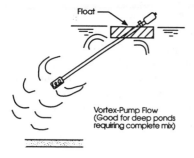

Float

Vortex-Pump Flow
(Good for deep ponds requiring complete mix)

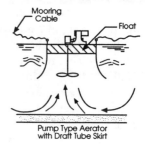

Mooring Cable

Float

Pump Type Aerator
with Draft Tube Skirt

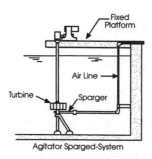

Fixed Platform

Air Line

Turbine

Sparger

Agitator Sparged-System

Fig. 8.12.   Air diffusion and mechanical aerators for aerobic reactors (aeration cells).

platform and walkway supports, which penetrate the water body, interfere with the uniform hydraulic circulation in the cell and provide locations for "dead spots" and solids accumulation in and around the support columns.

AEROBIC TREATMENT OPTIONS AND EQUIPMENT CHOICES

Figure 8.13 is a schematic depiction of an above-ground, aerobic treatment system. The below-ground system choices for treatment of winery process wastewater [≤ 1500–2500 gal (5679–9688 L) per day] for *small* wineries were discussed in a previous section and shown schematically in Fig. 8.8. It should be noted that the waste diversions suggested for small wineries are also applicable to large wineries, as shown in Fig. 8.13. For the small winery, the justification for waste diversion was for both a measurable reduction in solids loading on the treatment works and to prevent to the highest degree possible any carryover of colloidal, suspended solids to the soil absorption subsystem. For the large winery category, only the former reason is applicable. There are secondary benefits gained from the use of lees filters. The wine production benefits, for example, in gross lees filtration in large wineries are significant in the salvage of additional wine that would otherwise be lost with the lees. In *methodé champenoise* sparkling wineries, the disgorging lees are easily salvaged and separated in the frozen state. Sale of this alcohol-containing waste product to producers of industrial alcohols can offer a measurable cash return to the winery and offset somewhat waste treatment overhead charges. The "sludge return" feature shown in Fig. 8.13 is one method to stabilize and optimize the performance of the aerobic reactor, by supplying freshly settled sludge containing a large, active biomass. The sludge return system can be designed to meter some 8–10% of the total volume of settled sludge with the balance going to waste storage and eventual disposal in the vineyard or off site through a commercial septic tank pumping service.

Figure 8.14 illustrates a completed aeration cell (a) and paired settling cells (b) for an 85,000 case per year winery. The reclaimed wastewater from this facility was blended with natural runoff in a winter holdover storage reservoir for later seasonal use as a vineyard irrigation water supply.

Figure 8.15 shows a circular clarifier from concrete tank construction through completion. The finished system photograph shows the scraper drive motor (mounted on the railed catwalk) and the flash mixer and mixing tank for flocculent addition. A portion of the V-notched overflow weir and perimeter launder (trough) can also be seen in the interior of the tank under the catwalk.

The aerobic reactor and coupled clarifier subsystem can take other forms and utilize other proven bio-reduction treatment technologies. Table 8.10 is a summary of state-of-the-art methods with the author's assessment of their application for winery wastewater treatment. Alternatives to the aeration cell that was described in detail previously could be inserted into the process treatment schematic shown in Fig. 8.13 at the location

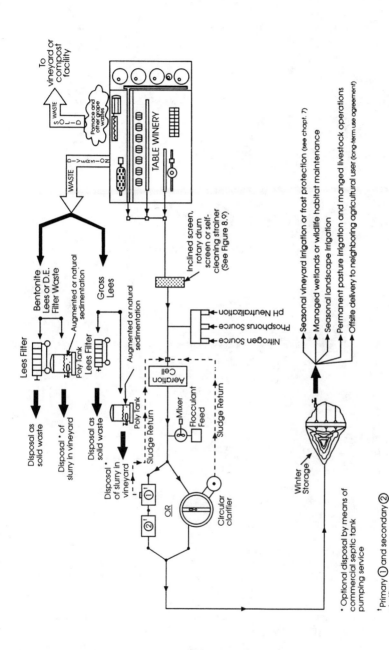

Fig. 8.13. Typical large winery process wastewater schematic with optional methods of disposal (less than 42,000 case production capacity)

243

Fig. 8.14.   (a) Poured concrete aeration cell for 85, 000 case annual production capacity winery ready for start-up and testing. (b) Paired settling cells that can be operated in series (primary and secondary clarification) or in parallel. Winter reclaimed wastewater storage pond (left center).

immediately downstream from pH neutralization and nutrient addition subsystem. Where the substitution of an alternative unit process would require a substantial change to the other components in the Fig. 8.13

Formed, but not poured, circular clarifier tanks showing waste sludge storage tank with access manhole riser (lower right)

Finished circular clarifier showing scraper drive motor ①, flocculant mixing tank ② with flash mixer ③ and interior weir plate and launder (effluent trough)

Fig. 8.15. Circular clarifier used for separating solids from aerobic reaction effluent for 180, 000 case annual production winery.

concept scheme, the addition or deletion of equipment or sub-processes required to create a workable assembly can be selected from the Table 8.10 choices by the system designer.

Clarification and Settling Aids

Removal of the bio-reduced solids which are in the effluent liquid–solid suspension leaving the aerobic reactor unit is necessary, so that the effluent

Table 8.10. Summary and Comparison of Above-Ground Treatment System Alternatives for Winery Wastewater

CONCEPT	SYMBOLOGY	REMARKS	REF.	ADVANTAGES	DISADVANTAGES	SUBJECTIVE RATING AND OVERALL SUITABILITY FOR WINERY WASTEWATER (Worst (1): Best (10))
A. Sequential batch		See Fig. 8.12 for O_2 Delivery Systems	(105)	Much like wine ferm. process. No discharge until proper level of treatment is achieved.	Labor intensive; biomass would have to be created for each batch.	(3)
B. Plug flow		Generally long, narrow tanks with step aeration; no longitudinal mixing.	(133)	Good level of treatment possible with careful management. Requires activated sludge return system because of cell wasting.	More space required; more elec. energy required; better suited for municipal waste treatment.	(3)
C. Complete mix		Continuous flow, contact, aeration, completely mixed.	{133}{134}	Straight aeration or activated sludge to improve performance or increase capacity. Good for ease of expansion and modular construction.	Requires careful management; fairly good buffering for shock loads.	(7)
D. Arbitrary flow		Partial mix, facultative pond.	{133}{134}	Good buffering for shock loads; if not properly managed, can produce odors from anaerobic zone.	More space required; good for large wineries.	(7)
E. Packed bed		Trickling filter fixed biomass (1)	(133)	Excellent for wastewater high in suspended solids; good for polishing primary effluent. Uses no elec. energy.	Not good for shock loads; surface is breeding place for filter flies and other aquatic insects; can produce odors if biomass destroyed by waste stream toxicity.	(5)
		Upflow anerobic filter with fixed biomass (2)	{135}{136}	Excellent for wastewater high in suspended solids; has ability to remove 90% suspended solids. Has not been tried by wine industry, but has good potential. Methane gas by-product. Uses very little space and can be easily added in modular units.	Not good for shock loads; biomass vulnerable to toxic waste stream (high or low pH etc). Not good for variable solids loading. Requires flare burnoff for methane or gas salvage.	(6)
F. Fixed film		Rotating biological contactor	(105)	Excellent for wastewater high in suspended solids; Units are easily added in modular units.	Elec. power consumption is moderate; equipment O & M is high; Not good for shock loads; Biomass vulnerable to toxic waste stream; Not good for variable flow or variable solids loading.	(4)

Notes: (1) Both packed bed alternative E(1) and E(2) and the fixed film system, "F," could be used as effluent polishing units downstream "A," "B," "C," or "D." They are also suitable as primary treatment units of winery wastewater. As with all system designs, it is the responsibility of the designer to select the system or combination of systems which will give the optimum performance for the least capital and annual operating cost.

will be suitable for reuse. The desirable limits of suspended solids and residual BOD were discussed previously in the subsection entitled Above-ground "Aerobic Waste Treatment System Performance Standards." Proper operation of the aeration cell will produce a mixture of agglomerated particles (consisting mostly of reduced solids, attached biomass and wasted cell material, and inorganic colloidal material) of various sizes and masses. Proper operation of an aerobic reactor will produce floc particles that will settle readily in either of the two principal clarifier types discussed previously (i.e., series- or parallel-connected settling cells or a circular clarifier with mechanical sludge collection and storage features). If the reclaimed wastewater is targeted for irrigation use, and either a properly sized, single-purpose storage reservoir or a dual-purpose blending/storage reservoir is used, additional suspended solids sedimentation will occur in the two-zone, facultative storage facilities. For wineries with large vineyard plantings, the dilution factor of reclaimed wastewater to impounded surface runoff or groundwater import is often in the 5:1 to 10:1 range; however, the storage facility should not be depended on as the "polishing unit" of the sedimentation/clarification works. Treatment plant upsets may produce less than the ideal effluent occasionally, and the storage reservoir can serve as an emergency buffer system. Routine treatment malfunctions with high concentrations of BOD and suspended solids that are conveyed to the storage reservoir, particularly with dilution ratios approaching 1:1, can cause deterioration in water quality in the reservoir from the following:

• Dissolved oxygen declines
• Nutrient imbalances which lead to algae blooms
• Loss of aesthetic values in terms of water transparency and general appearance

Figures 8.13 and 8.15 show respectively diagrammatically and photographically how a flocculation step (settling-aid injection system) can be incorporated into the process design. As suggested previously, the winery's waste treatment operator will, from experience and through effective two-way communication with the production supervisor, be able to anticipate the type of wine production operations that are occurring in the winery on any given day and, with the lead-time afforded by such knowledge, to fine-tune the flocculent injection unit to augment the sedimentation process, as required. In the first few years of small to medium size winery operations, the operator should periodically sample the aerobic reactor effluent to become familiar with the floc size and natural sedimentation characteristics associated with the various process events that occur in the winery. For very large wineries where many process steps are occurring simultaneously throughout the year, the wastestream will tend to be more uniform in character and the fluctuations in hydraulic and solids loading will be dampened by the repetitive nature of the events in the winery for each working day. The samples collected should be subjected to "jar tests", using replicate samples with a control sample (i.e., no flocculent addition). Several jars with various concentrations of the most commonly used flocculants, alum (aluminum sulfate) and ferric chloride, should be tested. Again, the rapidity with which the floc is formed, its size, and sedimentation rate should be observed and recorded, with the best results obtained used to guide the operator in achieving the optimum clarification for the wastestream, as sampled. As most chemical injection pumps can be adjusted through a fairly wide range of flow rates, a series of settings can be developed for the entire process year, taking much of the empiricism from the management of the wastewater system. Such recorded data are also invaluable when key utility personnel are absent for vacations or illness, or leave the winery entirely. These data should become part of the active section of the utility system operation and maintenance manual (O&M manual) as well as stored in the winery's computer data bank.

More efficient flocculants are available through water treatment chemical supply companies like Betz or Nalco, which offer a broad array of proprietary polyelectrolytes are designed to flocculate troublesome, suspended materials. The polyelectrolytes are considerably more expensive than the simple aluminum, and iron salts previously described, and their preparation ("aging," dilution, and injection) are more complex (129, 130).

SYSTEM EXPANSION AND EQUIPMENT CHOICES

Many wineries are built with planned, staged growth as markets and consumer acceptability and demand for the winery's products are strengthened over time. Winery development land use permits, for the United States at least, are generally obtained for the probable maximum production capacity anticipated, with local governmental jurisdictions generally requiring that utility systems be installed initially that will serve the ultimate wine production target. The disadvantage of this necessity of governmental compliance is that wastewater systems sometimes operate in an underutilized condition for 4–5 years, until full production is reached. It often takes some rather creative operational moves by the waste treatment system operator to get the best performance from the facility when solids and hydraulic loadings are below the design criteria values. [See earlier discussion on F/M ratios (i.e., food to microorganism ratios) for optimum aerobic treatment system performance.]

For wineries to return at some future date to the local planning body to increase the use-permitted production amount is generally an awkward and chancy proposition. Although it does mean that the winery is experiencing success in the marketplace, to increase the size of site utilities to accommodate the increment of growth is often very difficult if the initial master plan did not provide space and a modular design framework to allow the addition of treatment components without a major disruption of existing plumbing electrical and equipment units. Likewise, if the environmental document Environmental Impact Report, Environmental Impact Statement (EIR, EIS, or Environmental Assessment) was prepared for a fixed set of winery size parameters, it is possible that the planning body could require an entirely new environmental impact assessment to identify the incremental effect of a larger winery on traffic and natural resources. Planning for future growth should be at least in the mind of the utility system designer, so that future expansion/additions are as simple and straightforward as possible (137). More often than not, an undercapitalized winery will falter and the subsequent purchaser of the winery may have entirely different plans for production volume, methods, and wine varieties. For example, pump station structures, dry pits, or vaults can be enlarged slightly to accommodate future planned additions for a very small incremental cost, and the extra space, if not eventually utilized, allows the maintenance of the installed equipment to be performed by averaged size humans under reasonably comfortable circumstances. Oversizing the main electrical panel and buried conduits above what is actually required can save many valuable dollars in the future when additional conductors, signal cables, and circuit breakers are needed. Underground treatment systems (soil absorption) can be expanded if the space has been reserved in the initial placement and no permanent structures have been placed on the site. Figure 8.16 suggests how a conventional gravity system (no pump lifts required) could be added, if the master planning for the first installation had included some forethought on the hydraulics and plumbing for dividing the flow equally. In gravity sewers, where so-called open-channel flow hydraulic performance of pipe systems prevail, as designed, the sewers ideally flow at a one-half full condition. Thus, for a given pipe roughness and slope, as both cross-sectional area of flow and "hydraulic radius" increase, the quantity of flow "Q" increases in accordance with the open-channel flow relationships, as expressed in Manning's equation (138):

$$Q = 1.486/n \ R^{2/3} \ S^{1/2} \ A$$

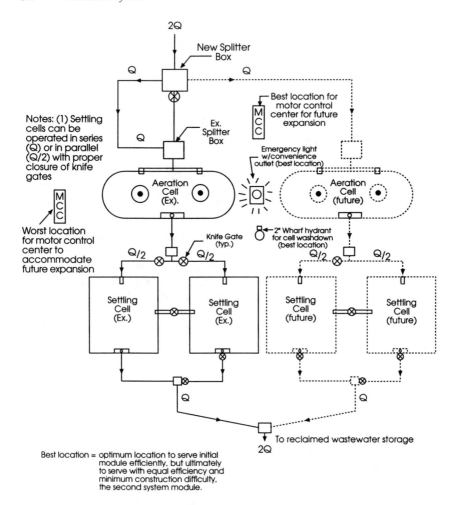

Fig. 8.16. Modular method for winery process wastewater system expansion to allow doubling of production capacity.

where *n* is Manning's roughness coefficient, *R* is the hydraulic radius (ft or m) = cross-sectional area ÷ wetted perimeter), *S* is the slope of the hydraulic gradient (ft/ft or m/m), *A* is the cross section of flow (ft^2 or m^2), and *Q* is the discharge (ft^3/s or m^2/sec). Therefore, for a 6-in. (150-mm)-diameter PVC sewer flowing one-half full with the slope of the hydraulic gradient (the slope of the water surface in this instance) at 0.01 or 1 ft (0.3 m) of fall for every 100 ft (31 m) of horizontal distance, the discharge would be

$$Q_{1/2} = \frac{1.485}{0.015} (0.124)^{2/3}(0.01)^{1/2}(0.098) \text{ for } n = 0.015$$

$$A = 0.098 \text{ ft}^2 \text{ (m}^2)$$
$$R = 0.098/0.79$$
$$= 0.124 \text{ ft (m)}$$

$$= 99.1(0.25)(0.1)(0.098)$$

$$Q_{1/2} = 0.25 \text{ cfs} = 113 \text{ gpm (7L/sec)}$$

Then for the same 6-in. (150-mm)-diameter gravity sewer flowing full, the discharge would be

$$Q_{full} = 1.486(0.125)^{2/3}(0.01)^{1/2}(0.097) \text{ for } A = 0.197 \text{ ft or m}^2$$

$$R = 0.097/1.57$$
$$= 0.125 \text{ ft (m)}$$

$$= 99.1(0.25)(0.1)(1.097)$$

$$= 0.49 \text{ cfs} = 221 \text{ gpm (14 L/sec)}$$

Thus, if the planned expansion of an existing winery doubled the maximum daily process wastewater discharge, the existing 6-in. (150-mm)-diameter sewer, as shown in the illustrative example, could accommodate the two-fold increment of flow hydraulically. In contrast, there are many gravity municipal sewers that, in fact, become pressure sewers because of infiltration and inflow from leaky pipe joints in areas of seasonal high groundwater (these same sewers exfiltrate in the summer when groundwater levels fall) and illegal roof drains and other stormwater connections to the sanitary sewer system. The manholes act as large manometers (in-line pressure gauges) with the water surface elevation in the manhole giving a reasonably good measurement of the magnitude of the pressure in the lines. For shallow manholes, where the pressure surcharge could not be accommodated, sewage overflow often occurs. This hydraulic overload condition should not be allowed to prevail in the winery's wastewater sewer system. The failure to divert the uncovered crusher/press pad stormwater drainage (post-harvest) could divert a large and unwanted volume of water to the sewer system and, hence, the waste treatment works, neither of which were designed for the overload situation. The design solution to the stormwater diversion problem was discussed subsection entitled "Wastewater Flows" and is shown graphically in Figs. 8.8 and 8.13.

Treatment concepts "B," "C," and "D" as shown in Table 8.10, which require mechanical or diffused aeration systems for successful operation, have several possibilities for expansion. If reasonable treatment performance has been achieved without activated sludge return, the inclusion of this subsystem could permit an additional increment of solids loading because of the increased bio-reduction efficiency that accompanies the activated sludge feature. If the winery has maintained good record of operating performance of its treatment systems, the designer can narrow the scope of the design and equipment additions to the barest minimum, as solids and hydraulic loadings will merely need to be extrapolated to the winery's expanded production capacity. (see also the subsection "Wastewater Flows"). The greater volume of oxygen required to reduce the incremental solids loading can be accomplished by the following:

(a) Increasing the power of existing floating mechanical aerators to provide more pounds of O_2 per hour
(b) Add other floating aerators, if aeration cell geometry and space permit
(c) And a submerged air diffusion subsystem to provide the dissolved oxygen increment

A more costly expansion option would require the splitting of the raw process wastestream into two equal parts. Figure 8.16 suggests schematically how a doubling of treatment capacity could be accomplished by cloning the existing system. If system expansion had been judiciously planned by the designer, the electrical and maintenance water supply service needs could be quite easily enlarged if

- The master motor control panel was located at the "best" location as shown in Fig. 8.16
- Grading, drainage, and maintenance access was considered for possible expansion in the initial design
- A 2-in. (50-mm) wharf hydrant is placed to allow its use with short hose runs for aeration and settling cell washdown and wave line scum/froth control
- The emergency light and convenience outlets for power electrical tools was placed prudently for use by maintenance staff for the existing and the expansion treatment units

For underground systems in small wineries, the reservation of an area for replacement of the soil absorption system in the event of its failure forces the designer to consider the best arrangement for simplified installation of future plumbing. If this area is not adjacent to the existing system [on the winery parcel, but remote i.e., greater than 500 ft (152 m) distant], there is very little that can be done to pre-plan an expansion or replacement. The additional treatment units and components (septic tanks, dosing siphons, and lift pumps) for an underground system, the same conceptual design philosophy illustrated in Fig. 8.16 can be employed to double the capacity of an existing process wastewater system.

Long pressure sewers should be oversized at the time of individual winery construction at least one pipe-size-increment greater than the apparent size dictated purely by system hydraulics. The incremental cost just for pipe material for the larger-diameter pipe is very low in comparison to the future cost of trenching, bedding, pipe laying and jointing, and trench backfilling to add additional pipeline capacity at some future time to increase capacity. (It is always comforting to remember that pipe carrying capacity is proportional to the square of the pipe diameter. Thus, a doubling of pipe size will yield four times the flow rate.)

Wastewater Metering

Potable water supply metering was discussed in considerable detail in Chapter 6 and it was described briefly in the subsection "Sewage Pump Stations" in this chapter, where dose counters were the economical metering method of choice. Unless a remote recorder in the winery is desired, the responsibility for flow monitoring should rest with the maintenance staff. Visually checking the dose counter requires that the observer visit the pump station or dosing chamber site, which in itself is a benefit of special importance. Daily visits to the subsystems of the treatment works, even if only for a short time, give the operator a visual, olfactory, and audio database for the operating system as it should be. Without resorting to extensive laboratory analysis or outside professional help, the operator can be forewarned of an impending malfunction, when a noisy pump bearing begins to indicate wear or imminent failure. A close look at the effluent leaving the aeration cell that displays obvious signs of "pinpoint" floc in lieu of the large, fluffy particles that are normally observed exiting the weir box tells the operator that system fine-tuning is in order. Remote data read-outs can be overdone at the expense of real-time observations; the dose counter helps to achieve that frequent personal observation objective.

There are a number of options for wastewater meters that are of the flow-through type (139) (i.e., no hydraulic restriction or meter component protrusion that could form the nucleus of jamming or blockage). The magnetic and sonic meters both fulfill that basic requirement. Using a standard turbine meter for treated effluent may sound practical, but something as simple as a leaf, length of thread, or grape stem segment can destroy what otherwise might be a complete and unbroken flow record if the foreign matter becomes enmeshed in the rotating turbine blades that provide the calibrated input for the flow record. For *raw* wastewater, the use of a flume and stage recorder in combination or the previously mentioned magnetic or sonic recorders all have an unobstructed, free flowing hydraulic geometry. Metering flumes, that are manufactured by Palmer-Bolus or Parshall, are constructed of fiberglass or PVC and are often cast into the base of a manhole.

The potable water supply metering described in Chapter 6 may be sufficient, and wastewater will not require separate metering if sensible subdivisions of flow are metered, as suggested in Fig. 6.9, so that landscape water, in-winery sanitary flushing, and other uses are separated from purely process use. Some governmental agencies are adamant about metering the raw wastewater inflow and outflow of reclaimed wastewater to create a very tight quantitative control over input and output. The cost to produce these data is often unnecessary and greater than the alleged pollution control benefit that is achieved by the enforcement action. For a winery that discharges to a municipal wastewater treatment system, wastewater input and output metering cannot be avoided. Charges to be levied against the winery for sewerage service will be based generally on flow records and BOD_5 or suspended solids loading.

Winter Storage Requirements and Water Balance

If total land confinement of the winery's treated waste effluent is to be accomplished annually without inadvertent spillage to surface waters, a storage facility capable of re-regulating the reclaimed wastewater production must be part of the complete treatment and disposal works. Figure 8.13 showed graphically the winter holdover storage facility with possible end uses for the water. Chapter 7 detailed the relationship between the reclaimed wastewater supply and the irrigation and frost protection requirements for the winery's vineyards. Table 7.2 is repeated here as Table 8.11 to illustrate the water accounting model format that is readily adaptable to any of several software spreadsheet programs. Two useful outputs are produced by the water balance model:

Table 8.11. Water Balance Simulation for Reclaimed Wastewater Reservoir Sizing

Month	V_{BM} (beginning Month Storage)	W_{IN} (wastewater inflow)	E_L (evaporation)	R_{IN} (rainfall)	I_{OUT} irrigation draught	S_{OUT} (reservoir spill)	V_{EM} (end of month storage)

1. The required volume of the winter storage reservoir is calculated for an assumed rainfall probability frequency that is acceptable to the local water pollution control agency, and
2. The volume of reclaimed wastewater available each month is determined, which can be either earmarked for the vineyard irrigation system or used to structure the acreage requirements for other non-vineyard beneficial uses such as permanent pasture or landscape irrigation.

Water balance equation:

$$V_{BM} + W_{IN} - E_L + R_{IN} - I_{OUT} - S_{OUT} = V_{EM}$$

Table 8.12 is an illustrative example of the water balance calculations to determine storage requirements to regulate the reclaimed process wastewater for a 50,000 case winery located in California's Napa Valley. The rainfall values shown in column 4 are for the wettest year of record (1983) for the St. Helena, California, NOAA weather station. The winery selected for this case study had no vineyards, but approximately 9 gross acres of native grassland was designated for conversion to permanent irrigated pasture. Three additional columns are shown in Table 8.12 that reconcile the volumes of reclaimed wastewater that must be used monthly, the required pasture acreage, and the pasture consumptive use [net plant use equals plant consumptive use minus irrigation losses, less available rainfall during the irrigation season (i.e., for northern California April 1–October 1 each year)]. The required volume of reservoir storage as seen from column 6 is approximately 693,000 gal (2,624,000 L) or, in conventional agricultural units of volume, 2.13 acre-ft (an area 1 acre in extent covered to a depth of 1 ft). From available State of California Department of Water Resources data, it was determined that the highest annual rainfall recorded for the St. Helena precipitation station had a statistical probability on a return frequency basis of 1 occurrence in 50 years. The 50-year annual rainfall does not necessarily occur only once in 50 years. It may be repeated several times in a 50-year span or not at all. So-called "Basin Plan" provisions for the Napa River Basin stipulate that reclaimed wastewater storage reservoir volume requirements need only accommodate, without spill, a 1 in 25-year return frequency rainfall year. Thus, the actual storage reservoir size was reduced. In some instances, for small wineries in very high rainfall zones and with limited space [60–70 in. (1524–1778 mm) per year average annual rainfall] removable reservoir covers have been designed to eliminate the extra volume of rainfall (column 4, Table 8.12). The covers are removed after the cessationof winter rain to maximize solar evaporation from the reservoir surface. Forlarge reservoirs, covering the surface to exclude

Table 8.12.   Water Balance Calculations for a 50, 000-Case Annual Production Winery for Sizing the Storage Reservoir and Irrigated Pasture

1	2	3	4	5	6	7	8	9	10	11
Month	Beg. Stg vol. (gal)	Wastewater inflow (gal)	Precip. (gal)	Reservoir evap. (gal)	E.O.M. storage (gal)	Pond excess to Irrig. (gal)	Spill (gal)	Irrig. Acres	Irrig. req. (gal & in./month) Inches	Gallons
Jan.	187,264	54,000	70,433	7,589	306,108	0	0		Prec.	—
Feb.	306,108	81,000	106,496	9,471	484,133	0	0		Prec.	—
Mar.	484,133	63,000	91,130	16,935	621,328	0	0		Prec.	—
Apr.	621,328	72,000	26,028	26,645	692,711[a]	187,829	0	1.83	3.77	187,829
May	504,883	81,000	3,387	31,485	557,785	250,106	0	1.83	5.02	250,106
June	307,679	63,000	2,885	38,572	334,992	138,970	0	0.83	6.15	138,970
July	196,022	63,000	0	42,210	216,812	152,077	0	0.83	6.73	152,077
Aug.	64,735	90,000	2,007	37,506	127,828	81,403	0	0.5	5.98	81,403
Sept.	46,425	99,000	3,010	29,917	118,518	64,932	0	0.5	4.77	64,932
Oct.	53,586	99,000	8,655	18,690	142,551	49,822	0	1.83	2.48	49,822
Nov.	92,729	63,000	78,587	8,843	142,016	0	0		Prec.	—
Dec.	142,016	72,000	83,165	5,444	291,737	0	0		Prec.	—
Totals		900,000	475,783	273,307		925,139	0			925,139

[a]692, 711 gal maximum reservoir storage required to prevent spill. (2, 624, 000 L.)
• See explanatory notes for source data in the pages following
• Management of irrigated pasture to give optimum growth; hence, optimum water use is described in the operation and maintenance subsection at the end of this chapter.
General Note: Volumetric units can be U.S. gallons, cubic meters or acre/feet.
Column No.:
  2 The beginning-of-month storage volume should be the total of the winery's reclaimed process wastewater for November and December plus the November and December rainfall for the year that will be incident on the reservoir surface less solar evaporation from the reservoir surface for the same period. (See Table 6.8 for estimated monthly distribution of total annual wastewater production for new wineries without metered flow records.) Total reservoir storage required is 693, 000 gal as per column 6.
  3 Monthly distribution of estimated or actual annual wastewater production (see Tables 6.7 and 6.8).
  4 See precipitation depth–duration–frequency curves for nearest weather station and select a year with annual rainfall that will have a minimum probability of return frequency of 1 in 25 years. [For the United States, weather records are found in National Oceanographic and Atmospheric Administration (NOAA) reports entitled "Climatological Data" and in several climate record, database programs from both state and federal sources.]
  5 Solar evaporation records are maintained by fully equipped weather stations in the NOAA network for the United States. California has developed monthly values of E_p (pan evaporation) plotted on a state map base. A California Department of Water Resources computer database can be accessed using a PC and modem that gives monthly estimates of evapotranspiration for various crops in various locales in the state. (Such data are required for columns 10 and 11 determination.)
  6 Columns 2 + 3 + 4 − 5; the desirable objective is to have the storage reservoir empty or nearly so at the end of the irrigation season each year.
  7 Column 7 in the illustrative example chosen the table is only relevant for the Northern Hemisphere, arid west, where the largest percentage of rainfall occurrence is generally November through March. Irrigation is required for permanent pasture and other perennial and annual crops generally from 1 April to 1 October each year. Thus, the irrigation schedule for other climate zones must reflect actual plant consumptive use less precipitation. [Monthly irrigation demand can be calculated using E_p (pan evaporation) and crop coefficients to correct the surface evaporation to plant evapotranspiration.] Crop coefficients are available for all regions of the world and have been compiled by the United Nations Food and Agriculture Organization (FAO) (140).
     For vines to produce grapes of superior quality, irrigation must be handled in quite a different manner than the scheduling of irrigation for pasture or landscape use. Thus, there can be less irrigation precision without end-product damage with most grasses and feed grain crops, whereas over-irrigation of vines to meet an unforeseen over-abundance of reclaimed wastewater might place one entire year's vintage at risk from over cropping or excessive cane growth and its attendant diversion of plant energy to plant structure and not fruit. There is greater water recycling operational flexibility and less risk to the winery venture in utilizing reclaimed wastewater for crop uses other than vines.
  8 Pond spill must be equal to zero.
  9 Acreage required = Monthly volume of irrigation water for dispersal (acre-ft)/[Applied irrigation water (ft)] − [Precipitation (ft)]
Note: Applied irrigation = plant consumptive use + irrigation efficiency. Acre-feet per acre per month = feet per month.
  10 Calculated monthly irrigation requirements for irrigated pasture.
  11 Repeat of column 7.

the rainfall component becomes physically and economically impossible. In locating reclaimed wastewater storage facilities, natural drainage courses should be avoided, even though the deeply incised channel can provide an ideal, naturally created storage space. Such sites are routinely sought after by dam builders whose principal goal is to maximize the capture of natural runoff (141). In order to economize on reservoir size, the opposite objective is desired by the designers of reclaimed wastewater storage reservoir. The storage and re-regulation of natural runoff is to be avoided unless local governmental regulations allow the commingling of natural runoff and reclaimed wastewater. (The methods normally used to allow conjunctive use of reclaimed and natural runoff were discussed in detail in Chapter 7.)

Water rights issues also complicate the on-stream location for a reclaimed wastewater storage facility. The diversion "to storage" and for direct diversion "to use" of natural runoff is allowed in most states only if a permit or license to do so is obtained under a rather lengthy administrative process. (This complex body of water law was discussed previously in Chapter 6 in the section entitled "Winery Site Water Resource Assessments.") If there are no alternative sites and an on-stream location must be utilized for reclaimed wastewater storage, a bypass channel can be constructed to exclude natural runoff. Although sometimes practical on secondary ephemeral streams, the likelihood of constructing a bypass on a perennial stream would have a very low probability of success. Even the most cursory environmental review would produce a large number of sound reasons not to disturb the stream and its aquatic resources.

In very arid regions where solar evaporation exceeds annual rainfall, winery process wastewater can be sent into the atmosphere as water vapor. Judicious design of the storage facilities which will exploit large reservoir surface area and shallow depths easily balances reclaimed water production against solar evaporation (see column 5 of Table 8.12).

Other Reclaimed Wastewater Reuse Options

As described in Table 8.12, the end use for illustrative purposes was permanent pasture irrigation. Other beneficial uses for reclaimed winery wastewater offer some advantages, but in most instances require more intensive management (142).

Aquaculture

In the last decade in the western United States and in other areas worldwide, fish farming has blossomed to meet an ever-expanding market for fresh fish and dietary protein, which the open-ocean fishery cannot readily supply. With the proper selection of fish species, both plankton and emergent aquatic plants in reservoirs can be controlled (101). Napa Valley's Chateau Montelena Winery has made a 2-acre pond the centerpiece of its visitor picnic area. Waterfowl and several fish species inhabit the pond and provide the visual accent and interest that make the perfect setting for enjoying their special wine offerings.

Because of nutrient additions to achieve winery wastewater treatability (see an earlier section on "Treatability of Winery Wastewater"), any excess nitrogen or phosphorus compounds can stimulate plankton (algae) growth. Although a modest plankton population is part of a healthy aquatic food chain, too much algae can deplete dissolved oxygen resources, destroying the rather fragile equilibrium among the pond's living organisms. An aquaculture-focused pond which uses reclaimed wastewater entirely will require more attention to maintain aquatic healthiness than say a pond with a 20–30% freshwater dilution ratio.

If there is an interest in commercial aquaculture, there are aquaculture companies who would, under some mutually agreeable arrangement, provide the fish stock and manage the pond in return for use of the reservoir system for rearing marketable fish. The closer the winery is to a large fresh fish market center, the more economically desirable the venture becomes.

Landscape Irrigation

Landscape irrigation was suggested previously as one possibility for utilizing a small percentage of the reclaimed wastewater production. The smaller the winery, the greater likelihood that the areal extent of landscaping will more closely match winery reclaimed wastewater output. Larger wineries will not necessarily have larger areas of landscape plant materials which will require supplemental irrigation. Thus, the disposal and reuse plan for large to very large wineries may involve the development of the several choices of beneficial water use to accommodate all of the winery's treated process wastewater. Decorative lawns, like the permanent pasture option previously described, require large volumes of water in arid, Mediterranean climates.

Fire Protection

Fire protection is another possible use for reclaimed wastewater. It is obviously a one-time use, unless, of course, a fire occurs. In open-storage facilities used for fire flow reserve, some additional water would be required for solar evaporation loss make-up. As will be described in Chapter 10, the National Fire Protection Association (NFPA) codes are very specific about the single-purpose use of water for fire fighting. Water quality is not an issue with NFPA, only that the water must be able to quench a structural or wildland fire by hose or interior sprinkler delivery. Small wineries protected with a single hydrant will, in most jurisdictions (County Fire Marshall or State Forestry Agency), require periodic testing and flushing of the hydrant to verify fire flow rate and residual pressure. If the stored fire protection water is allowed to stagnate and become anaerobic, the periodic testing chore will become odiferous and unpleasant. Also, a long-term period of anaerobicity will foster the growth of filamentous slimes, which could, if in abundance, partially block the outlet and, certainly, the small orifice sprinklers within the winery. Therefore, some attention should be given to the long-term storage effects of reclaimed wastewater, whether the intended use is for frost protection (strictly short-term seasonal use) or for fire protection (low probability of use except for periodic system testing). For the fire protection storage, the obvious and simplest method of maintaining water quality is the periodic flushing of the tank or reservoir contents and refilling with a fresh volume of reclaimed wastewater. As the NFPA code requires the maintenance of the requisite fire flow volume 100% of the time, the emptying and refilling of the tank might violate that code provision. A better approach would be a monthly check of dissolved oxygen concentration, and if below 2 mg/L, air should be sparged into the tank or hydrogen peroxide added to produce the needed O_2. A swimming pool algicide will keep unwanted growth from occurring.

Off-Site Use

Chapter 7 suggested that a use-market be sought by the winery if the size of the winery parcel or the lack of vineyards prevent on-site reuse of the reclaimed wastewater. In large grape growing areas, there are usually areas of both groundwater scarcity and abundance. If a vineyard owner can be found who is in need of water, a long-term (probably no less than 30 or 40 years) agreement should be drawn up between the parties with the responsibilities and obligations of each party clearly defined (i.e., Who pays for pumping energy, if required?; Is water delivery guaranteed?; Is water quality guaranteed?; Is water delivered at no cost to user?) The implementation of the reclaimed wastewater export plan becomes complex if utility easements from other property owners are required. This latter obstacle has prevented several off-site reuse schemes.

Export to an off-winery location is probably the least desirable of the options available, and because of the joint responsibilities which the use agreement will specify, the local regulatory agencies are less than enthusiastic about endorsing and approving such an arrangement.

Seasonal Wetlands and Wildlife Habitat Improvement

In the United States, the no-net-loss-of-wetlands concept is stipulated by the provisions of Section 404 of the Clean Water Act of 1974, as amended (143). There are often many opportunities for wineries in rural settings to improve wildlife habitat without interfering with vineyard operations or to create wetlands as the highest and best category of habitat (aquatic and terrestrial). Both the planning and execution of a created wetland project are complicated in the extreme and are beyond the scope of this brief discussion. The need for competent consultants to design wetland systems has been filled in the last decade, principally because many projects that have encroached on existing wetlands have had to mitigate the damage by creating new wetlands. Selection of the proper soils, plant mix, and micro-hydrologic regime is a part of the special wetland equation. Wetland creation should not be overlooked by wineries seeking a water reuse solution with a strong environmental statement.

All-Weather Access

A reasonably well-designed access road with no grades (slopes) exceeding 12% should provide vehicle access regardless of weather conditions. The need for access to the septic tank portion of underground systems was cited previously to allow septic tank pump trucks to position themselves within 20 or 30 ft (6–10 m) maximum of the sludge compartment manhole. For above-ground systems, year-round access is mandatory for not only routine maintenance and observation but also for repair and replacement if the equipment malfunctions. In wine growing regions where winter snowfall is likely to occur, special provisions must be made for roadway plowing and sanding, so that ingress and egress are possible at all times. The following are minimum access road requirements:

(a) Minimum paved width of 12 ft (3.7 m) with 2-ft (0.61 m) unpaved shoulders
(b) Maximum slope at 12%
(c) Loop geometry so that large trucks can negotiate ingress and egress easily
(d) For turns, minimum radius of curvature of 150 ft (46 m)
(e) V-ditch side drainage with 2% minimum cross-slope
(f) Cross-drains where required to direct accumulated flow to natural drainage courses

Fencing and Signage

Aboveground waste treatment facilities are attractive nuisances. The open ponds with steeply sloping sidewalls would be difficult to climb out of, particularly for children. Random vandalism of expensive electrical-mechanical equipment is also possible. Fencing can discourage children but will only delay vandals for as long as it takes to cut or scale the fence. Intrusion alarms that are now standard features in most wineries can be extended and installed to monitor the fenced, waste treatment system perimeter.

Warning signs are a legal requirement in most states in the United States, Central, eastern, and southern Europe do not seem to be as compulsive about hazard posting as the United States. Durable aluminum signs that can be attached to the perimeter fence at no more than 80-ft (24.4 m) intervals warn the viewer that wastewater treatment is being performed inside the fence and that only authorized employees are permitted to enter. Such signs may have a minimum impact on someone intent on entering the restricted area, but if tragic injury or death does occur inside the fence, both the winery's insurance company and legal counsel will be pleased that signs and fencing were in place, in the event that damage and liability actions against the winery are pursued by the illegal trespassers.

The need for emergency lights was identified in Fig. 8.16. This important illumination can be connected to the intrusion alarm system to flood the area with light and further discourage vandalism.

OPERATION AND MAINTENANCE OF PROCESS WASTEWATER TREATMENT AND DISPOSAL SYSTEMS

Operation and maintenance (O&M) manuals for utility systems have been highlighted in other chapters of this book and together with their use and importance in executing the designer's intent for the performance of the system (125). It would be a major breakthrough in the management of biological, mechanical, and electrical systems if after start-up, they could be tended by robotic units which utilize artificial intelligence based on fuzzy logic. The beginnings of non-human-intervention management can already be seen in the automotive industry and primary metals manufacture, where repetitive tasks are, in fact, very efficiently performed by programmable robots. Aerobic reactors for waste treatment, however, need not only human intervention but also frequent attention to measure and analyze the many variables and vital signs that give a picture of performance and general health.

Responsibility of Wastewater System Designer

The design package for professional services should include most if not all of the following responsibilities:

1. Planning preliminary design and environmental analysis
2. Final design (plans and specifications) and agency approvals and permits
3. Bidding assistance (pre-bid conference contract award details and precon-struction conference)
4. Construction services (approving equipment/material substitutions, check-ing shop drawings, monitoring construction progress, and inspecting criti-cal system elements)
5. Prepare record drawings (as-constructed) and prepare operation and main-tenance (O&M) manual

The O&M manual is like a cookbook, user manual, and history book all combined into one, and the designer is the best person to prepare the manual. The record drawings, cited as a part of the work product in item 5 above, are the reference for information concerning sizes and locations of all the components that make up the wastewater system. If changes are made to the system at some future time, the changes should be made to the record drawings so that it remains an up-to-date reference source.

Elements of an O&M Manual

Wastewater conveyance treatment and disposal systems for wineries vary considerably in order to meet different winery size and function and site and topographic limitations. The need for an O&M document for under-ground systems was described in an earlier section entitled "Operation and Maintenance of Sanitary Wastewater Systems" and detailed in Table 8.5. The O&M manual outline (Table 8.13) is all-inclusive for small wineries with either underground, soil absorption systems, or specially engineered systems (alternative systems). The manual becomes more simplified with the omission of sections that are irrelevant. The winery's Safety Manual becomes an important companion document to the O&M manual. It would not be improper to extract pertinent sections of the safety manual (once it has been prepared by winery management) and make specific references to personnel requirements, safety equipment, and emergency response procedures for the maintenance tasks of the wastewater system. This interrelationship of utility system operations and winery personnel health and safety is covered in detail in Chapter 12. The handling of hazardous chemicals, entering manholes and underground vaults, and general electrical safety are tasks for which cellar or utility staff should receive normal classroom training and familiarity exercises in the proto-type system. It is possible to combine all of the winery utility system O&M details into a single manual; however, the volume often becomes too bulky, and transport and use of the materials as a working document is less likely to occur with a 10-lb binder.

Table 8.13. Outline and Principal Subject Headings for a Typical Wastewater System Operation and Maintenance Manual

I. Introduction
    A. Summary of Design Data
    B. Description of System and Operation
    C. Personnel and Staffing Recommendations
    D. Relationship to Winery Safety Manual
    E. Preventative Maintenance
II. Monitoring and Record Keeping
    A. Hard-Copy and PC Data Base Duplication
        (1) Equipment History Records
    B. Sampling and Preservation of Samples
        (1) In-Winery Lab
        (2) Commercial Lab
    C. Sampling Schedule
    D. System Performance Standards
    E. Periodic Reports to Regulatory Agencies
        (1) Annual
        (2) Monthly
    F. Wine Processing Events Chronology
        (1) Flow Records by Process
        (2) Record BOD, COD, SS, and pH by Process
    G. References
III. Unit Waste Treatment Processes
    A. Description
    B. Operating Limits
        (1) Design
        (2) Overload Condition
    C. Collection Sewers
    D. Screening System
    E. Bio-Nutrient Additions and pH Neutralization Systems
    F. Aerobic Reactor
        (1) Aeration Requirements and Time Clock Settings, If Appropriate
    G. Clarification and Sludge Return
        (1) Flocculent Feed and Mixing Subsystems
    H. Pump Stations
    I. Electrical Controls and Instrumentation
    J. Waste Sludge Storage
    K. Reclaimed Wastewater Conveyance and Storage System
        (1) Algae, Aquatic Weeds, and Nuisance Insects
    L. Irrigation Subsystem
    M. Fencing, Locks, and Security
    N. Residual Solids Subsystems
        (1) Screenings (Handling and Disposal)
        (2) Waste Sludge (Handling and Disposal)
    O. Area Grounds Maintenance
        (1) Weed and Litter Control
        (2) Landscape Screens
    P. Valves and Slide Gates
APPENDICES
    A. Manufacturer's Equipment Recommended Maintenance and Lubrication Schedule
    B. Master Maintenance and Monitoring Schedule
    C. One-half Scale Set of Record Drawings
    D. Reclaimed Winery Wastewater Discharge Requirements as per Local Governmental Discharge Permit

Table 8.14.   Operating Data and Analysis.

```
SAMPLES:  Grab _____    DATE _____   TIME _____
          Composite _____   OPERATOR _____

WEATHER:
     Temp. _____Wind _____ Clear _____Cloudy _____

AVERAGE DAILY FLOW IN PLANT _____MGD

ANALYSIS & RESULTS (SEE FORM 0-3)
SETTLEABLE SOLIDS:                      SUSPENDED SOLIDS:
Influent _____ML/L  Influent _____MG/L _____LBS/D
Aerated Lagoon _____ML/L  MLSS _____MG/L _____LBS/D
Clarifier Effluent _____ML/L  Pol. Eff. _____MG/L _____LBS/D
Reclaimed wastewater
storage pond _____ML/L


BIOCHEMICAL OXYGEN DEMAND (5 DAY 20°C)
     Influent              TEMP _____ PH _____MG/L _____LBS/DAY _____
     Clarifier Effluent         _____    _____    _____        _____
     Reclaimed wastewater
     storage pond               _____    _____    _____        _____
                                _____    _____    _____        _____

Reduction _____%        Lagoon Loading _____

DISSOLVED OXYGEN
     Influent              TEMP _____ MG/L _____
     Mixed Liquor                 _____    _____
     Lagoon Effluent              _____    _____
     Clarifier Effluent           _____    _____

SLUDGE INDEXES
     Settleability _____%
     SVI _____
     Suspended Solids _____MG/L
          (see above)
```

Tables 8.14 through 8.22 are examples for the winery wastewater system O&M data sheets that illustrate the breadth and depth of the information required. As suggested in the O&M manual outline (Table 8.13), both hard-copy records and a PC database system are recommended. (An O&M software program which may be tailored for all of a winery's utility systems was previously cited on Chapter 5, in the section entitled "*Operation and Maintenance.*" The data collection format can be adapted to the winery's particular system and final reclaimed wastewater disposal method or methods.

Table 8.23 is a typical maintenance schedule for a large-winery above ground aerobic treatment and wastewater reclamation system. This checklist corresponds to the O&M manual outline, Table 8.13 and is shown in

Table 8.15.   Record of Aerated Lagoon Operating Parameters

MONTH _____   YEAR _____

DAY	FLOW INPUT					AERATED LAGOON				E A/S	
	Q FLOW RATE (mgd)	T LAGOON TEMP (°F)	S₀ BOD₅ INFLUENT (mg/L)	X₀ SS INFLUENT (mg/L)	PH INFLUENT	t AERATOR OPERATING TIME (HRS)	X₀ MLSS (mg/L)	D.O. (mg/L)	PH	EFFICIENCY OF PROCESS	
										% BOD₅	% SS
1											
2											
3											
4											
5											
6											
7											
8											
9											
10											
11											
12											
13											
14											
15											
16											
17											
18											
19											
20											
21											
22											
23											
24											
25											
26											
27											
28											
29											
30											
31											

Table 8.16.   Solid Handling Report

DATES	TYPE OF SOLIDS	POINT OF ORIGIN	FINAL DESTIN- ATION	HANDLING OPERATION DATA					FOLLOW UP MEASURES	REMARKS
				WHAT² DONE	EQUIPMENT	FLOW-RATE	TIME	QUANTITY		

1.   DISPOSAL IN VINEYARD, OFF-HAUL BY COMMERCIAL PUMPER TO SANITARY LANDFILL.
2.   TYPE OF OPERATION.

Table 8.17.   Pump Station Performance

DATE	WET WELL[1] WATER LEVEL SURFACE	SUCTION[2] INLET(S)	DISCHARGE PRESSURE	DISCHARGE[3] OBSERVED	CHECK VALVE OPERATION	LEAKS, NOISE, VIBRATION, ETC	CONTROL LEVELS	ALARM EQUIPMENT	REMARKS

1.   REMOVE SCUM & FLOATING DEBRIS PERIODICALLY.
2.   MUST BE KEPT CLEAR
3.   OBSERVE DISCHARGE AT END OF FORCEMAIN

Table 8.18.   Reclaimed Wastewater Storage Reservoir

DATE	UPSTREAM[1] FACE	DOWN-[1] STREAM FACE	NORTH[1] ABUT-MENT	SOUTH[1] ABUT-MENT	SEEPAGE (q p m)	SOLID GATE	SPILL-[2] WAY	RES. DEPTH (H)	AQUATIC WEEDS	MOSQUITO[3] CONTROL	REMARKS

1.   REPAIR EROSION QUICKLY TO PREVENT DAMAGE TO EMBANKMENT.
2.   SHOULD BE KEPT CLEAR AT ALL TIMES
3.   NOTE WHEN ATTENTION IS REQUIRED

equipment maintenance record format in Table 8.22. The checklist should be reproduced in poster size for wall mounting in the maintenance supervisor's office and in the utility maintenance staff shop, as constant reminders to both individuals of the vigilance that is needed in remaining current in the daily work tasks. Any veteran of equipment maintenance for any

Table 8.19. Irrigation Pump Station and System

DAY	CALCULATED ET CROP	TIME ON	TIME OFF	TOTAL HRS ON	FLOW METER READING (CUBIC FT.)	AMOUNT1 IRRIGATED (CUBIC FT.)	WEATHER	SOIL MOISTURE	VINEYARD NUMBER	PUMP DISCHARGE PRESSURE (p.s.i.)	REMARKS
1											
2											
3											
4											
5											
6											
7											
8											
9											
10											
11											
12											
13											
14											
15											
16											
17											
18											
19											
20											
21											
22											
23											
24											
25											
26											
27											
28											
29											
30											
31											
1.	SUBTRACT PREVIOUS DAYS READING FROM CURRENT READING.										

Table 8.20. Livestock Grazing Record

DATE	PASTURE NUMBER	SOIL MOISTURE	GRASS LENGTH	OVERALL CONDITION	STOCK WATER SYSTEM	NUMBER HEAD GRAZING	OPERATION PERFORMED	REMARKS

industrial endeavor, will be happy to report that deferring schedule maintenance inspections is the one very effective way of creating major plant shutdowns. Employing skilled, responsible maintenance staff and ensuring that they are continually trained in matters of maintenance protocol and personal safety, is one way that wineries can protect their plant capital

Table 8.21.   Log of Emergency Conditions

DATE	TIME	CONDITION ENCOUNTERED	HOW DETECTED	ACTION TAKEN	FOLLOW-UP	REPORT	REMARKS

investment and provide the highest statistical probability of trouble-free operation.

Staffing and Service Maintenance Contracts for Wastewater Systems

Item I.C of Table 8.13, O&M manual outline, provides the designer with the opportunity to suggest to the owner, the level of competence and training that the individual staff members who are given the hands-on responsibility for system maintenance should possess. In small wineries, it is often the assistant winemaker or the winemaker, who has collateral duties as maintenance staff. It does not take much imagination to conjure up a picture of a major wastewater system component failure during crush for a small winery with the harried and multitalented winemaker trying to effect repairs while overseeing the delicate pressing of over-ripe Chardonnay grapes.

Larger wineries *must* have dedicated staff whose primary job function is the maintenance of all mechanical and electrical equipment on the winery premises. As the size of the winery increases, more compartmentalization of maintenance work must occur, as the more important pieces of equipment (production and utilities) cannot be dealt with by only one staff person. Large wineries usually have skilled tradesmen from several fields (such as industrial electricians, electronic specialists, welders, pipefitters, and state-certified water and wastewater system operators).

For medium-sized wineries, the best approach for maintenance of waste-

Table 8.22.   Equipment Service Record

UNIT _____

LOCATION _____

NAME PLATE DATA   MAKE _____   _____

MODEL _____   _____

CAPACITY _____   _____

VOLTAGE _____   _____

REFERRED TO IN:

    EQUIPMENT MANUAL   ❏   VOL. _____   PAGES _____

    MASTER MAINT SCHEDULE

    O & M MANUAL   ❏   PART _____   PAGE _____

IINITIALS	DATE	TIME	PROBLEM NOTED	WORK DONE	TOTAL HOURS	REMARKS

water systems may be the contracting of specialized services from one of the local or nationwide industrial/municipal services companies (118). The scope of services can range from an operational oversight arrangement to the provision of a full-time staff operator (maybe just during crush/harvest). The service company employee can also be called on for troubleshooting and problem solving on an as-required basis. The experience and training that these Class IV and V Certified Waste Treatment Plant Oper-

Table 8.23. Typical Maintenance Schedule Checklist for Table Winery Process Wastewater System

Unit	Task	Frequency[a]	Remarks or O&M binder volume/section reference
A) Rotary drum screen	1) Debris box (empty when full). Debris box drain Debris box slab drain.	Daily during crush or as required. Clean daily as required	Stockpile and cover daily with soil to discourage fruit flies; incorporate into vineyard soils after crush.
	2) Two-grease fittings (each bearing of drum shaft)	Q	SAE 120 bearing grease
	3) Drive chain	Q	SAE 120 bearing grease
	4) Lift drum cover and wire brush any material lodged in the slots not removed by the spray bar	Daily during crush	
B) Parshall flume (ultrasonic flow meter)	1) Check flume throat to ensure that no debris accumulates and gives false flow readings	Daily during crush	—
C) Aerated lagoon splitter box	1) Check to see that flow split is about 50–50	M	—
	2) Wash down scum at waterline	D	—
D) Aerators	1) Vibration check by factory reps	Y	
	2) Lubricate zirc fittings		
	3) Check mooring line fittings and chafe protectors for wear	SY	
	4) Check amperage on each aerator	Q	
		M	Debris caught on propeller will show as higher than name plate amperage.
E) Clarifier mix basin	1) Check floc formation in basin	Daily during crush	Sludge return and fresh alum feed should be adjusted to give 1/4–3/8-in. floc particles or larger
	2) Check sludge from sludge return line	Daily during crush	Adjust timer for longer "off" interval if sludge content is low
mixer	1) Verify mixer operation; no lubrication required	Daily during crush	
F) Clarifier	1) Check influent inside settling trough and skirt to see if particles are settling adequately; open scum gate if scum layer exists	Daily during crush	
	2) Lubricate gear reducer, spur gear, and bearings	As per manuf. spec.; daily during crush	

Table 8.23.   Typical Maintenance Schedule Checklist for Table Winery Process Wastewater System (*Continued*)

Unit	Task	Frequency[a]	Remarks or O&M binder volume/section reference
	3) Check effluent on effluent through.		—
G) Sludge pump	1) Check drive belts	Weekly during crush; as per manuf. spec.	
	2) Lubricate bearings and seal cavity		
H) Effluent pumps	1) Check pump operation for 1 cycle and note time from start to finish; listen for excessive pump noise		—
	2) Lubrication	None required.	
I) Reclaimed wastewater holding pond	1) Note net volume increment daily as observed on staff gauge	Daily during crush and irrig. season (Apr.–July)	—
J) Reclaimed wastewater pumps	1) Check packing gland; should be 5–6 drops of water per minute leakage	Weekly during irrig. season when pumps operating	
	2) No lubrication required	—	
K) Chemical feed pumps	1) Preventive maintenance schedule	As per manuf. spec.	

[a]Frequency legend: Q = quarterly; SY = semiannually; M = monthly; Y = yearly; D = daily; W = winter season.

ators possess allow them to quickly analyze plant monitoring data and make appropriate adjustments in the dissolved oxygen level or pH to get optimum performance. A secondary benefit is the indirect training that the winery's staff person receives while working under the tutelage of the ultra-skilled consulting operator. As the winery's operator becomes more experienced, the maintenance service contract can be reduced in scope or terminated.

GLOSSARY

Activated sludge   The suspended solids in an aeration tank or at the bottom of a secondary clarifier in a sewage treatment plant, consisting mostly of living microorganisms.

Activated sludge process   A biological sewage treatment system in which living microbes, suspended in a mixture of sewage and air, absorb the organic pollutants and convert them to stable substances.

Advanced treatment   Purification processes used after or during secondary wastewater treatment to remove nutrients or additional solids and dissolved organics; also called tertiary treatment.

Aeration   A physical treatment process in which air is thoroughly mixed with water or wastewater for purification.

Aerobe   A microorganism that requires an aerobic environment to live and reproduce.

Aerobic   In the presence of air or available molecular oxygen.

Aerosol   A suspension of small solid or liquid particles in air.

Anaerobe   A microorganism that lives under anaerobic conditions, without free oxygen.

Anaerobic   In the absence of air or available molecular oxygen.

Aquatic organism   An organism that lives in water.

Autotrophic organisms   Self-nourishing green plants that obtain food from photosynthesis; the beginning link of the food chain.

Batch process   A process in which there is no inflow or outflow during the treatment phase.

Biochemical oxygen demand (BOD)   Measure of the concentration of organic impurities in wastewater. The amount of oxygen required by bacteria while stabilizing organic matter under aerobic conditions, expressed in mg/L, is determined entirely by the availability of material in the wastewater to be used as biological food and by the amount of oxygen utilized by the microorganisms during oxidation (usually a 5-day test at 20°C).

Biodegradable   Readily broken down or decomposed into simpler substances by biological action of microbes.

Capillary action   The movement of a liquid through a soil pore or retention by a solid surface, due to the interaction of adhesive, cohesive, and viscous forces.

Chemical oxygen demand (COD)   The amount of oxygen needed to oxidize all the organics in a wastewater sample; a measure of the level of organic pollution. A much more rapid test (3 h) than the 5-day BOD test. An equation can be developed to relate COD to BOD.

Chlorine residual   The total amount of chlorine (combined and free available chlorine) remaining in sewage or industrial wastes at the end of a specified contact period following chlorination.

Clarifiers   Settling tanks. The purpose of a clarifier is to remove settleable solids by gravity, or colloidal solids by coagulation.

Digestion   The decomposition of organic waste by microbes under controlled conditions in a sewage treatment plant.

Disinfection   The destruction of disease-causing microbes in water or sewage effluent, usually by the addition of chlorine or ozone.

Dissolved oxygen (DO)   The oxygen dissolved in water, wastewater, or other liquid, usually expressed in milligrams per liter (mg/L), parts per million (ppm), or percent of saturation.

Exothermic   Heat liberating.

Facultative autotrophs   Microorganisms that can use both carbon dioxide and organic compounds as carbon sources.

Facultative bacteria   Bacteria having the ability to live and grow both in the presence of and in the absence of free oxygen.

Flora   Plant life.

Food to microorganism ratio (F/M)   Amount of BOD applied to the activated sludge system per day per amount of MLSS in the aeration basin, expressed as pound of BOD per day per pound of MLSS. (See *Mixed liquid suspended solids.*)

Force main   A pipeline through which raw or treated sewage is pumped under pressure.

Gray water   Wastewater generated by water-using fixtures and appliances, excluding the toilet and possibly the garbage disposal.

Influent   Wastewater as received at a treatment facility.

Intermittent sand filter   A natural or artificial bed of sand or other fine-grained material to the surface of which wastewater is applied intermittently in flooding doses and through which it passes; opportunity is given for filtration and the maintenance of an aerobic condition.

Land application   A method for distributing partially or fully treated wastewater onto land where it receives further treatment by the soils and eventually reaches the groundwater. Land application can either be subsurface (see leachfields) or surface (irrigation, landspreading with a truck, etc.). In some cases, land treatment may be used as a treatment step alone, with underdrains to collect the effluent for disposal elsewhere. Generally considered alternative or innovative.

Leachfields   The most commonly used on-site disposal technique consisting of tiles or perforated PVC pipe, which distribute septic tank effluent for subsurface land application. Methods of subsurface waste effluent disposal, collectively, are called soil absorption systems.

Lagoon/stabilization/oxidation pond   An impoundment designed to enhance the natural purification of wastewater without major input of energy by allowing enough time and wind-induced or thermal mixing for biologic activity.

Limiting nutrient   That nutrient the of which concentration in the substrate limits the growth of the organism utilizing the substrate.

Manifold   A pipe fitting with numerous branches to convey fluids between a large pipe and several smaller pipes, or to permit choice of diverting flow from one of several sources or to one of several discharge points.

Metabolism   The chemical and physical processes continuously taking place in living organisms and cells, comprising those processes by which assimilated nutrient is built up into protoplasm (anabolism), and those by which protoplasm is used and broken down into simpler substances, with the release of energy (catabolism).

Mixed liquor suspended solids (MLSS)   Suspended solids in a mixture of activated sludge and organic matter undergoing activated sludge treatment in the aeration tank.

Nitrification   The conversion of ammonia into nitrates by bacterial action, causing a decrease in dissolved oxygen levels in water.

On-site disposal   Subsurface disposal of sewage at the location where it is generated, usually using a septic tank or leachfield.

On-site management district   A public entity authorized to operate and control groups of public or privately owned on-site systems.

On-site system   A self-contained system which provides both treatment and disposal of wastewater on an individual lot usually consisting of a septic tank and leachfield (also known as underground system).

Plankton   Small floating or weakly swimming animal and plant life in a body of water.

Potable   Drinkable.

Protoplasm   A semi-fluid colloidal material comprising the bulk of all animal and plant cells and consisting largely of water, proteins, fats, carbohydrates, and inorganic salts.

Protozoa   Microscopic unicellular animals.

Reserve area   An area set aside by the local sewerage authority to permit replication of a leachfield should clogging and failure occur.

Rotary drum screen   A physical treatment device in which water or wastewater flows through a revolving drum that is covered with a finely woven metal fabric that traps suspended solids.

Screening   A physical treatment process for wastewater in which large suspended particles are removed as the liquid passes through a wire mesh screen or strainer. The screenings are collected and disposed of as solid waste.

Secondary treatment   Biological treatment of wastewater designed to remove at least 85% of the suspended solids and biochemical oxygen demand.

Sedimentation   The slow settling and separation of suspended solids from a liquid under the force of gravity.

Septage   The solid and liquid material removed from a septic tank during pumping.

Settleable solids   That matter in wastewater which will not stay in suspension during a preselected settling period, such as 1 h, but either settles to the bottom or floats to the top.

Settling tank   A steel or concrete basin in which settleable solids are allowed to separate from water or wastewater under the force of gravity; also called a clarifier, settling cell, or sedimentation tank.

Specially engineered systems   On-site or underground systems that must be employed when soil, groundwater, or slope conditions do not permit the installation of a conventional septic tank and leachfield.

Suspended solids (SS)   Solids carried in water or sewage that would be retained on a fiberglass filter in a standard lab test.

Trickling filter   A biological sewage treatment unit in which dissolved organics are absorbed form the settled sewage as it flows over a fixed film of organisms attached to rock or other inert media.

Uniform Plumbing Code (UPC)   A code of practice frequently adopted by state regulatory authorities as the design standard for on-site systems.

Water balance   An accounting of all inputs and outputs to a reservoir to demonstrate that the volume of the storage facility is sufficient to control discharge (spill) under a given set of historical, hydrologic conditions (rainfall, runoff, evaporation, and seepage).

REFERENCES

102. GLOYNA, R. 1972. Wastewater recycling demonstration projects. *J. Water Pollut. Control Fed.* 13(4).

103. STORM, D.R. 1995. Cryptosporidium; a new drinking water threat. *Pract. Winery Vineyard* 15(6).

104. _____. 1980. *Design Manual: Onsite Wastewater Treatment and Disposal Systems.* Cincinnati, OH: USEPA, Mun. Environ. Res. Lab.

105. _____. 1979. *Wastewater Engineering, Treatment, Disposal and Reuse.* 2nd ed., Metcalf and Eddy. New York: McGraw-Hill.

106. _____. 1994. *Uniform Plumbing Code.* Walnut, CA: International Association of Plumbing and Mechanical Officials.

107. STORM, D.R. 1995. Research results on the effects of water softener brine discharges on septic tanks and leachfields. *Pract. Winery Vineyard* 15(6).

108. STORM, D.R. 1995. The best T.P. for your S.T. Winery Water and Waste. *Pract. Winery Vineyard* 11(4).

109. STORM, D.R. 1990. Choosing a winery site. Winery Water and Waste. *Pract. Winery Vineyard* 11(4).

110. STORM, D.R. 1984. Site planning for wineries: A sanitary engineer's view. Winery Water and Waste. *Pract. Winery Vineyard* 5(4).

111. STORM, D.R. 1987. Sewage pump stations: Operation, design decisions and equipment choices. Winery Water and Waste. *Pract. Winery Vineyard* 8(2).

112. STORM, D.R. 1985. Winery odors; achieving 24-hour protection, Part I. Winery Water and Waste. *Pract. Winery Vineyard* 6(4).

113. STORM, D.R. 1985. Winery odors; Achieving 24-hour protection, Part II. Winery Water and Waste. *Pract. Winery Vineyard* 6(5).

114. STORM, D.R. 1995. Something new in living sound walls and odor barriers. Winery Water and Waste. *Pract. Winery Vineyard* 16(1):80–81.

115. STORM, D.R. 1988. Controls and instrumentation. Winery Water and Waste. *Pract. Winery Vineyard* 8(6).

116. STORM, D.R. 1989. Waste treatment and biological activators. Winery Water and Waste. *Pract. Winery Vineyard* 10(3).

117. _____. 1978. *San Francisco Bay Drainage Area Basin Plan.* Sacramento, CA: State of California, State Water Resources Control Board.

118. PHILLIPS, G. 1995. Personal communication. Phillips & Associates Management and Technical Resources, Napa, CA.

119. CONVERSE, J.C. and OTIS, R.J. 1976. Alternate Designs for On-site Home Sewage Disposal. Small Scale Waste Management Project. Univ. of Wisconsin, Madison, WI.

120. INGHAM, A.T. 1980. *Guidelines for Mound Systems.* Sacramento, CA: State of California, Water Resources Control Board.

121. INGHAM, A.T. 1980. Sacramento, CA: *Guidelines for Evapotranspiration Systems.* State of California, Water Resources Control Board.

122. STORM, D.R. 1994. Trickle systems (aka: Drip); subsurface wastewater disposal with a new twist. Winery Water and Waste. *Pract. Winery Vineyard* 15(4).

123. STORM, D.R. 1979. *Stinson Beach County Water District, Wastewater Facilities Plan.* Davis, CA: Storm Engineering.

124. STORM, D.R. 1995. Gray-water recycling and reuse; laws begin to catch-up with technology. Winery Water and Waste. *Pract. Winery Vineyard* 16(3).

125. STORM, D.R. 1985. Water supply and wastewater treatment system operation and maintenance tips. Winery Water and Waste. *Pract. Winery Vineyard* 6(2).

126. STORM, D.R. 1991. New effluent control laws. Winery Water and Waste. *Pract. Winery Vineyard* 12(4).

127. STORM, D.R. 1993. Stormwater pollution prevention update. Winery Water and Waste. *Pract. Winery Vineyard* 14(2).

128. STORM, D.R. 1988. Process wastewater strainers and screens. Winery Water and Waste. *Pract. Winery Vineyard* 9(4).

129. STORM, D.R. 1992. Wastewater sedimentation and settling aids, Part I. Winery Water and Waste. *Pract. Winery Vineyard* 13(4).

130. STORM, D.R. 1993. Wastewater sedimentation and settling aids, Part II. Winery Water and Waste. *Pract. Winery Vineyard* 13(5).

131. _____. 1992. *Standard Methods for the Examination of Water and Wastewater,* 18th ed. Washington. DC: WPCP and APHA.

132. STORM, D.R. 1985. (Un)treatability of winery wastewater: Mysteries dispelled. Winery Water and Waste. *Pract. Winery Vineyard* 5(6).

133. _____. 1977. *Process Control Manual; Aerobic Biological Wastewater Treatment Facilities.* Washington, DC: U.S. Environmental Protection Agency.

134. _____. 1966. Advances toward understanding lagoon behavior. *Proc. of 3rd Sanitary Engr. Conf.* Univ. of Missouri Rept. No. 6.

135. SILVERIO, C.M. 1986. Anaerobic treatment of distillery slops using an upflow anaerobic filter reactor. *Process Biochem.* 21(6):192–195.

136. McCARTY, P.L. and Young, J.C. 1969. The anaerobic filter for waste treatment. *Journal of Water Pollution Control Federation* 41(5, Part 2):R160–R173.

137. STORM, D.R. 1986. Waste treatment system expansion: Where to begin. Winery Water and Waste. *Pract. Winery Vineyard.* 7(3).

138. VENNARD, J.K. and STREET, R.L. 1975. *Elementary Fluid Mechanics,* 5th ed. New York: John Wiley and Sons.

139. STORM, D.R. 1987. Flow measurement. Winery Water and Waste. *Pract. Winery Vineyard* 8(1).

140. DOORENBOS, J. and PRUITT, W.O. 1974. *Guidelines for the Prediction of Crop Water Requirements.* UN, Food And Agric. Org., Irrigation and Drainage Paper No. 25. Rome: FAO.

141. STORM, D.R. 1986. Small dams and impoundments. Winery Water and Waste. *Pract. Winery Vineyard* 7(4).

142. STORM, D.R. 1987. Land confinement and zero discharge. Winery Water and Waste. *Pract. Winery Vineyard* 8(4).

143. STORM, D.R. 1989. Coping with Section 404 of the Clean Water Act. Winery Water and Waste. *Pract. Winery Vineyard* 9(5).

CHAPTER 9

LIQUIFIED PETROLEUM GAS SYSTEMS

INTRODUCTION

There are several compelling reasons for including a discussion of liquified petroleum gas systems (LP gas or LPG) in this book.

First and foremost, as suggested in Chapter 4, section entitled "Boilers and Boiler Water Quality," the boiler fuel of choice for wineries in remote locations is propane, one of several commercially produced LP gases. In the United States, with the Clean Air Act Amendments of 1991 in force, the selection of propane as the boiler fuel almost guarantees that the exhaust emission requirements for the boiler will be met or exceeded without having to resort to expensive stack gas scrubbers and particulate entrapment devices.

The second reason for installing a propane storage (and dispensing) unit is to permit the use of LP-gas-fueled forklifts. The low-carbon-monoxide emissions produced in internal combustion engines that are fueled with LP gas, allow their use in enclosed winery spaces without hazard to the forklift operator or other cellar personnel.

A third, but less frequently used, application is propane fuel for standby

278

emergency generators. The ability of this fuel to remain uncontaminated (no moisture accretions or chemical breakdowns) over long periods of storage make it an ideal fuel for emergency equipment. The use of emergency standby generators was discussed previously in Chapter 6, in the section entitled "Water Storage." The design of fixed engines that are propane fueled is governed by National Fire Protection Association (NFPA) Standard 37, "Installation and Use of Stationary Combustion Engines and Gas Turbines."

A brief discussion of alternative boiler fuels was included in the Chapter 4 section entitled "Maintenance and Operation of Steam/Hot Water Systems." Large electric and natural gas public utilities are providing a limited supply of compressed natural gas (CNG), which may be more readily available than propane in some wine regions of the world.

GASEOUS FUEL CHARACTERISTICS

A comparison of the physical properties of the most commonly used gaseous fuels used in wineries clearly illustrates the following:

1. The fuel heat values in the vapor state
2. The specific gravity of the fuel in its vapor state [i.e., tendency to settle (heavier than air) or rise (lighter than air)]
3. Combustion product data

Table 9.1 is a summary of the physical and chemical characteristics of natural gas and propane, the two most common alternatives.

Please note in Table 9.1 that the fuel value for propane is greater than 400% of the fuel value of natural gas. The comparative unit costs of the two fuels, western United States, 1996 price levels are

Table 9.1. Gaseous Fuel Characteristics

	Natural gas	Propane (C_3H_8)
Heat value (BTU/ft^3)	551	2522
Specific gravity (air = 1.0)	0.47	1.52
Combustion product data $(ft^3/ft^3$ of gas)		
CO_2	0.56	3.0
H_2O	1.15	3.8
Nitrogen	3.77	18.5

Propane $1.35/gal (retail, volumes less than 100 gal)
($0.36/L)

Compressed natural gas: $0.61/gal ($0.17/L). The cost is based on conversion of cost per "therm" or 100, 000 BTU to its equivalent fuel value per gallon of 87 octane gasoline (145).

The relative densities of the two gases shows that the propane, if unconfined from tank or line leakage, will tend to seek low areas (sumps, dry pits, loading docks, and machinery vaults). The explosive potential of low-lying-area, pocketed propane gas is extremely high. Only 2.15% propane in air is required to produce an explosive mixture (106). Backyard Barbecues, recreational vehicles, and pleasure boat galley stoves and heaters that are fueled with propane and that have malfunctioned or been improperly operated have caused serious injuries and in some cases deaths. To prevent such accidents from LP gas in wineries, design guidelines set forth in National Fire Protection Association (NFPA) Standard 58 must be carefully followed by the winery's architect and mechanical/plumbing engineer (146).

DESIGN DETAILS

Because of the potentially explosive nature of the gas and its propensity to seek and settle in below-floor-level spaces, good sense and NFPA 58, Section 3-2.3, require that the fuel storage tank/forklift fueling facility installation be outside the winery building proper. The fuel complex can be under a weather shelter or roof, provided that the walled enclosure is not more than 50% of the roof perimeter. If the storage tank is to be dual purpose (fueling forklift trucks and for boiler fuel storage), a number of criteria must be met to accommodate both NFPA 58 safety provisions and forklift fueling convenience. By having the tank optimally located with the proximity of the tank to the steam boiler, boiler fuel pipe runs can be shorter and thus considerably less costly. If the cost of LP-gas fuel is important for accounting purposes, a meter should be installed on the filler hose. Figure 9.1 shows a commercial propane filling unit with a ring of protective bollards. Figure 9.2 shows the design details for LP-gas storage tank protective bollards.

Special provisions are required for storage tanks with volumes greater than 4000 gal (15144 L). They include ready access for emergency vehicles and the preparation of a fire safety analysis by a certified safety engineer, with the results of the analysis (safety recommendations) to be incorporated into the plan for the LP-gas storage tank subsystem.

Fig. 9.1.   Typical propane fueling tank installation.

The supplier of LP gas for the winery also has a statutory responsibility for the safe storage and periodic refilling of the fuel tank from a mobile, truck-mounted source. The winery's tank is usually leased on a long-term contractual basis with the supplier. Thus, the protective bollards, which remain as fixed features of the LP-gas system, are the responsibility of the winery owner to construct and maintain. If more competitive LP-gas fuel prices become available in the future from another gas supplier, the tank can be removed and the replacement tank installed inside the owner's bollard system. Storage tanks, up to several thousand gallons, are skid-mounted, with the tank saddles and supporting structural members all fabricated in accordance with NFPA 58. Maintenance (periodic painting, and replacement of pressure relief valves and tank replacement for mandated pressure vessel code testing and certification at 12-year intervals) (146) of the tanks, piping, and valves, up to the point where the fuel line to the building is connected, is the responsibility of the commercial tank supplier.

Any welding on LP-gas system piping must be performed by a certified pipeline welder as per the Uniform Plumbing Code (106). The welder must hold a certificate of competency from a certified welding school,

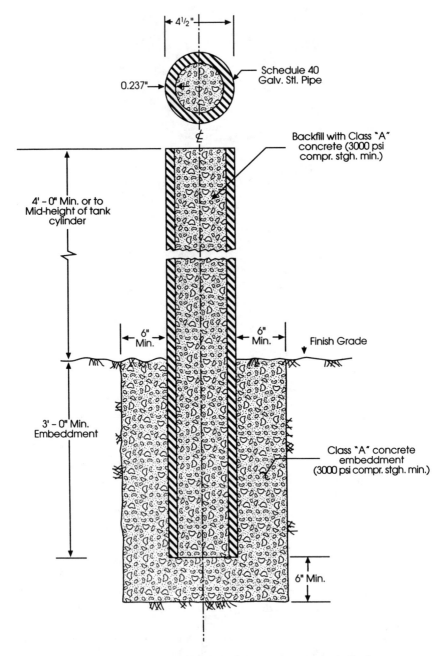

Fig. 9.2.   Typical design details for lp gas tank protective bollards.

which bases its curriculum and performance standards on the ASME Boiler and Pressure Vessel Code, Section IX.

Figure 9.3 is an illustration of a winery site plan which shows the setbacks from potential entrapment spaces or sources of spark ignition for a propane storage tank system with forklift fueling capability. A no-smoking zone for at least 50 ft (15.2m) around the propane tank should be designated and clearly posted.

The U.S. Occupational Safety and Hazard Administration (OSHA) in March 1995 established new regulations regarding the training and safety of industrial forklift operators (147). Although there is no specific mention of operator safety while fueling forklifts with LP gas, the winery should set up its own operating and training rules for forklift drivers that equals or exceeds the safety guidelines set forth in the new U.S. OSHA statute.

ODORIZED LP GAS, FORKLIFT GARAGING, AND GAS LEAK DETECTION

Commercial LP gas is required to have an odorant, which will act as a warning agent, if leaks in forklift or LP-gas storage or fueling system units should occur. The odorant is usually ethyl mercaptan. Having the odorant will not necessarily guarantee that an LP gas leak will be detected by forklift operators or other cellar workers. For example, because of the high specific gravity of the gas (1.52 vice 1.0 for air; see Table 9.1), the likelihood of the gas accumulating at the human nose level is quite remote, unless there is artificial turbulence in the enclosed space (warehouse or barrel room) from forced draft ventilation.

Garaging of forklifts then becomes very important if propane-fueled trucks are used. Forced ventilation in the space designated for forklift storage would be a necessity. The total number of complete air changes in the forklift garage space should be based on NFPA code requirements for the storage of vehicles. Outdoor garaging is another option if the winery's security system could offer the proper surveillance of the expensive lift-trucks. Gas leak detection systems are now manufactured for use in both large and small industrial facilities, where both toxic and explosive gas leaks require immediate detection (148). Wineries have at least two other gases that are a danger to cellar workers in relatively low concentrations. Sulfur dioxide (SO_2) is used as a biocide and preservative in wines, and carbon dioxide (CO_2) is evolved during the fermentation of wine grape juices and musts. Most "sniffer" detection units have the capability of identifying either toxic or combustible gases. The sensors are generally electrochemical for toxic gases and combustible gases use a catalytic bead

284

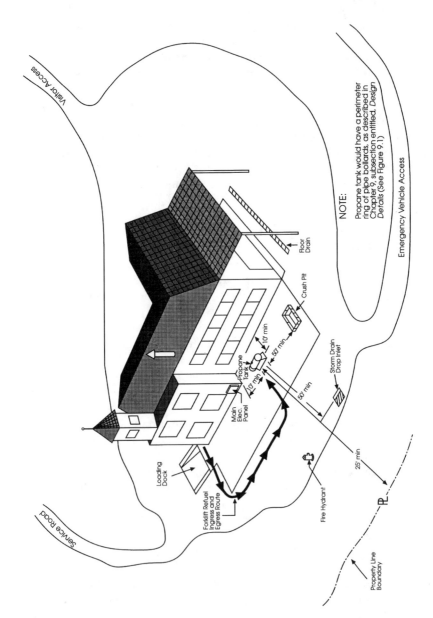

NOTE:

Propane tank would have a perimeter ring of pipe bollards, as described in Chapter 9, subsection entitled, *Design Details* (See Figure 9.1)

Fig. 9.3. LP-gas installation schematic showing setbacks and clearances. (Adapted from Ref. 145.)

system for determining combustible gas presence. A 4–20-mA detector output is amplified in the master controller to signal carrier power for subsequent processing (150). The "sniffer" alarm signals can be incorporated into the winery's master security alarm system in one way or another. The services of a 24-h security company dispatcher and/or an automatic telephone dialer would ensure that the gas leak information was transmitted to the local fire protection unit and safety officer for the winery.

PRESSURE RELIEF VALVES

All LP-gas cylinders must have an automatic pressure relief valve that releases the contents of the tank when tank rupture is imminent. The valves are generally spring loaded and set to release when the internal pressure reaches approximately 75% of the "test pressure" of the tank. Tank service pressures are usually 240 pounds per square inch gauge (psig) [1655 Pa] with test pressures twice service pressure. Therefore, the pressure relief valve setting is for 360 psig (2482 Pa). Propane is a non-corrosive gas, so the brass pressure relief valves have a service life of many years. The ASME Code for Unfired Pressure Vessels suggests pressure relief valve replacement after 10 years of service (149).

The same pressure vessel code recognized that small (8–10-gal or 30–40-L) propane tanks for forklift truck service might be subjected to higher tank pressures (usually due to higher temperatures) than larger stationary storage tanks. Thus, the code-approved service pressure for forklift LP-gas tanks is 312.5 psig (2155 Pa), to allow for higher temperatures from direct exposure to the sun (146).

Pressure vessels for Compressed Natural Gas must be certified for pressures nearly 10 times those for LP-gas cylinders. CNG pressures of 3400 psig (23443 Pa) are standard with 3000-psig (20685 Pa) injection pressures into engine combustion chambers. Thus, direct fuel conversion to high pressure CNG cannot be made without engine modifications.

OPERATION AND MAINTENANCE OF LP-GAS SYSTEMS

The relatively simple plumbing and vessel configuration of LP-gas systems makes routine operation and maintenance tasks important but not overly demanding in terms of timeliness or manpower. As described in the foregoing sections, safety in handling the fuel and its storage components is the most important of the O&M functions. The commercial supplier of the gas is responsible for the condition and preservation of the tank and associated plumbing and valves with the exception of the protective bollards, as described previously.

The fuel system for boilers that are fired with LP gas has been previously described in Chapter 4, including operation and maintenance guidelines.

The LP-gas-fueled forklift trucks serve as independent propane-powered subsystems that require the periodic maintenance care that all winery rolling stock should receive. If the winery has a service contract with the forklift supplier or is leasing them, then there is only the periodic refueling tasks that are performed by the cellar staff.

OTHER FORKLIFT OPTIONS

If electrical/battery-powered forklifts are the lift truck of choice for the winery, the space and the electrical requirements for a battery charging and maintenance center create an entirely different set of design problems that must be dealt with by the winery architect and electrical engineer, who must be familiar with the special needs of direct-current power systems design. A detailed description of this optional choice for forklift prime movers is beyond the scope of this reference book.

GLOSSARY

Compressed natural gas   A substantial reduction in volume and increase in pressure to allow economic nonpipeline transport and/or use of the high methane content gas as a clean-burning fuel for internal combustion engines.

Dry pits   Recessed, concrete-lined below-ground spaces used for processing equipment in wineries and for ease of raw material handling in the case of wineries. (See *Machinery vaults.*)

Embeddment   To place within, as with the embeddment of anchor bolts to secure machinery to its foundation.

Fuel value   An expression of the potential heat content of a fuel usually expressed in BTU (British Thermal Units) per pound or kilocalories per kilogram.

Machinery vaults   Underground concrete spaces, generally with movable covers or access manholes, which contain electromechanical equipment and associated controls and instrumentation.

Moisture accretions   Moisture acquired incidentally from changes in ambient environmental conditions or properties of a substance (pressure, temperature, or volume).

Protective bollards   Traffic-proof, structural members designed to prevent incidental vehicular damage to explosive gas storage tanks, electrical switch gear (transformers, control panels), or critical fire protection plumbing (backflow preventers and post indicator valves).

Rolling stock Colloquial expression for all types of vehicular, wheeled equipment [forklifts, pickup trucks, all-terrain vehicles (ATVs)] generally contained in the inventory of a winery.

Specific gravity A dimensionless ratio of the density of a substance to the density of distilled water at a given temperature (the specific gravity of water being recorded as 1.0).

Stack gas scrubbers Combustion emission gas cleaning devices that utilize particle ionization entrapment from electrostatic screens, water spray flushing, and activated carbon absorption to remove undesirable combustion products.

Sumps In drainage nomenclature, sumps serve as the terminus and accumulator for drain water to provide (a) sedimentation and entrapment of suspended material, (b) a quiescent pool for skimming and removal of oil and grease, and (c) as a wet pit for a pump or metered, gravity flow outlet for conveyance to a wastewater treatment system.

REFERENCES

144. MARKS, L. S. 1989 *Mechanical Engineers Handbook,* 14th ed. New York: McGraw-Hill Book Co.

145. GARCIA, J. 1995. *Personal communication.* CNG Vehicle Division, Pacific Gas and Electric Co., San Francisco, CA.

146. _____. 1992. *Standards for the Storage and Handling of Liquified Petroleum Gases.* Quincy, MA: National Fire Protection Association.

147. _____. 1995. *Fed. Reg.* March 14, 1995.

148. _____. 1995. *Gas Monitoring System for Hazardous Gases.* Milpitas, CA: Sierra Monitor Corp.

149. _____. 1980. *ASME Code for Unfired Pressure Vessels,* 1980 ed. Philadelphia, PA: ASME.

150. ANDERSON, W. 1995. *Gas Monitoring Detectors for Toxic and Combustible Gases.* Milpitas, CA: Sierra Monitor Corp.

CHAPTER 10

FIRE PROTECTION SYSTEMS

WINERY FIRE HAZARD AND OCCUPANCY CLASS

Every analysis of fire protection and prevention, whether it be a winery or fireworks factory, has two main objectives in common:

1. safety for the occupants of the building
2. protection for the building and contents from fire damage or total loss

Urban wineries have the same problems as wineries located at rural sites, but with the former, the solutions are considerably less complex and less costly. Only the rural winery setting will be discussed in this chapter, although the statutory requirements are the same for both rural and urban locations. The winery structure must be planned and detailed by the winery architect to meet Building Code and Uniform Fire Code requirements. The building designer cannot make a winery completely inflammable or fireproof, but he/she can, with judicious selection of construction materials and the control of draft and ventilation within the building envelope, reduce the risk of catastrophic fire and prevent its spread, should one occur. Multiple-option egress points should be detailed by the architect and approved by the local building official or fire protection agency representative.

What makes a winery a moderate-hazard occupancy classification? The flammable properties of wine are shown in Table 10.1. The fiberboard cartons in which finished wines are packaged are also highly combustible, as are the wooden pallets on which the cased wine is stacked. Paper in all forms is classified as a Class "A" combustible (not to be confused with the "Group" alphabetic designation given in Table 10.1 for flammable materials). An additional sub-classification is given to so-called "High-Piled Combustible Storage" as contained in Volume I of the Uniform Fire Code (152). Wine case goods have a commodity Class I (i.e., cased alcoholic beverages not exceeding 20% alcohol). Fire protection requirements for that portion of the winery that is devoted to case storage only, are detailed in the Uniform Fire Code. Fire protection requirements are subdivided into high-pile-storage gross floor area size categories from 500 square feet (s.f.) [47 m^2] to 500, 000 square feet (s.f.) [47000 m^2]. For example, a non-public-access wine case goods storage area between 2501 s.f. and 12, 000 s.f. (233 m^2 and 1115 m^2) could have the allowable combinations of fire extinguishment and fire detection equipment presented in Tables 10.2 and 10.3.

Wineries that are producing fortified wines or who distill wine for high-proof spirits are in an entirely different category than table wineries. No attempt will be made in this discussion of winery fire protection systems to cover all possible wine product categories and fire risks. The codes that guide the winery's building design team will allow for differences in building type, products produced, and degrees of protection that are achievable. The design team also has the responsibility to ensure that the design is in conformance with the codes and balanced optionally between allowable fire safe structural features and the best fire suppression technology.

Table 10.1. Flammable Properties of Wine (≤ 14% Ethanol)

Liquid	Flash point (closed cup) [°F]	Water solubility	Suitable extinguishing agents	Hazard
Wine	130–150	Soluble in all proportions	Water Alcohol foam Carbon dioxide or dry chemical	• Group B[a] • Group D[b]

[a]Group B is combustive when in contact with oxidizing materials; requires sprinkler protection and segregation from oxidizing materials.

[b]Group D is hazardous when heated; may explode or give off flammable or oxidizing decomposition vapors when moderately heated.

Table 10.2. Fire Protection and Life Safety Requirements for Warehouse Spaces

Commodity class	Size of high-piled storage area (s.f.)	Automatic fire extinguishing system	Fire detection system	Building access for fire apparatus[a]	Smoke and heat removal	Curtain boards[b]	Small hose valves and stations
I Wine case storage	2, 501–12,000 [Option I]	Yes	NR	NR	NR	NR	NR
I Wine case storage	2, 501–12,000 [Option 2]	NR[c]	Yes	Yes	Yes	Yes	Yes

[a]Fire apparatus access to space not greater than 150 ft and fire apparatus access road with all-weather, unobstructed width 20 ft or greater and overhead clearance of 13.5 ft or greater.

[b]Curtain boards = a ceiling-hung barrier to prevent the spread of smoke and heat along the ceiling (aka: draft curtains)

[c]NR = not required.

Table 10.3. Pallatized Storage Limits

Commodity class	Size of high-piled storage area (s.f.)	Maximum allowable basal pile dimension (ft)	Maximum allowable storage height (ft)	Maximum allowable pile volume (c.f.)
I Wine Case Storage	2, 501–12, 000 [Option 1]	100 × 100	40	400,000
I Wine Case Storage	2501–12,000 [Option 2]	100 × 100	30	200,000

Note: Wine warehouse storage space illustrative example based on allowable design criteria in Uniform Fire Code.

FIRE THREATS

Fire threats for wineries can be of two general types:

1. wildland fires
2. structural fires

Wildland fire risks are probably greatest in the Mediterranean-type climates of the world's wine producing regions. Hot, dry summers create the ideal ingredients for seasonal wildfires. For example, on the steeply sloping mountains that bound both the east and west sides of California's famed Napa Valley, some 28 wineries occupy heavily wooded sites, surrounded generally by conifer species on the west (Mayacamas Mountains) and with oak-brush-grassland vegetative complexes on the east (Vaca Mountains). Napa's 1981 Atlas Peak fire destroyed some 25, 000 acres (10163 Ha) of mixed oak, brush, and grassland on the east side of the valley before being contained. With sites for new wineries on the valley floor virtually non-existent, upslopes on both sub-alpine margins of the valley have put many of the new facilities in a high-hazard area for wildfires. Although open-space gaps can provide natural fire stops, a winery's extreme remoteness from the nearest organized fire protection station, however, requires that external wildland fire protection be installed, including, as a minimum, an adequate number of hydrants to protect all sides of the winery structure with hydrant spacing no greater than that specified in the Uniform Fire Code (UFC), App. III, Table A-III-B-1. The number of hydrants required is about one for every 1000 gallons per minute (gpm) [63 L/sec] of fire flow, as calculated for winery building type as defined by the Uniform Building Code, and the "fire area" (as defined in the UFC) as floor area in square feet. To more clearly illustrate the use of the UFC and UBC on a prototype winery example, fire flow and fire flow storage will be calculated on a step-by-step basis in a subsection following. As these design criteria will form the basis for the size and capacity of the major features of the winery's fire protection system, the package submitted for design review will receive very close examination by the responsible fire protection administrative authority for the region. As the cost for a remote winery's protection basics can represent as much as 25% of the winery construction site utility budget (153), the planning inputs and overall fire safety objectives should be carefully prepared with balanced inputs by the winery owners and the winery design team. For particularly complex systems, a state-certified fire protection engineer can serve as the primary designer or in a review capacity for the architect, mechanical engineer, and site utilities engineer collaborative design team to verify assumptions and to pre-certify that the plan will pass administrative and technical review. There is nothing more frustrating than having the fire protection system plans and specifications rejected as not being responsive to code requirements. Unnecessary delays can wreak havoc on a tight construction schedule and impair the advance ordering of specialty fire protection components, some of which require a 6-month lead-time for fabrication.

Structural fires in wineries, as in other industrial facilities, can originate in a number of fire-prone areas. National legislation for tobacco smoking prohibition in buildings has eliminated at least one source of accidental fires in administrative/office spaces (i.e., the wastepaper basket fire that historically was ignited by a still burning match or pipe smoker's cinders emptied carelessly into the highly combustible wastepaper). In the 4-year period 1983–1987, fires attributed to smoking materials (cigarettes, pipes, and cigars) dropped an impressive 17%, roughly paralleling the reduction in tobacco use nationwide (155). Paper and other solid waste recycling in general has also probably contributed to the reduction in the number of fires in waste containers. With nearly universal segregation of paper, glass, and metals and the total prohibition of toxic and/or inflammable materials from the solid waste stream, the probabilities of solid waste associated fires has been greatly reduced.

Structural fires are possible even in wineries constructed of reinforced concrete. Electrical fires can occur in the best designed wineries. Electric motors that overheat (and wineries have an abundance of electric prime movers) and ignite insulation materials produce toxic smoke that can be hazardous to cellar workers. All electric motors of fairly recent origin have so-called heat strips or high-temperature activated relays, which automatically terminate the power supply, when the design high-temperature is reached. (The loss of Halon 1301 and 1211 as effective suppressants for electrical fires and their non-ozone-depleting substitutes, will be discussed in the subsection entitled "Portable Fire Extinguishers"). There are instances when the motor high-temperature protection has failed, so there is no guarantee that there will not be an electric-motor-associated fire in a winery. The fire hazard associated with the winery's warehouse spaces has been previously discussed. The very valuable electronics that make up a large part of contemporary equipment in any winery administrative office (FAX, PC, copier, and printer) can produce fires on a statistical basis at about the same frequency as electronic appliances do in household fires or about 18.4% nationally (155) of all causes of fire.

Office furnishings are combustible and could augment the fuel for an administrative office fire. Fireproof materials are not available for office furnishings, but fire-resistant furniture, rugs, and window coverings can be produced from material that produces no toxic smoke during combustion. Fire-resistant paints and interior partitions also assist in slowing a fire's progress, once ignited.

Other sources of explosion/fire from liquid petroleum gas (LP gas), which is often used in remote winery locations for steam boilers and many for forklift truck fuel, has been described previously in Chapter 9. The special explosion/fire prevention measures in winery building design that must be taken in the use of this fuel were also detailed in Chapter 9.

Wineries with kitchens are as vulnerable as their residential equivalents, with fires caused by cooking activities causing some 24% of all fires nationally (155). Ventilation flue fires from accumulated grease and dust are second only to stove fires from overheated cooking oils in the restaurant category. Exhaust hoods for commercial-class ranges will require sprinkler installation for fire suppression, supplemented with portable extinguishers, as will be described in a subsequent subsection.

Additions or repairs to stainless steel process piping or tanks, within the winery proper, are welded with so-called shielded-arc welding units. There are no effective, preventable measures to minimize the fire threat from welding. Certified welders brought to the winery for specialty welding are well aware of the hazards and often come equipped with carbon dioxide extinguishers. The designated winery safety officer should always be formally notified that welding is to take place so that the proper fire precautions can be taken.

Every industrial building that is designed for human occupancy must be provided with sufficient exits to facilitate the rapid escape of occupants in case of fire or other life-endangering emergency. No locks or other fastenings shall be installed to hinder egress from a building in any way except approved "panic hardware." Review and approval of detailed winery evacuation plans are made part of the permit process that is administered by the local fire district or county fire marshal. The Uniform Building code, the Uniform Fire Code, and, more recently, the Americans with Disabilities Act (ADA) each set forth requirements that must be met. For example, under the ADA, individuals without sight must be provided with audio fire signals of some type that can allow a blind person to exit a structure in the event of a fire. The Americans with Disabilities Act Design Guidelines (ADADG) were prepared by the responsible U.S. agency which administers and enforces the provisions of the ADA statute. The Department of Justice has pared the ADADG; however, there apparently are ambiguities in the recently issued code, and design professionals are often forced to retain ADA interpretive specialists in order to avoid future professional liability actions (156). The winery's architect will be fully aware of the overlapping nature of the total fire protection code packages and should be able to demonstrate that all requirements have been met or exceeded.

Stairs can well serve a person who must exit a multistory building who is not physically impaired, whereas the construction of interior ramps to serve the disabled for emergency egress can add 15–20% to the exterior dimensions of the building envelope. Some designers opt to create horizontal exits (horizontal movement of occupants on the same floors to exterior stair or ramp systems) which allow emergency decent outside of the building proper. This method of evacuation has historical significance and was first codified by the National Fire Protection Association in 1916 as "Outside Stairs for Fire Exits" to allow the retrofitting of multistory dwelling units in densely populated metropolitan areas for fire exit safety (154). The current edition of the NFPA *Life Safety Code Handbook* (157) details the minimum exit/evacuation provisions that must be made for various categories of building structure and occupancy.

The classification of structures by "occupancy" to which the code requirements are applied is related to the maximum number of people within the structure at any one time and the fire hazard that activities within the structure present. Wineries are classed for occupancy purposes as Group F, Division 1, alcoholic beverage production (152). A comparison of industries with Fire Code occupancy risks that are grouped together along with wineries are as follows:

- Auto manufacturing
- Carpet and rug cleaning
- Clothing
- Dry cleaning and dyeing
- Food processing
- Paper mills and paper products

The generic risk classification under "occupancy" for wineries and others in the Group F category would be *moderate hazard*. High-hazard industries would include paint shops, explosive manufacture, chemical production, oil refineries, and tents or air-supported structures. The occupancy classification and hazard grouping are the criteria by which fire flow rates and fire flow durations are determined. An illustrative example of determining fire flow and duration for a winery structure will be covered in a subsection following.

Structural fires in wineries can involve the main and subservice electrical systems. The local agency with regulating responsibility for fire protection will review the architect's plans for emergency egress, assuming that the lighting system is inoperative. Obviously, emergency lighting systems also have an important role during other power outage situations (i.e., earthquake, tornado, and electrical service system carrier failure). [The winery's health and safety officer should periodically conduct fire drills with electrical power "off"—oftentimes exercises, such as fire drills shortly after occupancy is taken, reveal deficiencies in the emergency lighting system (i.e., dark areas in stairwells, etc.).] Emergency lighting systems can include exit signs, incandescent and fluorescent battery lamps (wall or ceiling mounted), and standby electrical generators. The required illumination for exit routes including corridors and stairways is detailed in Appendix I-B of the Uniform Fire Code (152). Performance standards for emergency lighting are contained in the National Electrical Code. Some of the important provisions of the emergency lighting design criteria are as follows:

1. Average initial illumination of 1 footcandle along the entire route of egress.
2. Emergency lighting must operate at the 1-footcandle power for a minimum period of 1.5 h.

3. Transfer from primary source electrical power to emergency power must occur within 10 s.
4. Emergency lighting illumination should overlap sufficiently so that the failure of one lamp will still allow safe passage along the exit route.
5. No point along the exit route should be more than 100 ft from the nearest visible sign (changes in exit route direction shall be marked with an arrow symbol).
6. Emergency lighting must be tested every 30 days manually using the test switches mounted on each battery-powered unit (manual testing must be performed even though self-diagnostic test systems are a feature of the emergency lights).

STRUCTURAL FEATURES FOR FIRE PREVENTION AND SPREAD CONTAINMENT

Containment is a universal design practice for limiting the spread of building fires. The logic behind this design concept is to slow the spread of fire while suppression systems, such as sprinklers, perform the extinguishment. Walls, ceilings, floors, door, and windows are "fire rated" based on the number of hours they can be expected to contain a fire. Just as shipbuilders embraced the practice of compartmentalization to contain fires and flooding on seagoing vessels, so did building designers, to limit the spread of both smoke and flames. Holes through fire-rated building walls, ceilings, or floors, called penetrations in architectural parlance, are necessary to carry pipes, electrical conduits, and heating, ventilating, and air conditioning (HVAC) ducts from building zone to zone. Properly designed "fire-stops" can restore the incised structural unit to full pre-penetration fire rating. ("Fire rating" is the time-index used to measure the ability of a material or materials to retard the transfer of heat and to prevent the direct exposure of flames to the area immediately on the other side of the fire-rated material or structural unit.) There are a number of different codes (generally the requirements are nearly equal) which specify fire-stop requirements. It is necessary for the winery's architect and structural engineer to determine which code has been adopted by the local building officials for the winery location. The codes and their corresponding firestop section designations are as follows:

Code	Appropriate Firestop Section
1. National Building Code	915.6.1
2. Standard Building Code	1001.3.5 and 1001.3.6
3. Uniform Building Code	4304
4. National Fire Code	NFPA 101 and 90A

Typical wall or floor penetration fire-stops are shown in Figs. 10.1 and 10.2. Figure 10.3 is a state-of-the-art heating/air conditioning or ventilation duct automatic closure device (heat activated) that prevents the dispersal of smoke or fire through the winery building envelope.

The caulking compounds that are used singly or in combination with mechanical collars will expand to approximately 10 times their volume to seal even the smallest opening. The high temperatures associated with a structural fire produce the swelling of the so-called intumescent compounds, which prevents even the smallest amount of smoke to pass through the wall, ceiling, or floor. The code(s) are unanimous in their requirement of a positive, fire-stop device that will close the penetration completely, if the pipe, for example, is plastic (resin) and the pipe material will burn.

The type of wall, ceiling, or floor, through which the opening passes is also a variable that must be dealt with by the architect in achieving fire-stop code compliance. A frequently referenced design guide is the Gypsum Association's *Fire Resistance Design Manual* (158). While extolling the marvelous properties of gypsum (calcium sulfate and water), the manual does contain nearly 75 fire-rated wall, ceiling, and floor systems construction details for 1-h up to 4-h ratings, using various combinations of gypsum wallboard. (For those readers with a fertile imagination, there is a special subsection in Ref. 158 on so-called Type X gypsum board or in fire protection nomenclature, x-rated gypsum board; no one under age 18 need research this section.)

Fire resistance ratings for other wall and ceiling systems using plaster as well as gypsum materials are contained in Chapter 6 of the *Handbook of Industrial Loss Prevention* prepared by Factory Mutual Engineering Corporation (151). An extremely thorough coverage of almost all combinations of building construction from wood and steel through pre-tensioned and post-tensioned concrete is contained in this informative reference. Factory Mutual Engineering Corporation, or FM, is a fire equipment testing institute, whose seal appears on approved materials and equipment, usually accompanied by the UL stamp of the nearly identical twin testing institute, Underwriters Laboratories. The role of these testing institutes in rating fire protection equipment for use in the United States and elsewhere will be discussed in a subsequent subsection.

FIRE PROTECTION SYSTEM TESTING INSTITUTES

A wise plan reviewer for California's State Fire Marshal once told me that my career as a designer of fire protection systems will be greatly simplified if I remembered the following dictum: ". . . never specify any equipment

System No. CAJ1081
(Formerly System No. 449)
F Rating—2 and 3 Hr (See Item 3)
T Rating—0 Hr

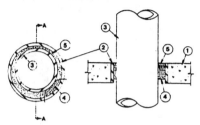

SECTION A-A

1. Floor or Wall Assembly—Min 4-1/2 in. thick reinforced lightweight or normal weight (100-150 pcf) concrete floor or min 5 in. thick reinforced lightweight or normal weight concrete wall. Wall may also be constructed of any UL Classified Concrete blocks*. Max diam of opening is 16 in.
   See Concrete Block (CAZT) category in the Fire Resistance Directory for names of manufacturers.
2. Metallic Sleeve—(Optional)—Nom 16 in. diam (or smaller) steel pipe, conduit or steel EMT cast or grouted into floor or wall assembly, flush with floor or wall surfaces.
3. Through Penetrants—One metallic pipe, conduit or tubing to be installed either concentrically or eccentrically within the firestop system. Pipe, conduit or tubing to be rigidly supported on both sides of floor or wall assembly. The following types and sizes of metallic pipes, conduits or tubing may be used.
   A. Steel Pipe—Nom 10 in. diam (or smaller) Schedule 10 (or heavier) steel pipe. The annular space shall be min 1/4 in. to max 2-1/4 in.
   B. Steel Pipe—Nom 1/4 in. diam (or smaller) Schedule 10 (or heavier) steel pipe. The annular space shall be min 0 in. (point contact) to max 1-15/16 in. When nom 12 or 14 in. diam steel pipe is used and/or when min annular space is less than 1/4 in., the F Rating is 2 hr.
   C. Conduit—Nom 6 in. diam (or smaller) steel conduit or nom 4 in. diam (or smaller) electrical metallic tubing. The annular shall be min 1/4 in. to max 3-1/4 in.
   D. Copper Tube—Nom 6 in. diam (or smaller) Type L (or heavier) copper tube. The annular space shall be min 0 in. (point contact) to max 1-13/16 in. When copper tube is used, F Rating is 2 hr.
4. Forming Material*—Min 3 in. thickness of min 3.5 pcf mineral fiber insulation firmly packed into opening as a permanent form. Forming material to be recessed from top surface of floor or from both surfaces of wall as required to accommodate the required thickness of fill material.
   USG Interiors Inc.—Type SAF
5. Fill, Void or Cavity Material*—Sealant—Min 1 in. thickness of fill material applied within the annulus, flush with top surface of floor or with both surfaces of wall. Dry mix material mixed with water at a rate of 2.1 parts dry mix to 1 part water by weight in accordance with the accompanying installation instructions.
   United States Gypsum Co.—Type FC
*Bearing the UL Classification Marking

F = fire rating
T = temperature rating
intumescent = material that expands with increase in temperature

NOTE:
The Material was extracted by David R. Storm from UL's 1995 *Fire Resistance Directory.*

Fig. 10.1.   Typical fire stop and/or temperature and smoke attenuator detail for wall or floor penetration by a pipeline.

System No. CAJ6001
(Formerly System No. 99)
F Ratings—2 and 3 Hr (See Items 1 and 2)
T Rating—0 Hr
L Rating At Ambient—2 CFM/sq ft (See Item 5)
L Rating At 400 F—less than 1 CFM/sq ft (See Item 5)

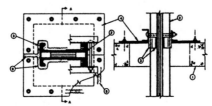

SECTION A-A

1. Floor or Wall Assembly—Lightweight or normal weight (100-150 pcf) concrete. Min thickness of concrete floor or wall assembly is 2-1/2 in. for 2 hr F Rating and 4-1/2 in. for 3 hr F Rating. Wall may also be constructed of any UL Classified Concrete Blocks*. The size of the rectangular through opening shall be such that the annular space between the outside corners of the busway (Item 2) and the periphery of the through opening is in the range of 1/2 in. to 3 in.
   See Concrete Blocks (CAZT) category in Fire Resistance Directory for names of manufacturers.

2. Busway—For 3 hr F Rating, the busway shall consist of a nom 9 in. wide (or smaller) by 4-1/2 in. deep (or smaller) "I"-shaped steel and aluminum enclosure containing factory-mounted copper bars rated for max 600V, 1600 A. For 2 hr F Rating, the busway shall consist of a nom 23 in. wide (or smaller) by 4-1/2 in. deep (or smaller) "I"-shaped aluminum or steel and aluminum enclosure containing factory-mounted copper bars rated for max 600 V, 4000 A or aluminum bars rated for max 600V, 3000A. The busway shall bear the UL Listing Mark and shall be installed in accordance with all provisions of Article 364 or the National Electrical Code, NFPA No. 70. Busway to be installed with min 1/2 in. to max 3 in. clearance from its outside corners to the sides of the rectangular floor or wall opening and shall be rigidly supported on both sides of floor or wall assembly.
   See Busways and Associated Fittings (CWFT) category in the Electrical Construction Materials Directory for names of manufacturers.

3. Fill, Void or Cavity Materials*—Wrap Strip—Nom 1/4 in. thick intumescent elastomeric material faced on one side with aluminum foil, supplied in strips. Min 2 in. wide strip formed to follow the contours around the entire periphery of the busway (foil side exposed) and secured in place with steel wire. Busway wrap to extend 1 in. above and 1 in. below the top plane of the floor assembly. For wall assemblies, the busway wrap is to be installed in the same manner used for floor assemblies, but it shall be installed symmetrically on both sides of the wall assembly. When steel cover flange (Item 7) provided by the busway manufacturer is used, the busway wrap is not used.
   Minnesota Mining & Mfg. Co.—Types FS-195, FS-195+

4. Fill, Void or Cavity Materials*—Intumescent Sheet—Rigid aluminum foil-faced sheets with galv steel sheet backer. Sheet cut to tightly-follow the contours of the busway and with a min lap of 2 in. on all sides of the through opening. Sheets to be installed with the galv steel sheet backer exposed (aluminum foil facing against floor or wall surface). Sheet secured to top surface of floor and to both surfaces of wall using min 1/4 in. diam by 1-1/2 in. long steel expansion bolts, or equivalent, in conjunction with min 1-1/4 in diam steel fender. Max spacing of sheet fasteners not to exceed 5 in. OC When steel cover flange (Item 7) supplied by the busway manufacturer is used, the sheet(s) shall be secured using the same fasteners used to secure the steel cover flange sections. Prior to installation of the sheet(s), a nom 1/4 in. diam bead of caulk (Item 5) shall be applied to the top surface of the floor and to both surfaces of the wall around the perimeter of the through opening.
   Minnesota Mining & Mfg. Co.—Types CS-195, CS-195+

5. Fill, Void or Cavity Materials*—Caulk—Generous application of caulk to be applied around the base of the busway or the busway wrap at its egress from the intumescent sheet(s) in addition to completely covering the busway wrap up to the interface(s) with the busway. Nom 1/4 in. diam bead of caulk to be applied as a gasket between the intumescent sheet(s) and the floor or wall surfaces. All seams in the intumescent sheet(s) made to accommodate the busway to be covered with a generous bead of caulk.
   Minnesota Mining & Mfg. Co.—Types CP-25 S/L, CP-25 N/S, CP-25 WB, CP-25 WB+. (Note: L Ratings apply only when Type CP-25 WB+ caulk is used.)

6. Steel Cover Strip—(Not Shown)—Min 2 in. wide strip of min 0.020 in. thick (No. 26 gauge) galv steel centered over entire length of each slit made in intumescent sheet (Item 4) to permit installation about the busway. Prior to installation of the steel stip, the seam or slit in the intumescent sheet shall be covered with a nom 1/4 in. diam bead of caulk (Item 5). Steel cover strip secured to galv steel sheet backer of intumescent sheet with steel sheet metal screws or steel rivets spaced max 3 in. O.C. on each side of seam or slit.

7. Steel Cover Flange—(Optional, Not Shown)—Four-piece interlocking cover flange formed from nom 1/8 in. thick steel, supplied by busway manufacturer. Steel cover flange to lap min 2 in. on floor or wall surface on all sides of the through opening and secured to top surface of floor and to both surfaces of wall using min 1/4 in. diam by 2 in. long steel expansion bolts, or equivalent. Sheet fasteners installed in factory-drilled and field-drilled holes in cover flange sections, through intumescent sheet: and spaced max 5 in. O.C. When steel cover flange is used, a generous application of caulk (Item 5) shall be applied around the base of the busway at its egress from the cover flange(s) and around the perimeter of the cover flange to cover the edges of the cover flange and the intumescent sheet.

*Bearing the UL Classification Marking

F = fire rating
T = temperature rating
intumescent = material that expands with increase in temperature

NOTE:
The Material was extracted by David R. Storm from UL's 1995 *Fire Resistance Directory.*

Fig. 10.2.   Typical fire stop and/or temperature and smoke attenuator detail for wall or floor penetration by a busway.

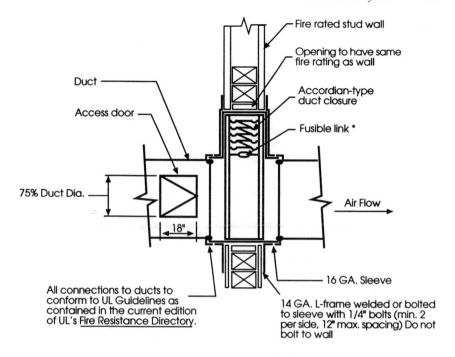

Fire rated stud wall

Opening to have same fire rating as wall

Accordian-type duct closure

Fusible link *

Duct

Access door

75% Duct Dia.

18"

Air Flow

16 GA. Sleeve

All connections to ducts to conform to UL Guidelines as contained in the current edition of UL's Fire Resistance Directory.

14 GA. L-frame welded or bolted to sleeve with 1/4" bolts (min. 2 per side, 12" max. spacing) Do not bolt to wall

* Automatic closure of duct when high temperature melts fusible link and prevents the spread of smoke through the winery's heating/air conditioning or ventilation system.

Fig. 10.3.   Typical smoke and fire damper detail for HVAC duct penetrating a fire-rated curtain wall.

for fire protection systems unless it is UL rated or FM approved and is painted red. . . .'' UL and FM are the fire protection system equivalents of NSF (National Sanitation Foundation) and ANSI (American National Standards Institute) for plumbing and sanitation fixtures and equipment, respectively. No one quarrels with the thoroughness with which UL and FM conduct their testing and approval of equipment items. Both institutes publish ''directories of approved equipment'' of approved equipment (159, 160). If the winery's fire protection system designer finds an equipment item that is particularly suited to the needs of the project, but it is only listed in the UL directory, and the winery's insurance risk is being assumed by one of the nine-odd insurance companies which endorse and financially support the FM testing group, the UL equipment item would not likely receive approval for installation. It is preferable to have the

designer find a suitable substitute for the equipment in question, so that
FM approval can be unconditional.

Factory Mutual System, of which Factory Mutual Engineering Corporation is the testing arm, is supported by nine large insurance companies
(eight American and one British). Wineries, which are insured by one of
the eight FM risk underwriters, would have their fire protection plans
reviewed for code compliance on a contractual consulting basis. FM would
also ensure that only FM approved equipment and materials are used
exclusively for construction purposes. The nine insurance companies are
as follows (161):

1. Arkwright Mutual Insurance Co., Boston, MA
2. Blackstone Mutual Insurance Co., Providence, RI
3. Boston Manufacturers Mutual Insurance Co., Waltham, MA
4. Fireman's Mutual Insurance Company, Providence, RI
5. Manufacturers Mutual Fire Insurance Company, Philadelphia, PA
6. Protection Mutual Insurance Co., Park Ridge, IL
7. Affiliated F.M. Insurance Co., Providence, RI
8. Factory Mutual Engineering Corp., Norwood, MA
9. F.M. Insurance Company Ltd., London, England

Underwriters Laboratories was founded in 1894 and for the first 22 years
of its existence, financial support came from the insurance industry (161).
UL is now a completely independent laboratory and operates on a non-
profit basis on revenues from material suppliers and manufacturers who
require the label and certification if their products are to be specified for
use in fire protection systems. As with the FM System, field inspections and
follow-up evaluations of previously certified equipment and materials are
also made by the UL staff from one of five testing stations in the United
States:

- Chicago, IL
- Northbrook, IL
- New York, NY
- Santa Clara, CA
- Tampa, FL

One further note should be highlighted in the relationship of insurance
carriers and their role in the design and approval process for fire protection systems. Insurance companies may "advise" on plans and specifications prepared by a licensed design professional, but their advisory recommendations do not have the force of law. For the designer to ignore the
insurance company's recommendations would be imprudent and most
assuredly result in higher insurance premiums for the winery. Reducing
fire risks to some acceptable minimum is one of the important roles of the
insurance carrier. It is the role of the system designer to balance the

incremental costs of more system sophistication or component redundancy against the benefits in terms of life safety, fire damage reduction, and reduced fire insurance premiums. A clear and concise economic analysis should be prepared for the winery owner's consideration and decision making.

FIRE PROTECTION WATER SUPPLY SYSTEMS

General Considerations

Water supply systems for fire extinguishment must be single-purpose in function. Some "maybe" and some definitely "won't suffice" water storage options will be discussed. A decorative landscape pond could be a "maybe" source of supply, if:

- The volume of the pond can be maintained equal to the calculated fire flow storage requirement 100% of the time
- The management of the pond will be such that plankton, aquatic weeds, and suspended sediment would not impair the hydraulics of the hydrant and/or sprinkler systems
- The pond would have to be protected from freezing in cold climates.

A sprinkler frost protection water supply storage facility would not meet the single-purpose criterion unless the drawdown was limited to the volume of storage above some minimum pool level elevation and the elevation of the frost protection supply pump was fixed at or above the "full" elevation level for the fire protection storage volume. In other words, the pump intake for the fire protection pump would have to be located below the "full" elevation requirement for minimum fire protection storage. The same requirements for a firm and unfailing supply of water and pond management to protect water quality and freezing would also be mandated for the frost protection pond. A swimming pool that can be operated continuously, year-round, without freezing can be a candidate for a fire protection water supply reservoir. Even if the swimming pool was not hydraulically connected to the winery's hydrant and/or sprinkler system but a fire pumper truck is able to gain emergency access to the pool, under certain conditions, the swimming pool's stored volume of water could be credited to the winery's gross rate and duration of fire flow, subject to the local fire marshal's opinion.

Fire Flow Rate and Duration

Insurance companies with a recent history of large claims from devastating fires (Sands Hotel, Las Vegas, and Oakland Hills fire, Oakland, California) have narrowed the choices for wineries, which can be equated to risk and loss or damage by fire. The premiums that must be assumed by the winery for protection from fire may not be the most costly component of the total winery insurance package, but it is significant. The choices for systems that the architect and engineers must make are not necessarily absolute and dictated by a single set of fire code requirements. As suggested previously in this chapter, fire protection means, first and foremost, safety for winery employees and visitors. Winery assets including structure, inventory, and equipment must be protected to some minimum level (highest insurance premiums) to the maximum possible level of fire protection (lowest insurance premiums). Somewhere in between the least expensive (just meeting the code) and having a full-time fire department manned and with the necessary equipment on the winery parcel (exceeding minimum code requirements) is the optimum system that is economically feasible to construct, with insurance premiums that equate to some level of risk that winery ownership is willing to assume.

Rate of flow, duration of flow, and dynamic pressure (i.e., pressure in the pipe network when the system is activated) are the hydraulic criteria by which fire protection water supply systems are judged. For sprinkler systems, so-called *primary demand* is the rate in gallons per minute at which water will be required to supply those sprinklers that may be open and discharging initially until other sources become available (i.e., pump truck connections to sprinkler risers when organized fire-fighting units arrive on the scene). *Primary supply* is the rate in gallons per minute at which water is instantly available at the pressure required for effective sprinkler operation. *Total demand* is the rate in gallons per minute at which water will be required to supply *all* the sprinklers in the winery that may ultimately be activated. *Total supply* is the rate in gallons per minute which can serve all sprinklers at the design pressure over and above any flow that is required for concurrent use of exterior hydrants or interior hose stations.

Fortunately, the rate and duration of flow requirements for sprinklers, exterior hydrants, and interior hose stations are not independent and additive in satisfying the code's flow requirements. For example, sprinklers would be activated in an untended winery in the event of a pre-determined temperature rise in the building space in question (i.e., warehouse, office, or kitchen). Hydrants would not be activated until fire fighters arrived on the scene. In a normal sequence of fire extinguishment events for structural fire, the sprinklers might quench the flames and prevent the spread of fire. The firemen on the scene would likely inactivate the sprinklers and complete the extinguishment with hose streams. Thus, there is rarely a time when sprinkler and hydrant flow requirements would have to be met concurrently.

California's Napa Valley will again be used as the venue for a case study to illustrate the derivation of rate of fire flow and duration of flow for a remotely located winery. [*Author's Note:* Even though the 1994 edition of the

Fire Code has been published, the adoption of the revised code by states and local governments often takes 1–2 years. Thus, the only recent major change made to the Napa County code governing fire protection is the mandatory provision that all wineries be equipped with automatic sprinkler systems. Most assuredly, other more stringent fire protection requirements will be adopted by Napa and other jurisdictions throughout the United States as the re-codification process slowly occurs over the next few years.]

In the subsection of this chapter entitled "Five Threats," the classification of wineries as an industry group was previously reported as Group F, Division 1, alcoholic beverage production (152). The risk classification was for Group F industries, *moderate hazard*. The latter classification is the first step in the fire flow duration analysis. To simplify the step-by-step process, the Napa County, California guidelines have been reproduced as Table 10.4. Explanatory notes in the right column of Table 10.4 provide thetechnical support logic for a particular judgment decision or technical choice. Further, to assess what difference the 1994 Uniform Fire Code (152) might make in the magnitude of fire flows and duration of fire flows for the same winery, the design criteria for fire flow determination as contained in Appendix III-A, Table A-III-A-1 will be used.

As suggested previously, the base fire flow value for the hypothetical winery used for illustrative purposes in Table 10.4 might be greater or more conservative if the 1994 Uniform Fire Code design guidelines were used.

Table A-III-A-1, Minimum Required Flow and Flow Duration for Buildings (152) would produce the following value:

> 420, 000 ft^3 was assumed for the hypothetical winery. If an average ceiling height of 12 ft is assumed (higher in warehouse and tank rooms and lower in other production and administrative spaces) the floor area or "fire area," the nomenclature used in the Fire Code, is 35, 000 ft^2 with "ordinary construction," as previously defined in Table 10.4 is the Building Code equivalent of Type IV-H.T. with 1-h ratings. A base fire flow value of 3250 gpm is required for a minimum duration of 3 h. To make the comparison to the fire flow and fire storage values derived in Table 10.4 on the same basis, assume that the base fire flow amount can be reduced by at least 40% as was done under the Napa County, CA fire protection regulations:

> | | |
> |---|---|
> | Allowed fire flow reduction: | 3250 × 0.40 = 1300 gpm |
> | Net required fire flow: | 3750 − 1300 = 2450 gpm |
> | Required fire flow storage: | 2450 × 60 × 3 = 441,000 gal |

Thus, the more conservative 1994 Uniform Fire Code would require in Napa County, CA nearly 200% of the current code requirement for rate of flow and approximately a 1000% increase in the fire flow storage requirement. The designer must take a very close look at incremental costs for full sprinkler

Table 10.4. Fire Flow and Duration of Fire Flow Determination for a Hypothetical Winery in Napa County, CA, as Calculated from Their Design Guidelines

Design guidelines	Author's explanatory notes

IV. Water Systems and Access Requirements for Light, Moderate, and Heavy Occupancy Groups

1. Definitions: Reference: 1991 Uniform Building Code & Uniform Fire Code
   a. Light Hazard Occupancy

Apartments	Institutions
Condominiums	Libraries (except
Churches	large stack room area)
Colleges	Museums
Clubs	Nursing, convalescent &
Dormitories	care homes
Dwellings	Office buildings
Hospitals	Rooming houses
Compressed inert gas	Schools
storage	

   b. Moderate Hazard Occupancies

Asylums	Warehousing of normal	
Hotels	combustibles	
Prisons	**Wineries**	Moderate hazard
Sawmills	Welding Shops	classification for
Gas stations	Furniture stripping works	wineries (see
		item (IV,3)
Lumber yards	Kiln drying systems	
Compressed		
Inert gas		
manufacturing		

   c. High hazard occupancies
   Aircraft hangars/shops
   Chemical works or storage
   Cotton picker & opening operations
   Explosives & pyrotechnics manufacturing
   High-piled combustible storage in excess of 21 ft high
   Linoleum and oil cloth manufacturing
   Linseed oil mills, storage or processing oil refineries
   Flammable liquids bulk plants or storage (in excess of 500 gal)
   Paint shops
   Pyroxylin plastic manufacturing & processing
   Pesticides manufacturing, storage, and shipping
   Shade cloth manufacturing
   Solvent extricating
   Tents and air-supported structures
   Varnish works or application of varathanes
   Warehousing of combustibles/flammables *and* other occupancies involving processing, mixing, storage, and dispensing flammable and/or combustible liquids

(*continued*)

Table 10.4. Fire Flow and Duration of Fire Flow Determination for a Hypothetical Winery in Napa County, CA, as Calculated from Their Design Guidelines (*Continued*)

Design guidelines	Author's explanatory notes

IV. Water Systems and Access Requirements for Light, Moderate, and Heavy Occupancy Groups

For the purpose of classifying the types of constructions which apply to a particular building, the following will be used:
  a. Wood construction
  b. **Ordinary construction**
  c. Fire resistive
  d. Noncombustible
	Ordinary construction assumed (wood frame and doors; stucco or masonry exterior siding with gypsum board interior walls)

3. *Formula to Determine Required Fire Flow per Hour (gph) @ 20 psi Dynamic minimum*

		Author's notes
Light hazard $\dfrac{\text{cu ft}}{600}$ = Required Flow for 1 h (gpm)		Volume of winery structure assumed to be 420,000 cu ft *or*
Moderate hazard $\dfrac{\text{cu ft}}{400}$ = Required Flow for 1 h (gpm)		$\dfrac{420{,}000}{400} = 1050\ gpm$
High hazard $\dfrac{\text{cu ft}}{200}$ = Required Flow for 2 h (gpm)		

The fire flow as determined above need not exceed:
5000 gpm for wood construction
5000 gpm for ordinary construction
3500 gpm for fire-resistive construction
3000 gpm for noncombustible construction
except that for a normal one-story building of any construction type, the fire flow need not exceed 3,500 gpm.

4. The Required Fire Flow—shall not be less than 150 gpm @ 20 psi dynamic

5. All Flow Requirements Stated—must be certified by a licensed engineer. The location and number of hydrants shall be approved by the Napa County Fire Warden.

6. The net adjusted fire flow credit/surcharge shall not exceed 75% of base fire flow.

7. In all cases where a Fire Department connection is used as part of an approved sprinkler system, a fire hydrant as approved by the Napa County Fire Warden shall be installed within 50 ft of the Fire Department connection.

    A Fire Department connection will be defined in the section "Sprinkler Systems."

8. Building access shall be provided by a "Knox Box" lock device approved by the Napa County Fire Warden. Any cases gates shall be equipped with a "Knox Box" lock device to provide single-key access to Fire Service Personnel.

Table 10.4. Fire Flow and Duration of Fire Flow Determination for a Hypothetical Winery in Napa County, CA, as Calculated from Their Design Guidelines (*Continued*)

Design guidelines	Author's explanatory notes

IV. Water Systems and Access Requirements for Light, Moderate, and Heavy Occupancy Groups

    9. For the protection of life and property, access road requirements to the commercial site and around the buildings therein shall be approved by the Napa County fire Warden.

Work Sheet

This form is designed to allow the developer to determine the quantity and quality of fire defense measures required for the building(s) proposed for construction.

1. From page 6, number 3 of the guide, determine if building will be classed as Light, Moderate or High: <u>Moderate</u>
2. Using the most current Uniform Building Code publication (as adopted by Napa County), determine if proposed building construction is defined as wood, ordinary, fire resistive or noncombustible: <u>Ordinary</u>
3. From the working plans for the building, determine the total cubic volume: <u>420,000</u> cubic feet
4. Based on occupancy type (light, moderate, high), divide the number of cubic feet determined in (3) above by the factor: 600, light; 400, moderate; 200, high; shown in (A) on the "Brief Recap" sheet attached to the package. This will determine the base fire flow. Round off your answer to the nearest 50: <u>1050</u> gpm.

Design guidelines	Author's explanatory notes

IV. Water Systems and Access Requirements for Light, Moderate, and Heavy

5. Multiply the gpm figure determined in (4) by 60 for light/moderate hazard occupancies, and by 120 for high hazard occupancies. This will give you the <u>base figure</u> for required fire protection water storage capacity: 63,000 gallons.
6. Fire Protection water storage is independent of domestic or other use. Domestic and other use shall be in addition to Fire Protection. All pipes or valves required for domestic or other use shall be installed on the tank above the level designated for Fire Protection purposes only.
7. Credits/surcharges:
   A. From (4) above, enter base fire flow: <u>1050</u> gpm. Now, from the table below, circle the figure which best describes the building construction type/hazard class as proposed.

(*continued*)

Table 10.4. Fire Flow and Duration of Fire Flow Determination for a Hypothetical Winery in Napa County, CA, as Calculated from Their Design Guidelines (*Continued*)

Design guidelines	Author's explanatory notes

IV. Water Systems and Access Requirements for Light, Moderate, and Heavy

		Light	**Moderate**	High
1.	Wood Construction	0%	0%	+10%
2.	**Ordinary**	−5%	−5%	+10%
3.	Non-Combustible	−25%	−25%	−25%
4.	Fire Resistive	−25%	−25%	−25%

Add or subtract the percentage allowed due to construction type/hazard class. T/H adjusted gpm <u>1000</u>.

B. Heat or smoke detectors (installed in accordance with NFPA Note: Do not use (B) if (C) is used: Deduct 5% from T/H adjusted gpm: _____

C. Heat or smoke detectors installed and maintained as a functioning part of a NFPA approved 24-h monitored Fire Alarm System: 1. Deduct 15%: <u>158</u> gpm or if alarm terminates at fire department communications center, 2. deduct 20%: _____ gpm.

> Choice "C"-1 is recommended as a minimum, or 892 gpm.
>
> Choice "D" is now a "policy" mandate for wineries in Napa County, CA. (163). Partial sprinkler coverage in critical building areas is probably the most economical option.

D. Proposed occupancy to be completely covered by NFPA approved automatic Sprinkler System, complete with audible flow alarm which is monitored 24 hours per day at the Napa County Fire Department Emergency Command Center. Deduct 40% _____ gpm or occupancy partially covered by NFPA approved sprinkler system that is 24-h monitored at the Napa County Fire Department Emergency Command Center: Deduct 15%: <u>−134 gpm or 758 gpm.</u>

E. Entire roof surface of nonflammable materials as defined by UBC and NFPA: Deduct 5%: <u>−38 gpm or 720 gpm.</u>

> Cement shakes are architecturally attractive, durable, and fireproof. Required in high-hazard fire areas.

Table 10.4.   Fire Flow and Duration of Fire Flow Determination for a Hypothetical Winery in Napa County, CA, as Calculated from Their Design Guidelines (*Continued*)

Design guidelines	Author's explanatory notes

IV. Water Systems and Access Requirements for Light, Moderate, and Heavy

F. Lack of adequate separation between buildings—see table below:

Single building has been assumed.

Distance between structures	Light hazard		Moderate hazard		High hazard	
	W/O	FRNC	W/O	FRNC	W/O	FRNC
0–10 ft	+10%	+ 5%	+10%	+5%	+20%	+5%
11–20 ft	+ 5%	+ 2.5%	+ 5%	+2.5%	+15%	+2.5%
21–30 ft	+ 2.5%		+ 2.5%		+10%	

w/o = ordinary construction
FRNC = fire resistant or noncombustible
Enter appropriate separation credit or surcharge here: N.A.

Net totals: This is the net adjusted gpm required.   700 gpm
Multiply × 60 (1 hour storage) = storage 42,000 gallons[a]
Multiply × 120 (2 hour storage) = storage _____ gallons[a]

See item IV-3

The net adjusted fire flow credit/surcharge shall not exceed 75% of Base Fire Flow.[a]

40% deductions < 75% ∴ OK

[a]If a municipal or private water system serves the proposed site and cannot supply required fire flow and storage, "on site" storage will be required.

coverage of the building with high-tech smoke alarms and remote 24-h alarm monitoring at the nearest fire station to obtain a larger reduction in flow requirements for additional system sophistication and reliability. Under no circumstances will the newest addition of the Fire Code permit a fire flow below 1500 gpm (152). Therefore, with the maximum credit taken for alarms and monitoring, a winery of the building type and size assumed for this example would require as a minimum nearly 200% more fire fighting hydraulic capability than is required under the pre-1994 Fire Code Edition adoption era.

The fire protection system construction and operating cost implications to wineries that are in the planning or expansion mode of the more rigid fire protection hydraulic requirements to be brought about eventually by the universal local adoption of the 1994 Uniform Fire Code are going to be significant. Possibly, there will be concurrent reductions in insurance premiums from the winery loss risk reduction that the more conservative fire protection regulations create.

Fire Protection Storage Facilities

The single-purpose use mandated by the Fire Code for fire storage ponds and reservoirs requires that the utilities system designer produce some very imaginative solutions to satisfy the 100% availability of the necessary fire flow.

There are similarities between the creation of storage facilities for winery potable water supply systems and fire protection systems. As mentioned several times in a previous section, the winery classification under discussion is "remote" without access to publically owned water supply or sewerage systems. For potable water supply storage facilities, storage volumes may fluctuate according to water supply and demand parameters. Maintenance of the requisite fire flow storage to offset solar evaporation in open reservoirs and seepage in unlined reservoirs can be accomplished by several methods:

(a) Refrigeration evaporative-cooled condenser water blow-down—This water supply can amount to 4–5 gpm for a 40-ton refrigeration unit. Blow-down in evaporative coolers occurs when the salinity of the cooling water reaches a pre-determined level of salt concentration. To blend the saline water with the stored volume of fire protection water would not impair its usefulness for fire extinguishment. However, if the fire pond also serves the winery as a landscape amenity and/or habitat for waterfowl or fish species, the salinity balance of the reservoir contents would have to be periodically monitored. (The mere fact that solar evaporation is taking place tends to increase the concentration of the dissolved mineral salts in the reservoir; unless rainfall accretions are significant and annual flushing and dilution of salts occurs, water quality parameters must be carefully monitored to maintain a healthy aquatic environment.)

(b) Rainfall and solar evaporation are seldom in harmony, at least in Mediterranean-type climates, where winters are wet and summers are hot and dry. Needed make-up water for solar evaporation depletions cannot be guaranteed by relying on the hydrologic cycle. For some grape growing regions in the eastern United States and Europe, almost daily summer rainfall is a normal occurrence, and evaporation or seepage loss make-up water for the fire storage pond can take place naturally.

(c) Reclaimed process wastewater is also a possibility for both the primary fire protection water supply and for reservoir evaporation/seepage loss makeup. (This recycling opportunity was suggested previously in Chapter 8 in the subsection entitled "Other Reclaimed Wastewater Reuse Options.") The reclaimed wastewater storage facility can often be divided into two cells by means of concrete or timber bulkheads. A polyethylene pond liner could be utilized to create an impervious basin for the fire storage element of the reservoir. If the major portion of the annual reclaimed wastewater production is used for vineyard or other irrigation, solar evaporation loss make-up could be provided to the fire flow storage pond by one of the

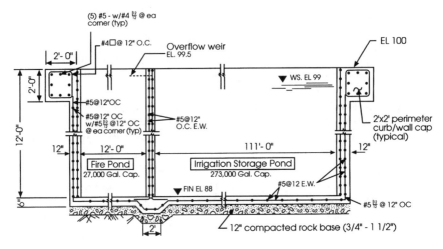

Typical structural details of a combined fire protection storage and irrigation storage pond. (Not to scale)

As-constructed irrigation and fire protection storage pond combination.

Fig. 10.4.   Fire protection storage constructed as part of reclaimed process wastewater winter hold-over irrigation storage.

methods described in items (a), (b), or (d). Figure 10.4 is illustrative of a reclaimed process wastewater storage facility being used for both winter storage of irrigation water (fluctuating water level) and fire protection storage (non-fluctuating water level). As designed, the reclaimed process wastewater is discharged to the fire pond. When full, the fire pond storage excess flows over a weir-slot in the transverse wall separating the fire pond from the irrigation storage facility. Irrigation draughts on the irrigation pond allow that storage volume to fluctuate while the fire pond remains full.

(d) A well utilizing stored groundwater is an acceptable alternative to the more commonly utilized surface supplies. Make-up water for solar evaporation and/or seepage would not be required under a groundwater utilization scheme as the primary source for fire protection water. The aquifer that is used for primary fire protection source water should have groundwater level stability and water table elevations that are not too deep. For example, in the fire flow calculation that employed the criteria, as contained in the 1994 edition of the Uniform Fire Code, a discharge winter hold-over irrigation storage requirement of 3250 gpm was developed. (See Table 10.4 and discussion presented previously in this chapter.) If the active groundwater level was 700–800 ft below natural ground surface, the horsepower requirement for a single pump to serve as the primary fire pump would be extraordinarily large. A more economical arrangement might be the use of a smaller-horsepower pump (same lift requirement of 700–800 ft but flow rate of only 300 gpm) to supply a surface fire storage facility with a booster fire pump, that would have hydrodynamic requirements much lower than the high lift and high flow rate pump needed for a combined groundwater extraction and fire flow rate maintenance. Also, with a mechanical/electrical failure of a single-use pump, the required flow rate and duration could still be met with, for example, a standby, emergency diesel-powered pump.

Visual inspections to verify that the required fire flow storage is, in fact, available at all times can be augmented with the addition of a level sensor in the pond that will activate when the contents of the reservoir fall below a preset value. The principal purpose of the alarm is to alert winery personnel of a rapid loss of reservoir storage from structural collapse (embankment failure from seismic event or overtopping from excessively high rainfall and runoff) or hydraulic malfunction from fire hydrant pipe network failure, which would cause a rapid loss of storage pond contents.

For open reservoirs, the local fire official may require the installation of a so-called staff gauge at a location in the reservoir that is easily seen. The staff gauge can be calibrated in metric units (which is advisable for U.S. wineries, as metrication is no longer an option, but the law of the land). The Leupold & Stevens Company of Beaverton, Oregon manufacture staff gauges with durable porcelain enamel paint lettering on non-corrosive metal staffs (164). The winery's utility systems designer should provide an area-capacity curve that combines the essential reservoir geometry into a useful, graphic form. Figure 10.5 is a typical area–capacity curve for a fire protection storage pond in conventional formatting for such data. Figure 10.6 is a spillway rating curve that permits design reviewers to verify that the spillway will pass the agreed-upon spillway design flood without overtopping the dam structure.

The contents of reservoirs that are constructed with tributary, native watersheds will receive a measured volume of suspended sediment over the

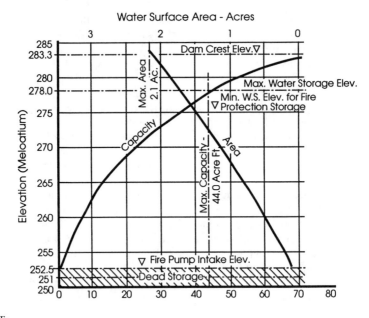

NOTE:
(1) 1 acre-foot ~ 326,000 gallons (U.S.).
(2) Standard dam safety design practice is to require at least five(5) feet of freeboard from maximum water surface elevation to the dam crest.

Fig. 10.5.   Typical area–capacity curve for fire protection storage pond of non-uniform geometry.

service life of the storage facility. A reasonably accurate estimate can be made of the annual sediment accretion and an additional increment of reservoir volume calculated to store the sediment. In other words, a volume of "dead storage" is set aside strictly for capturing and retaining watershed-produced sediment. (See Fig. 10.5, which illustrates the "dead storage" concept.) The enactment of the Federal Water Pollution Control Act and its various amendments in the 1970s along with the 52 state models that add local reinforcement for the protection of water resources, the discharge of sediment is identified as controllable and preventable. Prior to recognition of sediment as a stream contaminant, so-called sluice pipes were often installed in dams at their deepest point to permit the periodic "sluicing" of accumulated sediment. Other sediment management options for fire storage ponds are as follows:

- Periodic dredging of bottom sediments is accomplished using a portable suction dredge. (See discussion of this subject matter in Chapter 6.) Care must be taken in selecting the location and method of disposal for the dredged materials. A dredging permit will be required in most all govern mental jurisdictions.

(ft)	(ft)	(cfs)
W	H	Q
21	1	65.1
21	0.5	23.02
21	2.0	184.13
21	30	338.27

W = Width of weir (feet)
H = Depth of water over weir crest (feet)
Q = Spillway discharge (cubic feet per second)

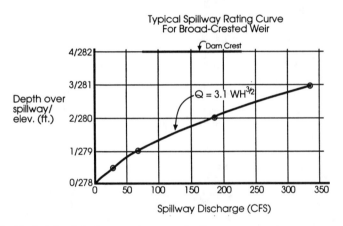

Fig. 10.6. Typical flood-discharge rating curve for fire storage pond emergency spillway.

- Installing a sedimentation basin upstream from the fire storage reservoir is a common practice in the far western United States, where regional soil conservation programs have been in place for 40 years. The entire purpose of the basin is to entrap the sediment so that periodic harvest and disposal of the fine-grained soil can be easily accomplished. Technical assistance may be available to wineries that are in close proximity to a field office of the U.S. Department of Agriculture (USDA), Soil Conservation Service. (Financial support was also historically available for soil conservation projects on private lands through Public Law 550, the Watershed Protection and Conservation Act. Budget constraints at all levels of government generally prevent outright monetary grants, although low-interest-rate loans are certainly a possibility.)

One additional and important feature of fire protection storage ponds is the intake structure. Placing the intake at the bottom of the storage structure creates the potential for clogging the inlet screen or, at best, requiring the fire pump during its weekly test cycling to ingest and bypass through the test piping a quantity of very abrasive inorganic sediment and organic detritus (leaves and dead aquatic plants) that may have settled on or near the intake port. It is far better to raise the intake from the reservoir bottom a vertical distance of 1.5–2.0 ft by placing the intake on concrete pipe pedestals or saddles. (The designer must take into account the loss of volume below the elevated intake structure.) Figure 10.7 depicts an intake

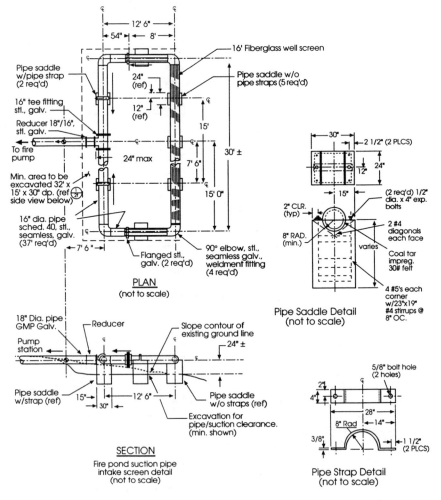

Fig. 10.7.   Typical pump intake system for fire protection storage pond.

system constructed of a large-diameter, fiberglass well screen. The hydraulics of the intake screen must be such that the maximum velocity through the well screen openings must be equal to or less than 0.5 ft/s. The low inlet velocities will prevent the transport of unwanted solids through the screen. Another design feature that can be added to the intake screen system, for a very modest incremental cost, is an air sparging system to permit the periodic cleaning of the intake screen slots. Even with the best fire pond management practices (algae and sediment control program), partial clogging of the intake may eventually take place. The air sparge system requires a shut-off valve at the extreme upstream end of the suction pipe (local fire official may or may not allow the valving of the suction line; as a minimum, a tamper-proof valve would be required, so that the suction line could not be inadvertently closed). Air from a trailer-mounted air compressor or portable tanks can then be introduced through a pipeline connected to the suction pipe, just downstream from the tamper-proof valve. The air bubble introduced into the suction line will reverse flow and clear the screen slots very effectively.

Fire Flow Pressure

The designer for the winery's fire protection water supply system must by code provide a minimum of 20 psi (1.4 kg/cm^2) at the hydrants and 10 psi (0.7 kg/cm^2) minimum at the highest sprinkler level in the winery building (152). Both of the minimum pressure requirements are for dynamic pressures or pressures when the hydrants and/or sprinklers are discharging [the dynamic pressure will be approximately the static pressure, (i.e., hydrants and sprinklers in the shut-off condition, less the friction loss in the pipe network)]. The local fire official may place other dynamic pressure requirements on the system at his/her discretion to meet the fire safety requirements for the particular winery installation. The 20 psi (1.4 kg/cm^2) minimum dynamic pressure does not seem to be adequate for fire extinguishment from a hydrant unless the fire truck booster pump is considered. A fire company crew, upon arrival at the fire scene, will attach a suction hose to the hydrant [hydrants may also have a 4-in. (102 mm) so-called "steamer or pumper connection"] so that the truck-mounted fire pump can boost the hydrant line pressure to 150 psi (10.5 kg/cm^2) or more for servicing several 2½-in. (64 mm) hoses carried by the pumper truck units. Some insurance carriers do not permit the installation of hydrants with 4-in. (102 mm) pumper connections because if the water supply is public or municipal, a significant pressure drop in the pipe network could be created by the pumper truck connection, impairing flow, which could result in less than adequate fire protection for nearby structures if a local fire spreads to adjacent properties (165).

"Jockey pumps" are a necessary feature of the primary fire pump systems, which require the primary pump to provide the required sprinkler and hydrant flow at the design pressure. Jockey pumps prevent the main pump from cycling if hydrant or other minor system leakage allows a pre-determined system pressure to fall below a specified point. The jockey pump maintains the system at a low but constant pressure until:

(a)   a sprinkler sector is activated or
(b)   a hydrant is opened causing a precipitous drop in the line pressure

With a real-time fire event, the jockey pump is inactivated through a booster controller (main pump and jockey pump) and the main fire pump is activated, bringing the system to full fire flow discharge and pressure in a matter of seconds.

Fire flow storage that is located vertically above the winery site is still the best and most reliable hydraulic arrangement (see previous discussion in Chapter 6 on gravity flow advantages). Elevated tanks in topographically flat regions of the wine world have historically fulfilled the gravity requirement and they have provided pump-independent, reliable fire flow in municipalities throughout the United States and elsewhere. Initial cost and annual maintenance of such a structure is an obstacle that is hard to overcome in this era of lean winery construction budgets and the fundamental business necessity of reducing operating overhead. Finally, elevated tanks have never won any awards for architectural beauty. Even the oblate spheroids, the so-called "Hortonspheres" that look like overgrown mushrooms dominating many skylines in the U.S. midwestern states, still look like tanks. Contemporary environmental law requires that new projects meet some minimum test of aesthetic values. Elevated tanks would not appear to stand the acceptability test that an EIR/EIS would require.

Fire Pumps

Fire pumps have evolved from the steam-driven, reciprocating type in the late 19th and early 20th century to the electric motor and finally diesel-powered centrifugal or turbine pumps. Diesel-powered centrifugal pumps are considered to be the most reliable and dependable contemporary prime movers (161). As described in the earlier subsections of this chapter and other chapters of this book, most all wineries have a requirement for utility system independence because of their rural and generally remote locations. A wholly self-sufficient fire protection water supply system is certainly one of the most important of those utility complexes. Fire pumps have several different subcategories [i.e., booster pumps to add energy in the form of pressure for otherwise volumetrically adequate water supplies, primary fire pumps which deliver the fire flow at the correct rate and pressure (typical of winery application with surface supply source), and vertical shaft, turbine-type pump as the primary fire pump system]. As emphasized previously, the pumping equipment and controls must be certified for fire protection use by Underwriters Laboratories (UL) or Factory Mutual Company (FM) testing institutes. The fire protection pump system designer must prepare a thorough analysis of local electrical power reliability and dependability and, during the preliminary design stage, offer the rationale to the winery owners and fire officials on the best apparent prime mover choice (i.e., electric motor or diesel, compression-ignition motor). Electrical power utility companies are less than enthusiastic about releasing power outage statistics for a particular sector of their power distribution network. (This very important aspect of the winery's electrical power supply was previously discussed in Chapter 2.) Wineries that are located at the end of a non-looped power grid are generally the most vulnerable to power interruptions. Long extensions of electrical service to very remote winery locations places the power dependability factor in the very low category.

National Fire Protection Association publication No. 20, Centrifugal Fire Pumps, provides the designer with all of the technical guidelines which govern the design, materials of construction, and performance standards for fire pumps that will be acceptable for fire protection use.

A typical vertical shaft turbine-type fire pump system arrangement is shown in Fig. 10.8.

Figure 10.9 shows a diesel-powered fire pump system with the discharge bypass carrying the full output of the fire pump. The local fire official is rating the system prior to acceptance. During the test phase, the diesel pump must successfully undergo at least 10 automatic and 10 manual starts, each time reaching the pump manufacturer's specifications for discharge pressure and rate of flow. The local building official cannot issue even a temporary occupancy permit to the winery until the fire protection system has been thoroughly inspected and tested. Any hydrants in the system are also rated individually by the fire official to again demonstrate that the design fire flow and the minimum residual dynamic pressure at the hydrant of 20 psi (1.4 kg/cm^2) are both achieved. Sprinkler system testing will be discussed in a subsequent section of this chapter.

The diesel fire pump controller is a very sophisticated, expensive, programmable testing, starting, and record-keeping system to give the maximum control reliability to the pump and its appurtenant equipment. These essential pieces of equipment can perform the following functions although they are not necessarily limited to those functions:

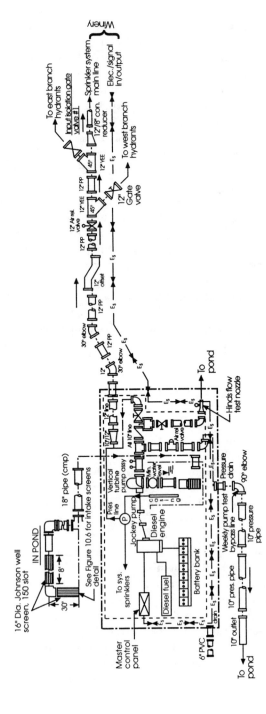

Fig. 10.8. Schematic arrangement of a diesel-powered primary fire pump system for a 190, 000-case annual production winery (vertical-shaft, turbine-type pump).

318

Fig. 10.9. Fire pump system discharge rating using a calibrated, "hinds" test nozzle.

Monitor and record

- Engine speed; alarm signal for overspeed
- Low oil pressure
- High cooling system temperature
- Engine fail-to-start on test cycle alarm
- Low battery bank charge
- Low fuel tank supply
- Low pump room temperature (failure of space heater)
- Low source water supply level
- Low suction pressure
- Jockey pump failure

On a "safety shutdown" sequence, the controller will inactivate the engine if it is in a test cycle and low oil pressure or low battery bank charge occurs. An alarm annunciator in the winery can alert winery personnel of the problem so that the malfunction can be investigated. However, during this period, if there is a pressure drop in the system from an in-winery sprinkler activation, due to smoke or fire, the controller overrides the malfunction signal and starts the fire pump.

At the end of each weekly retest cycle, the engine/pump running data are compiled by the controller and printed in hard-copy format. These data are important for the winery to review and file and to submit to the local fire official during periodic inspections, so that the performance of the fire pump since the last visit can be easily and quickly evaluated by the inspector. Even though the fire pump can test and evaluate all of its vital system parameters without human intervention, it is strongly suggested that someone from the winery maintenance staff also personally witness the weekly start-up and testing cycle.

A pump system adjunct that is often required by the local fire official when the winery's primary source of fire protection water supply is surface in nature is a so-called fire department draft pipe and pumper head connection. The logic that supports the installation of a secondary intake and discharge from the pond supply is quite simple and adds a reliability dimension to the overall fire flow delivery system that is well worth the relatively modest incremental cost. A schematic layout for both the primary and secondary fire flow delivery system is shown in Fig. 10.10. Figure 10.11 depicts a hypothetical, auxiliary draft pipe and pumper head and its associated design details. In the event of a primary fire pump malfunction during an actual fire emergency, the answering fire crew can connect a pumper truck to the draft tube. With two 2½-in. (64 mm) hoses connected to the discharge side of the truck's pump apparatus and the nearby two-head hose connection, the primary fire flow pipe system can be charged to 50% or 60% of its design capacity, allowing at least nominal performance of the winery's interior sprinkler system and possibly a single hydrant. The only additional important requirement of the back-up fire flow supply subsystem is that the fire truck access road be designed for all-weather and surfaced (side ditches, cross-drains, and sufficient room for the large fire trucks to maneuver with celerity).

Pump capacity in gallons per minute is based on both the requirement for automatic sprinkler discharge and the number of hydrants that must be serviced simultaneously. The interior sprinkler system is often a "design-construct" contractual arrangement. Thus, there is often a need for design collaboration between the designer of the primary fire pump system and the sprinkler system engineer. (Sprinkler systems will be discussed in detail

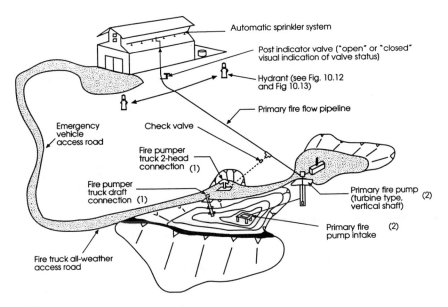

Automatic sprinkler system

Post indicator valve ("open" or "closed" visual indication of valve status)

Hydrant (see Fig. 10.12 and Fig 10.13)

Primary fire flow pipeline

Emergency vehicle access road

Check valve

Fire pumper truck 2-head connection (1)

Fire pumper truck draft connection (1)

Primary fire pump (turbine type, vertical shaft) (2)

Primary fire pump intake (2)

Fire truck all-weather access road

NOTES:

(1) Design details for secondary fire department pumper truck connection and pumper truck draft pipe connection are shown in Figure 10.11.

(2) Details for a typical diesel powered primary firepump system were shown in Figures 10.7, 10.8, and 10.9.

Fig. 10.10. Schematic layout of typical diesel-powered primary fire pump and secondary fire department connection.

in a later section.) Obviously, the primary fire pump system cannot be designed until the sprinkler requirements are known, so the winery design team coordinator (usually the winery architect) must arrange to have the sprinkler system designed as soon as architectural working drawings of the spaces are completed. Other interior plumbing and piping systems (potable water, steam, electrical conduits, telephone, etc.) must share chase and cableway overhead space with the sprinkler piping; thus, the architect and his mechanical and electrical engineer specialists must assume the responsibility for ensuring that wall, ceiling, and floor penetrations meet the fire rating standards described previously in the section entitled, "Structural Features for Fire Prevention and Spread Containment."

One issue that cannot be overlooked for diesel-powered primary fire pump systems is combustion air, particularly important when the engine, pump and controls are housed in a protective structure. For example, a 420-horsepower Caterpillar diesel operating at 1750 rpm would require 855 cubic feet per minute (cfm) of air to sustain combustion (166). Lou-

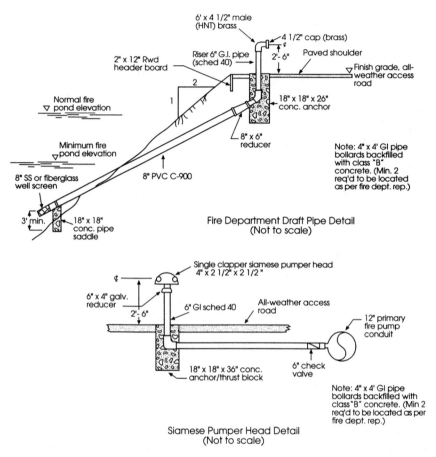

Fire Department Draft Pipe Detail
(Not to scale)

Siamese Pumper Head Detail
(Not to scale)

NOTE:

4″ × 4′ Gl pipe bollards backfilled with class "B" concrete. (Min 2 req'd to be located as per fire dept. rep.)

Fig. 10.11.   Secondary fire department connection fire flow augmentation.

vered and screened vents with a sufficient cross-sectional area opening to prevent the ingress of flying insects (150-mesh screen) must be provided.

Ambient temperatures within the pump housing must be kept at 40°F or greater. Auxiliary electric heaters are generally installed within the pump enclosure to prevent unduly low temperatures and condensation, which could damage sensitive electrical and electronic components. The master fire pump controller often has an optional resistance heater installed in the master control panel box to prevent freezing.

Other considerations for the fire pump that is diesel engine driven are as follows (167):

(a) Noise muffling and silencing: Large-horsepower pumps that may be required to meet flow, lift, and residual pressure requirements for a particular winery are started on a test cycle at least weekly. Unless properly silenced, the sound level can be not only annoying to winery visitors but also damaging to hearing systems of winery utility personnel who must observe the fire pump test cycle (168).

(b) Diesel fuel tank: Under current aboveground, fuel storage tank design and construction standards, the tank must be either double walled or a containment basin must be constructed around the tank equal to the volume of the storage vessel.

(c) Fire pump controller: This very important ancillary piece of equipment was depicted in Fig. 10.9 and described in detail in connection with the narrative, which described the basic fire pump system.

FIRE ALARMS AND INSTRUMENTATION

The so-called fire detectors are the devices that react to the products of combustion. The signals from those detectors are transmitted by wire to a master processor that can be custom-programmed to fit the space, occupancy, and fire threat parameters for any winery large or small. As will be described later in this section, the rapidity with which that signal reaches the fire department station house may be the difference between a little and a great deal of fire damage. The detectors can be classified by type, and the winery's fire protection engineer will have a fairly wide range of choices that will meet Uniform Fire Code and National Fire Protection Association standards. There are five basic types of fire detectors, as shown in Table 10.5 (174).

Combination detectors, using both UV and IR microwave sensors, are commercially available. The marriage of these two basic types adds an element of reliability to the fire signal processing step. The combination units will not activate a fire alarm signal until both units detect the presence of the specific radiation source. Some of the more sophisticated units have a feature installed, which calibrates the opacity of the lens. If a dirty lens is recorded, a special audio-visual alarm is transmitted to the master fire alarm panel, alerting supervisory personnel that a detector malfunction has occurred (174). The latter feature does not eliminate lens dusting, but it does give an early warning if a spider has built a web over the lens since the last cleaning and inspection. The winery will pay extra for these features.

The era of microelectronics and instrument miniaturization has significantly advanced fire protection technology, making a wide array of systems available that allow the ready detection of fire for each of the four stages of combustion (165):

Table 10.5. Fire Detectors: Their Capabilities and Limitations

Type	Description
A. Thermal	
(a) Fixed temperature	Signal given when temperature reaches a preset value
(b) rate of rise	Signal given when temperatures rise exceeds a pre-determined allowable rate.
B. Smoke	Unit contains a source lamp which is beamed across an open space to a photo-electric cell. If the calibrated cell detects less light from the source, the alarm signal is activated.
C. Ionization	A small, self-contained chamber has a small quantity of radioactive material, which ionizes the air between two electrically charged disks. Any combustion products (vapors, smoke, or aerosols) adsorb the ions, reducing the current between the disks, which activates the alarm signal. These units must be strategically placed as the combustion materials must arrive at the sensor by natural air currents that exist within the building space.
D. Infrared (IR)	These detectors measure microwave radiation in the infrared sector of the spectrum. They are true optical devices and must be in direct line-of-sight with the sources of combustion or they fail in their function. Special filters must be employed so that false alarms (hot engines, sunshine, incandescent lights, and mirror reflections) do not occur. Lenses must be frequently inspected and cleaned.
E. Ultraviolet (UV)	Similar to type D and the sensor in the UV part of the microwave spectrum is also line-of-sight. Any physical obstruction disables the UV unit. Inspection and cleaning of the lenses is an important maintenance task.

- Incipient stage: the production of invisible combustion products
- Smoldering stage: smoke produced, but no flames or heat in evidence
- Flame stage: clean flame sometimes without smoke (very short time between flame stage and subsequent stage)
- Heat stage: all of the elements of combustion in evidence, including visible smoke, invisible vapors, flame, and heat

Automatic fire detection systems have no direct effect on the fire, as only extinguishment can perform that function. What early detection does afford is the opportunity to fight the fire while it is still of a manageable size. Early detection of a fire can usually prevent the fire from becoming a large, seriously damaging fire by at least a factor of 2 (170). Further, of the fires discovered and reported within 5 min of ignition, only 5% spread beyond the space of fire origin, and 17% spread beyond the initial room of ignition when detection times exceeded 5 min (170). Additionally, statistics related to direct line alarms, which alert the fire department station house or dispatcher, indicate that damage is reduced by at least 85% and it is only reduced by 40% with local alarms (171).

Automatic sprinklers give the potential for reducing the probability of large damage loss to an industrial building by a factor of 6 (172). All pertinent references that discuss the effectiveness of early detection of fire in reducing damage agree that in spaces where people are present and an alarm is activated immediately by a person or persons (for wineries that equates to the standard working day and some evening and nights when special testing and/or promotional events are taking place), the damage reduction from fire is about equal to the fast response of an electronic fire detection system connected to a commercial security dispatcher or directly to the fire department station house. Other statistics related to fire causes and probabilities were discussed in section of this entitled "Fire Threats."

The fire detection system package has several basic elements. Although it might appear to be prudent to combine state-of-the-art building management systems (BMS) with the fire alarm system, there is a reluctance on the part of many fire officials to approve the combination (173). Building management systems are computer- or PC (microprocessor)-based master controllers, which are capable of operating all of the winery's mechanical systems at their optimum efficiencies. Because the BMS operates ventilation and heating/air conditioning dampers, it was a logical engineering follow-on for the BMS system manufacturers to try and consolidate the fire safety and building energy management functions into one economical unit. Although the firm position of fire officials against BMS–fire safety and detection system combinations may soften in the future, the apparent duplication of some building system functions and the mandated independent, stand-alone requirement for the fire safety system is not necessarily the potential construction cost impact for wineries that it appears to be. The electronic components for the systems continue to be down-sized in dimension and cost. Anyone involved in the design-end of fire safety will probably agree that duplication of fire alarm system components and functions is not necessarily bad. For example, if a ventilation duct damper is automatically closed or opened by the fire safety system, and the Building Management System indicates on its LED display that vent duct damper No. 2 is closed or open, that increases the verification of that action by 100%.

The fire protection engineer must evaluate the various potential sources of combustion in the winery, so that the selection of the detection devices will have the range and sensitivity to allow the earliest possible discovery of a fire event.

The simplest and most reliable thermal detection system is the fusible-link, which is contained in the sprinkler head in those winery spaces that have the highest potential for fire. Winery warehouse spaces and their potential fire hazard and relationship to the applicable section of the Uniform Fire Code were discussed earlier in the section entitled "Winery Fire Hazard and Occupancy Class." It is likely that, as a minimum, winery warehouse administrative and food service spaces will require sprinklers. [*Author's Note:* An important issue to consider by winery ownership is that most all UL- and FM-approved systems can comply with the Uniform Fire Code requirements. It is the special role of the designer to see that the best system is tailor-made for the winery and its special conditions.] How those sprinklers will be activated and controlled can be accomplished by at least four system types or combinations (165, 169):

1. Conventional
   (a) Fusible-link, self-activating sprinkler heads.
   (b) Master control system with multiple detectors [based on current measurement at the master panel (i.e., too little current gives a trouble signal or too much current gives an alarm signal for the building zone where problem is occurring)]. This system can contain a so-called "pilot head" in each building zone or subzone. The pilot head is a fusible-link plug in a small (½-in. pipe) pipe filled with compressed air. A high temperature which causes the link to melt releases the air in the pipe, which, in turn, activates the valve controlling only the water supply to the set of sprinklers that are located in the fire-affected zone.
2. Addressable
   (a) This system uses conventional detectors, such as those described previously in Table 10.5. These advanced units employ integrated-circuit computer chips and digital communication technology. The master panel communicates with each building space for which detectors are installed and can receive one of three messages:
       (1) OK
       (2) Trouble
       (3) Alarm
   These units have the ability to poll each detector and annunciate the results. More supervisory control is required with addressable systems; it is probably only justified for large or very large wineries.
3. Addressable analog
   (a) Similar in construction and function to the addressable system except in the nature of the information transmitted to the master panel. For example, the sensor can send a digitized smoke signal to the master panel that is proportional to the exact amount of smoke that it senses. The master panel analog then compares the smoke data and con-

cludes whether there is a fire. Polling of the sensors takes place every few seconds. The most significant advantage of this system is that it allows the master panel to test the sensitivity of each detection device in the network individually, thus eliminating many false alarms from faulty detectors. This degree of sophistication is probably not justified for a winery with a moderate hazard category.

There are other issues associated with the design of the detection system that must be addressed by the winery's fire protection engineer:

- Degree of detection system supervision
- Methods to employ for alarm verification and the establishment of detector sensitivity (balance between good fire safety and frequent false alarms)
- Battery back-up systems for the master control panel
- Designation of the receiver for the alarm signal (i.e., fire station directly or 24-h commercial fire alarm/security communication center and dispatcher)

The total fire protection package is a very important element of the winery's site utility system and winery owners and general managers should educate themselves sufficiently to render sound decisions, when faced with the wide range of choices for equipment function and reliability.

SPRINKLER SYSTEMS

Overview

Sprinklers and their role in fire suppression and occupant safety have been referenced in previous sections of this chapter. Sprinklers have been used in commercial and industrial buildings since 1896. The current edition of National Fire Protection Association (NFPA) Standard 13, "Standard for the Installation of Sprinkler Systems" (1994 edition) has been revised and modified some 50 times since its initial publication at the end of the 19th century (175). Just as with the 1994 edition of the National Fire Code (NFC), referenced the subsection entitled "Winery Free Hazard and Occupancy Class," where illustrative comparisons were made to identify the increases in fire flows for wineries and other industrial activities over the 1991 Edition of the NFC, the NFPA 13 standard is likewise more conservative than its predecessor 1991 version.

The impact of the 1994 edition over the more liberal 1991 edition can be observed in several requirements that could directly effect new winery design or winery expansion/retrofit plans. Table 10.6 covers a number of the relevant changes to NFPA 13. For example, there are major changes to

Table 10.6.   A Comparison of Selected NFPA 13 Standards for the 1991 and 1994 Current Edition (*"Standards for the Installation of Sprinkler Systems"*)

Item	1991 NFPA 13 requirement	1994 NFPA 13 requirement
Sprinkler pipe materials (176)	Copper or Steel	Copper Steel CPVC (chlorinated polyvinyl chloride, ASTM F442) PB (polybutylene, ASTM D3309)
Elevator hoistways (176) and machine rooms	No requirement	Sidewall sprinklers shall be installed within 2 ft of the hoistway bottom. Ordinary sprinklers (135–170°F) or intermediate sprinklers (175–225°F) shall be installed in elevator machinery rooms.
Electrical equipment rooms (175)	No requirement	If electrical equipment room has any other use beside housing switch boxes, control panels, starters, breakers, etc., sprinklers shall be required. No sprinklers shall be required for electrical equipment rooms if: (a) room contains only electrical equipment (b) the electrical equipment is of the dry type (c) the room has a minimum 2-h fire resistance rating (d) the room cannot contain any combustible material storage
Fire Department (175) Connection for Automatic Sprinkler Systems[a]	Fire department connection required for systems having 20 or more sprinkler heads unless the local fire authority waives the requirement	Fire department connection required for all automatic sprinkler systems unless: (a) winery is not serviced by an organized fire department (b) sprinkler is of the deluge type that exceeds the hydraulic capacity of the fire department's equipment.

[a]A fire department connection is shown in Figs. 10.11 and 10.14.

The sprinkler system designer must include sufficient safeguards in the construction specifications to prevent soldering flux and plastic pipe primer and solvent detritus from copper and CPVC pipe use, respectively, from clogging and interfering with the proper operation of the sprinkler heads.

storage area/warehousing portion of a winery structure that more clearly defines whether NFPA 13 provisions are really applicable to the particular method and mode of, say, finished wine storage. Thus, the 1994 version of the code states that if empty wooden pallets are stacked up to 6 ft in height with four such stacks together with aisle separation of 8 ft between four-stack sets, NFPA 13 does apply. If storage of finished palletized wine exceeds 12 ft in height, NFPA 13 does not apply, but NFPA 231, "Standard for General Storage" does apply. Also, for pallet stacks which exceed 15 ft in height, the use of the special early-suppression fast-response (ESFR) sprinklers can provide the protection specified in NFPA 231 for stacks to 25 ft. [*Author's Note:* There are exceptions to the high-stack sprinkler system fire safety compliance with ESFR sprinklers, which may not necessarily apply to the normal warehousing practices in wineries. If large quantities of plastic materials are included in the cartonized products (styrofoam wine shippers, for example) or if plastic pallets were used instead of standard wood construction, so-called "in-stack" or "in-rack" sprinkler arrays would have to be employed (178).]

Sprinkler heads can also be a problem for both the designer and the contractor. In known "hot spots" in a sprinklered winery warehouse space, administrative office or food service area, proximity to steam lines, high intensity luminaires, or other high temperature sources, special sprinklers of the high temperature type must be installed in those locations. The insurance carrier's rating agent will mandate the temperature rating of the special sprinkler heads. If a false sprinkler system activation and deluge occurs after construction of the winery, because the "hot spot" was somehow overlooked, both the designer and the sprinkler contractor could be liable for any damages and certainly would carry the burden for the replacement of the low temperature sprinkler heads with the specialty type.

Sprinkler Design and Selection Considerations

The foregoing subsection and the previous subsection entitled "Fire Flow Rate and Duration" makes it patently clear that wineries with their Moderate Hazard classification and Group "F" occupancy class, Division 1, will generally require automatic sprinkler coverage of some type and extent. Having the winery's design professional merely meet the several code requirements, that have been previously identified, may not necessarily provide the best available protection for the winery at the least cost. It must be remembered by winery management and designer alike that the process of cost minimization and fire protection optimization is often a trade-off among structural fire resistance, sprinkler coverage, and the responsiveness of alarm and master control systems. It has been previously demonstrated how fire flow and duration "credits" can be gained by incorporating certain features into the winery's master fire protection plan [i.e., (1) use of non-combustible construction materials, (2) smoke detectors and the means of transmitting the alarm signal, and (3) nonflammable roofing materials.]

The selection of sprinkler type is a multifaceted decision with aesthetics,

hydraulics, and NFPA 13 installation protocol requirements, all combined
into a matrix of cost and fire protection effectiveness. It is not considered
necessary in this book to explore in elaborate detail the advantages and
disadvantages of the six or seven hydraulic and functional styles that are
currently available in the marketplace.

In terms of location and intended use, sprinklers can be classed by the
intended direction of the active spray pattern:

• Upright

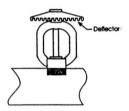

• Pendant

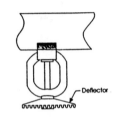

• Sidewall

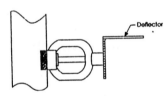

Note: Sprinklers shown are generic, standard type with fusible link.

Sprinklers can with certain limitations be recessed and hidden architectur-
ally with chrome, enamel (any color) cover plates or decorative rosettes,
which blend with the ceiling or wall finish in conference, tasting, or re-
ception room areas.

A further subdivision of sprinklers by "response" gives three additional
categories and one more that is still under a testing and product development program by NFPA (177).

   • Quick-response
   • Early suppression-fast response
   • Quick-response extended coverage
   • Quick response early suppression (undergoing testing and product devel-
     opment)

Although "standard" sprinklers were functionally designed to control fires by preventing their spread, the new class of sprinklers are designed to suppress or extinguish fires by discharging a large volume of water directly onto the burning surface (178).

Standard sprinklers discharge at a rate that is proportional to orifice diameter, the orifice constant, and the dynamic pressure, when the system is operational. The equation, which relates the three orifice discharge parameter, is

$$Q = K \sqrt{P} \qquad \text{(Ref. 151)}$$

where P is pressure in psi, K is a constant which is (for 1/2 in. = 5.7, for 17/32 in. = 8.0, and for 3/8 in. = 2.85.

Table 10.7 lists calculated flows for a standard sprinkler head with a ½-in. orifice.

As described previously, the sprinklers both standard and "new age," can be activated with a fusible link that melts at a specific high temperature. For wineries, the warehouse spaces would have the highest potential for heat release; thus, for a temperature–environment controlled space to protect the finished bottled wine with ambient ceiling temperatures probably never exceeding 80°F, the recommended activation temperature for fusible link devices would be either 175°F or 212°F. Other winery spaces such as offices, laboratories, and food service centers (except electrical equipment and elevator shaft pits/headspaces and hoist machinery equipment rooms) will have a lower potential rate of heat release. The designer could still meet NFPA 13 compliance if the activation temperature rating for the sprinklers in those spaces was reduced to 135–165°F (151). The very real problem of installing sprinklers with uniform temperature activation in a winery space with a high temperature abnormality (near heat registers, steam pipes, or high temperature luminaires) has been previously described. The caution is repeated here for emphasis to have all potential "hot spots" carefully located and identified by the winery architect or mechanical engineer and have them communicated to the sprinkler system

Table 10.7. Discharge for a Standard Sprinkler with a ½-in. Orifice Under Various Discharge Pressures

Pressure (psi)	K	Discharge (gpm)
10	5.7	18.0
20	5.7	25.5
35	5.7	34.0
50	5.7	40.0
75	5.7	49.5

designer for his information and action in planning and layout of the final sprinkler array.

The NFPA *Automatic Sprinkler Systems Handbook,* 6th edition 1994, provides an excellent reference for designers who may be confused by the abundance of available sprinkler types and performance standards. Tables 10.8 and 10.9 from the sprinkler systems handbook can be used to compare the fast-response-type sprinklers both in terms of hydraulic performance, temperature ratings, and color code. For example, an extra-large-orifice quick-response sprinkler, which has a 5/8-in. (0.625-in.) orifice with a K of 11.3, produces a discharge approximately 200% greater than the ½-in. orifice under various pressures, as shown in Table 10.7.

The "glass bulbs" listed in Table 10.9 refer to a second method of sprinkler activation (fusible links have been previously described), wherein a glass bulb, acting as an orifice plug, is filled with low-boiling-point alcohol. When a predetermined temperature is reached, the glass bulb, which is sealing the orifice, bursts from the internal pressure of the boiling alcohol, thus releasing the water through the sprinkler orifice.

Table 10.8. Sprinkler Discharge Classification and Identification

Orifice size (in.)	Orifice type	K factor[a]	Nominal ½-in. discharge	Thread type	Pintle	Orifice size marked on frame
1/4	Small	1.3–1.5	25	1/2-in. NPT	Yes	Yes
5/16	Small	1.8–2.0	33.3	1/2-in. NPT	Yes	Yes
3/8	Small	2.6–2.9	50	1/2-in. NPT	Yes	Yes
7/18	Small	4.0–4.4	75	1/2-in. NPT	Yes	Yes
1/2	Standard	5.3–5.8	100	1/2-in. NPT	No	No
17/32	Large	7.4–8.2	140	3/4-in. NPT or 1/2-in. NPT	No / Yes	No / Yes
5/8	Extra large	11.0–11.5	200	1/2-in. NPT or 3/4-in. NPT	Yes / Yes	Yes / Yes
3/4	Very extra large	13.5–14.5	250	3/4-in. NPT	Yes / Yes	Yes / Yes
5/8	Large drop	11.0–11.5	200	1/2-in. NPT or 3/4-in. NPT	Yes	Yes
5/8	ESFR	11.0–11.5	200	3/4-in. NPT	Yes	Yes
3/4	ESFR	13.5–14.5	250	3/4-in. NPT	Yes	Yes

[a]K factor is the constant in the formula $Q = K\sqrt{P}$, where Q is the flow in gpm and p is the pressure in psi. For SI Units: $Q_m = K_m\sqrt{P_m}$ where Q_m is the flow in L/min, P_m is the pressure in bars and $K_m = 14\ K$.

Notes: (a) The system shall be hydraulically calculated. (See Chapter 6.) (b) Small-orifice sprinklers shall be installed in wet systems only. *Exception:* Small-orifice outside sprinklers for protection from exposure fires installed in conformance with Section 3-7 shall be permitted.

Source: Reprinted with permission from NFPA 13, *Installation of Sprinkler Systems* Copyright © 1994, National Fire Protection Association, Quincy, MA 02269. This reprinted material is not the complete and official position of the National Fire Protection Association, on the referenced subject, which is represented only by the standard in its entirety.

Table 10.9.   Sprinkler Temperature Ratings, Classifications, and Color Codings

Max. Ceiling Temp.		Temperature Rating		Temperature classification	Color code	Glass-bulb colors
°F	°C	°F	°C			
100	38	135–170	57–77	Ordinary	Orange or red	[a]
150	66	175–225	79–107	Intermediate	Yellow or green	
225	107	250–300	121–149	High	Blue	
300	149	325–375	163–191	Extra high	Purple	
375	191	400–475	204–246	Very extra high	Black	
475	246	500–575	260–302	Ultrahigh	Black	
625	329	650	343	Ultrahigh	Black	

[a]A large number of glass-bulb sprinklers have their frame arms coated for aesthetic reasons. This makes frame arm color coding confusing at best. Therefore, the temperature rating of glass-bulb sprinklers is determined by the color of the encased liquid, which is nothing more than a low-boiling-point alcohol. The temperature rating of this type of element is controlled by the size of a small air bubble that is trapped in the glass tube of the device. The relative size of this bubble establishes the temperature rating of the sprinkler.

Source: Reprinted with permission of NFPA 13, *Installation of Sprinkler Systems Copyright* ©1994, National Fire Protection Association, Quincy, MA 02269. This reprinted material is not the complete and official position of the National Fire Protection Association, on the referenced subject, which is represented only by the standard in its entirety.

Figure 10.12 depicts the major plumbing components for a typical industrial automatic sprinkler system in an urban environment. It is different than the systems shown in Figs. 10.10 and 10.11, in that the urban installation is serviced by a municipal water system, whose thousands of residential, commercial, and industrial users must be protected from back-flow into the municipal pipe network if a partial vacuum is ever created in the water main to which the Fig. 10.12 sprinkler system is connected. The key plumbing feature shown, that is a mandatory requirement of the Uniform Plumbing Code 1994 edition is the so-called back-flow-preventer, which is the above-ground unit consisting of two swing check valves in series and an air-gap or air inlet valve that will break the partial vacuum to atmospheric pressure, before water in the building's sprinkler system can be drawn into the municipal network. Homeowners with lawn sprinkler systems have a simpler installation to prevent surface contaminated water from entering the interior residential pipe network should, for any reason, a partial vacuum be produced. The small residential sprinkler system air inlet valve operates in principle just like the larger valve on the industrial system.

Also seen on Fig. 10.12 are flow switches that are connected either directly to the local fire department dispatchers switchboard or to a master fire alarm controller panel in the building. The fire alarm master controller can also initiate a signal to a commercial security/alarm dispatcher or to the fire department directly. The flow switch informs the receiver of the signal that the sprinkler system is in operation and that a fire is in progress.

B.   Position indicator valve locked in the "open" position. (Flow switch sends signal to security/alarm system central dispatcher signifying operation of the building sprinkler system)

A.   2¹/₂ inch siamese fire department connection (See Figure 10.11)

C.   Backflow preventer showing dual check valves and isolation valves with tamper-proof swtiches. (Valves in the "open" position are unlocked)

Fig. 10.12.   Typical industrial hydrant and sprinkler plumbing array (fire flow service from a municipality).

Other important hydraulic and functional features of sprinkler systems that will only be briefly described are as follows:

- Pipehangers—Various types of devices for attaching pipes to ceilings and walls to allow permanent connections of the pipe network to resist dead-load, hydraulic thrust, seismic, and temperature (expansion or contraction) forces.
- Position Indicator Valves (PIV)—Installed at the fire protection water supply piping connection to the winery building, they show that the sprinkler system is charged and give ready access to fire department personnel or winery maintenance staff. The valves are locked open and keyed to a master lock system used by the local fire department. A shut-off condition may be required after a fire has been controlled (fire department) or to repair a leaking or malfunctioning sprinkler. The PIV has a windowed, alphabetic indicator showing "open" or "closed." (See Fig. 10.12.)
- Flushing and Test Connections—As suggested by their labels, these fittings, which are part of any sprinkler system, can be either separate units or combined. They are necessary for start-up testing and periodic performance tests that are made by the local fire jurisdiction, under the provisions of NFPA Standard 25, "Inspection, Testing and Maintenance of Fire Protection Systems."

The footnote in Table 10.8 which specifies a strainer for those installations, which have sprinklers with orifices smaller than ⅜ in., would not apply to a winery installation where minimum orifice sizes would be ½ in. Also, the reference to "wet systems" for the installation of small-orifice sprinklers refers to those systems that are fully charged all the time. In cold climates, dry-pipe systems are often installed to prevent freezing of the sprinklers pipe network. (Dry and wet barrel fire hydrants will be discussed in a subsequent subsection, where cold weather considerations make dry barrel hydrants a necessity.) For many reasons, it is often desirable for the winery to retain an architect/engineer team of design professionals, who are very familiar with the climate constraints for a particular winery location [e.g., a warm-weather architect/engineer from San Diego, CA (where it never freezes) might find it awkward to efficiently plan and design a structure and site utilities for a winery to be constructed in Ohio, on the shores of Lake Erie, where winter freezes are commonplace and snow is only something that the Southern California architect reads about in ski magazines.]

MISCELLANEOUS FIRE PROTECTION EQUIPMENT AND APPARATUS

Fire Hydrants

Fire hydrants are the second line of defense in an interior, structural, electrical, or warehoused product fire. Automatic sprinklers are still the first and most important fire protection system in the winery. The 2½-in. hoses which connect to hydrants are for use by professional fire fighters only. Hose racks or hose reels placed strategically in the winery are generally sized for 1½-in. hose, which can be used to control or extinguish interior fires by winery employees with only a minimum of training. Local fire department personnel are usually very accommodating and will assist in periodic training sessions on the use of small diameter hoses and other portable extinguisher equipment.

Hydrants outside the winery are generally located around the perimeter of the winery building, with the local fire marshal making the final decision on location and type of hydrant required to be compatible with local fire department equipment (hose connections, suction hose connections, and accessibility 100% of the time with no exceptions). Hydrant spacing for wineries may be 200–250 ft apart, which minimizes pressure losses through the hose and allows a quicker response to a fire if the fire fighting team does not have to connect excessively long runs of hose before water can be played upon the fire source. Protection of the hydrants from damage by routine winery truck traffic or visitors unfamiliar with the flow of vehicles in and around the winery is essential, particularly during the grape harvest season. This is most easily accomplished by the installation of pipe bollards that will prevent accidental damage but not interfere with fire crews making a connection during an emergency. (See Figs. 10.13 and 10.14.) Emergency vehicles must have unobstructed access to the winery with super-master pass keys and key pad combinations for security gates, so that there will be no delays in reaching the winery and entering the structure during non-working hours. Emergency medical vehicles should have a prescribed route if the winery is some distance off the traveled secondary roads. If service roads and visitor access roads are separate, road signage should clearly indicate the preferred route. Wineries with abundant real estate should also have their architects locate an open area (not necessarily paved), but with easy access to the winery. This area would be dedicated to life-flight helicopters for emergency landings. All of the gate, building lock, and preferred emergency route details should be formalized and documented and provided to the local fire jurisdiction and the fire company, which would be the most likely to answer a medical or fire alert. Many of the emergency access items described are logged into hard-copy records by the fire department representatives, who periodically visit the winery for alarm, sprinkler system, and primary fire pump inspections. For complicated multi-level winery structures with difficult access (i.e., stairs, ramps, etc.), the fire department will prepare a "fire fighting plan" to which the winery production and administrative staff may contribute.

Hydrants are produced in two basic types:

- Dry-barrel
- Wet-barrel

The wet-barrel type is often referred to as the California type and is installed in climatic zones where freezing does not occur. (See subsection on "Sprinkler Design and Selection Considerations" for dry-pipe and wet-pipe sprinkler systems.) Figures 10.13 and 10.14 show the principal design features of the wet and dry barrel types, respectively. Note that both types have a breakaway feature that minimizes damage to the bury section, valves, or operating stem in the dry-type hydrant. The operating stem breaks cleanly at the location shown as "valve stem coupling" in Fig. 10.14. Figure 10.13 shows an optional spring-operated check valve that closes when the hydrant shears at the location shown as "break-off riser." Repairs to hydrants can usually be made quickly and the fire protection system put back

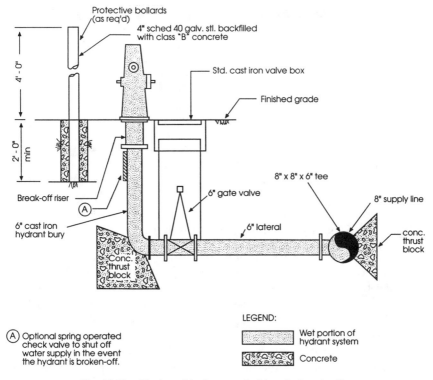

Fig. 10.13.   Wet-barrel hydrant; typical installation detail.

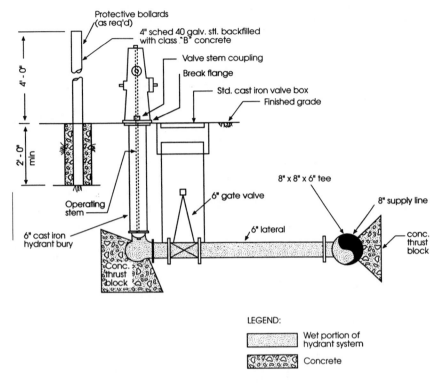

Fig. 10.14.   Dry-barrel hydrant; typical installation detail.

in operation in the shortest time with the least loss of fire flow storage and repair manpower and cost. The dry-type hydrant does not require the optional check valve, as the operating stem below the "break flange" remains intact.

Interior and Exterior Fittings and Equipment

The fittings that are associated with interior hose stations are of two classes. Hose stations with 1½ in. (38 mm) hose on reels or racks are designed to be used by winery employees during the work-day to control blazes that may be outside sprinklered areas until the professional fire-fighting team arrives. These small hose units, together with the portable fire-fighting equipment, constitute the only "hands on" apparatus that the winery staff can use. Larger valves, 2½ in. (64 mm) with hose connections are often installed in pairs to give the fire fighters additional sources of water both inside and outside the winery, away from the exterior hydrants. Figure 10.15 shows the typical fittings and apparatus that are commonplace items

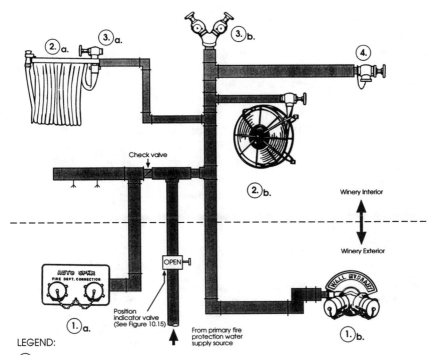

LEGEND:

(1.) **a.** Sprinkler inlet: 2 1/2 inch

    b. Hose connection (outlet): 2 1/2 inch

(2.) **a.** Hose rack

    b. Hose reel

(3.) **a.** Interior valve (single)

    b. Interior valve hose connection (dual)

(4.) Class I : 2 1/2 inch hose valve (interior). For fire dept. use only.

NOTE:

Fire fighting equipment items 2a, 2b, 3a and 3b are associated with 1½ inch fittings and hoses, which may be used by winery occupants/employees.

Firefighting equipment items 1a, 1b, and 4 are for 2½ inch hose size, which only trained fire department personnel can safely use.

Fig. 10.15 Interior and exterior fire-fighting equipment and fittings commonly used in wineries and other industrial buildings.

in wineries worldwide. The set of dual fittings for exterior use in augmenting or providing alternative fire flow to sprinklers, if the primary fire pump has malfunctioned during an emergency, can be specified in a wide array of architectural treatments from colored epoxy enamel through the more conventional chrome and brass finishes.

Another related fire protection system complication can be avoided if the design engineer is firm in not permitting the contractor to use different manufacturers of the same gate valve, air release valve, or hydrant. If "mix and match" components are allowed, at some future time when the winery needs to order replacement parts, there would be no commonality in the ordering process, parts list, or necessary costs, which will create a small nightmare for the utility systems maintenance supervisor, his crew, and the winery's purchasing agent.

Portable Fire Extinguishers

Portable fire extinguishers are inexpensive devices that are placed where fire danger is great and localized. Winery forklifts, vineyard equipment, propane fueling stations (see Chapter 9), food service kitchens, and office spaces.

Clearly stamped on each extinguisher is the type of fire for which the unit is designed. The classification of fires is as follows:

Fire Class	Material
A	Wood, textiles, paper, or trash
B	Oil, gasoline, grease, or paints
C	Electrical or electronic equipment including motors, control panels, computers, and copy machines
D	Flammable metals such as magnesium, phosphorus, powdered aluminum, or zinc

The preferred material for extinguishment of the four classes of fires is in a state of flux because of the ozone-depletion potential of some of the most effective fire suppression substances. The complete elimination of Halon 1301 from the list of effective extinguishments by year 2005, as required by the Montreal Protocols (previously discussed in Chapter 5 in the section entitled "Refrigerants"), has created some problems for the fire protection industry to find substitutes which are suitable (179). Fires in the "C" classification are directly affected by the loss of Halon 1301. In addition to electronic/electrical equipment fires, Halon 1301 was also historically used in commercial and military aircraft fire suppression systems. The following is a list of fire types versus the most often used materials for extinguishment:

Fire Class	Materials	Type of Fire Suppressant
A	Wood, textiles, paper	Water, glycol, CO_2, soda-acid, or Halon 1301*
B	Oil, gasoline, grease, or paint	CO_2, sodium or potassium bicarbonate, or foam†
C	Electronic/electrical equipment	CO_2, sodium or potassium bicarbonate, or Halon 1301
D	Flammable metals	Proprietary dry powder extinguishments (dry sand and sawdust–talc mixtures have been successfully used). Material used must exclude oxygen.

*Halon 1301 has often been called the "magic gas" because of its ability to extinguish electronic fires without damage to sensitive equipment. It is in a liquid state when charged into a portable extinguisher, but gasifies upon release to normal atmospheric pressure. Halon 1301 is colorless, odorless, and non-toxic. The bromine contained in Halon 1301 is dissociated when it reaches the atmosphere, and it is believed to strip the third atom of oxygen from ozone (180).

†"Foam" is produced in portable extinguishers when a solution of sodium bicarbonate and an aluminum sulfate activator are combined. The resultant foam with its entrained CO_2 bubbles smothers the fire by excluding oxygen from the burning Class "B" substance.

The most important aspect of portable fire extinguishers is the unswerving necessity to have them inspected and recharged once annually by a professional fire extinguisher service company. Most service companies have a mobile unit, so that the extinguishers are never removed from their locations for very long periods. NFPA 10 "Portable Fire Extinguishers" also provides the interval between hydrostatic pressure tests of the cylinders, which is normally 5 years. An inspection tag attached to each extinguisher unit helps the winery staff in keeping the cylinder recharging current. It is always better for the winery to take the responsibility for keeping the extinguishers maintained in an operable condition, rather than having the local fire jurisdiction representative appear for an unscheduled fire protection system inspection and find that the extinguishers have not been serviced for 3 years.

Spacing of extinguishers is generally the responsibility of the local fire marshal, but for planning purposes, a winery can expect to have one unit for every 2000–3000 ft^2 (186–279 m^2) of interior space, with travel distances to extinguishers no greater than 75 ft (23 m), as recommended by NFPA 10.

OPERATION AND MAINTENANCE OF FIRE PROTECTION SYSTEMS

The principal reference for the testing and maintenance of fire protection systems is contained in NFPA Standard 25.

Wineries constructed in the last 5 years will have the advantage of the sophisticated controllers and fire detection systems that are virtually self-monitoring.

Starting at the fire protection water supply system, the main source of water, whether it be an open reservoir or groundwater, should be under constant surveillance through a system of water level alarms that will transmit a signal for both "low"- and "high"-water conditions. A visual check of reservoir contents is also possible with daily reading and recording of water level from a staff gauge or electrically driven stage recorder that gives a hard copy record of reservoir contents usually for a 7-day period. The need for evaporation or pond seepage make-up water can be predicted if the minimum volume of fire storage water is examined weekly in light of the rate of water loss, so that measures can be taken to replenish the supply before the low water alarm is activated.

The pump intake screen cannot be visually inspected unless a diver or low-light-level underwater video camera is used. Mineral encrustation of the intake screen may eventually interfere with the hydraulics of the system. If designed properly, the screen will have an excess of opening space 20–30% above that actually required to prevent excessive head (energy) losses at the intake. A well drilling company which is familiar with the cleaning of encrustations on well screens and well casings might have the equipment and expertise to clean the fire pump intake screen. The well-equipped companies also usually have small, underwater video cameras for well casing inspection that could be adapted for intake screen inspections in an open reservoir. (See the subsection "Well Casings" in Chapter 6.)

The fire pump and fire pump controller (either diesel or electric motor driven), if of fairly recent vintage, will have a weekly test cycle programmed into the controller logic that will ensure the successful start up and test run of the fire pump with a time interval of no longer than 7 days. A printed or digitized record of the weekly tests is also provided by most controllers for weekly review by the winery utility systems maintenance supervisor, and by the local fire jurisdiction inspector during his periodic inspections of the system. The maintenance document provided with the fire pump details the daily, weekly, quarterly, and yearly maintenance tasks that must be accomplished. With the availability of computer software that can encompass all winery equipment preventative maintenance procedures, as described in previous chapters, it is not difficult to imagine how easily it would be to modify the program to include the fire pump, jockey pump, diesel/electric prime mover, fuel systems, heaters, and other major components of the fire protection water supply system.

For the open reservoir system, the same aquatic system maintenance program that was detailed in Chapters 6 and 7 must be strictly followed. Excessive growth of plankton and periphyton (algae and emergent aquatic weeds on the shoreline) will decrease the quality of the water supply to the point where clogging of the intake screen system with organic detritus might occur.

If reclaimed winery process wastewater is stored for fire protection, there is an additional element of care required to prevent eutrophication (aging) of the pond from over-enrichment by bio-nutrients from the residual carbonaceous and nitrogenous waste material remaining in the treated effluent.

The NFPA requires that the sprinkler and hydrant system be inspected and tested at a specified time interval. Commercial service companies are available to perform the inspection and testing. If possible, the contractor who installed the system would be the logical choice to perform the maintenance and testing tasks. If the work load and available staff of the local fire jurisdiction permits, the winery should notify the fire marshal of a scheduled sprinkler inspection and test, so that it could be witnessed, possibly in conjunction with inspections of other units of the overall fire protection system.

Finally, the fire alarm system must be monitored and tested frequently to ensure its unblemished operation, if and when it is needed. Fire drills and controlled alarm activations should be routinely scheduled by the winery, so that orderly evacuations of the production and administrative spaces may be made without hurry or panic. If tasting rooms have difficult ingress and egress routes (multistory buildings or long circuitous stairways or ramps), the tasting room staff should be trained to assist the visitors, who might be unfamiliar with the evacuation routes to safety in the event of a fire.

Special events at the winery deserve even more attention when it comes to fire safety and evacuation. Fire safety planning for large gatherings should be accomplished in concert with the local fire jurisdiction, who will provide the "extra dimension" of professional fire safety training and knowledge for the event planning.

The operations and maintenance of winery fire protection systems is almost self-correcting, because if the winery forgets some maintenance or record keeping task, the local fire marshal will be sure to remind you of your omissions on his/her next inspection visit and follow-up letter of admonishment, listing the oversights.

GLOSSARY

Across-the-line electric motor starting   The electric drive motor of a fire pump that starts at full voltage up to speed almost instantly.

Aqueous film-forming foam (AFFF)   A quick knock-down foam that spreads an aqueous film over the surface of the flammable or combustible liquid.

Branch lines   The pipe that supplies the sprinkler heads.

Butterfly valve   A very thin control valve. Is either open or closed. Flow cannot be metered with partial opening. Some butterfly valves are not both UL list and FM approved.

Check valve   A valve used to prevent the reversal of flow. Valve plate has unidirectional movement.

Cold soldering   When sprinkler heads are improperly installed so that one head discharge will cover another adjacent head, the discharge will keep the adjacent head cool and prevent it from fusing. Sprinkler heads cannot be placed closer than 6 ft in order to prevent cold soldering.

Corroproof heads   Sprinkler heads or nozzles that are coated with a wax material or lead to prevent corrosion when the heads or nozzles are installed in a corrosive atmosphere or in an area where they are exposed to weather.

Coupling guard   The cover over the coupling on the shaft between the driver and the fire pump to prevent personnel injury should someone not realize that the shaft is rotating at high speed.

Cross main   Pipe that supplies lateral sprinkler lines.

Deflector   The water discharging from the orifice of a sprinkler head impacts on the deflector to form the discharge pattern of the head. The deflector of a spray sprinkler upright is large, that of a spray sprinkler pendent is small.

Dry powder   A dry powder agent that is used to suppress combustion involving combustible metals.

Escutcheon   The ceiling plate that covers the hole in the suspended ceiling where the drop nipple penetrates the ceiling to supply a pendent-type sprinkler.

Eutetic solder (fusible link)   The fusible element of the sprinkler head that is a compound of material designed to become soft, or melt, at a specific temperature, providing the temperature rating of the head. The most common temperature ratings are 165°F, 212°F, and 286°F, but nearly every manufacturer supplies a vast range of heads with different temperature ratings.

Extinguishers   A portable device used by an individual acting alone to suppress a fire. Extinguishers are classified according to the type of combustion they best extinguish:

> "A": Ordinary combustibles such as wood, paper, etc.
> "B": Flammable liquids.
> "C": The agent in the extinguisher can be safely applied to energized electrical wiring and equipment.
> "C": For use with combustible metals.

Factory Mutual (FM)   An independent testing organization. Any equipment, device, or system manufacturer who wishes to display the FM label submits samples to the organization for extensive testing.

Fire watch   The appointment of an individual to remain in an area unprotected by the existing sprinkler system to watch for any fires. Used when it is necessary to shut down an existing sprinkler system to make repairs, remodel, or make a connection.

Flammable liquid   A liquid having a flash point below 100°F. There are several definitions of a flammable liquid based on their flash points and boiling points as described in NFPA Standard 30.

Flash point   The minimum temperature at which a liquid will give off sufficient vapor concentration at the surface of the liquid to form an ignitable mixture with air.

Flow control valve   A deluge valve that can be opened and closed automatically. Controls the discharge regardless of the supply pressure.

Heat actuated device (HAD)   A pneumatic rate of rise detector. A hollow, ellipsoidal-shaped device. When the combustion heat rises at least 15°F/min, the air in the device expands and activates an alarm device.

Hot start   The device that maintains adequate temperature in the diesel drive of a fire pump when it is not operating, so that the engine will start quickly when called upon to activate.

Infrared detectors   Flame detectors that are activated by infrared wavelength radiation.

Inspector's test connection   A test connection to test the operation of all the system alarm devices. When the inspector opens the test connection valve, the water must flow within 60 s. If this flow does not occur in 60 s, it indicates that the dry pipe system will require some device to exhaust the air pressure, or a device to admit air to the intermediate chamber of the dry valve.

Jockey pump   A pump that maintains a constant pressure on fire protection water supply system piping and is sometimes referred to as a pressure maintenance pump.

National pipe threat   A standard that indicates the size, dimension, and number of threads per inch for a pipe thread. When ordering a fire department pumper connection, a siamese connection, the specification must always read: "siamese connection threads shall comply with local fire department specification."

Orifice   The discharge opening of a sprinkler head or nozzle. A standard sprinkler head has ½-in. orifice, a large sprinkler head has a 17/32-in. orifice.

Outside screw and yoke valve (OS&Y)   A control valve which, when open, has the stem extending several inches above the wheel so its condition (open or shut) can be determined at a glance.

Post indicator valve   A post that extends a few feet above grade and is attached to a nonrising stem gate valve in the underground pipe. The post has a window that indicates whether it is "open" or "shut."

Post indicator valve assembly   A post indicator is attached to a gate valve installed underground. The post indicator valve assembly is a one-piece unit, whereas the post indicator is two—a post indicator and a gate valve. The post of the post indicator valve assembly has the indicating window (open or closed).

Reduced voltage electric motor starting   When the power source to the electric drive motor of a fire pump cannot accommodate the tremendous surge of power to an across-the-line motor, a reduced voltage motor starts in stages so as not to place a massive drain in the electric power source.

Reducing coupling (RC)   A pipe fitting used to connect two pieces of pipe with different diameters or sizes.

Riser   The pipe that rises vertically from the valve to the bulk main or system piping. The pipe rises from floor to floor to supply the sprinkler or standpipe systems on each floor.

Riser nipple   If the cross-main is at a lower elevation than branch lines, the riser nipple is the pipe from the cross-main to the elevation of the branch line. If this piece of pipe is greater than 6 in. in length, it is referred to as a *riser pipe.*

Sectional or isolation valves   A control valve that sections off an isolated zone of underground piping. This facilitates closing off a piping of the underground system for repairs while leaving the remainder of the underground supply operational.

Spray sprinkler pendant (SSP)   A sprinkler head used in the pendant position.

Spray sprinkler upright (SSU)   A sprinkler head used in the upright position.

Tamper switch   A switch on a system control valve that prevents tampering with the valve.

Test blanks   When a portion of a sprinkler system is to be tested without testing the entire piping system, a test blank is installed to isolate the area to be tested from the remainder of the system. These test blanks are identified by red painted lugs that protrude beyond the flange to clearly indicate their presence.

Thrust blocks   A calculated mass of concrete poured against a change in underground pipe direction, hydrant, or valve to prevent the change in momentum reaction of the water in the piping from causing failure of the tee, ell, valve, or hydrant.

Ultraviolet detectors   Flame detectors that are activated by the ultraviolet wavelength radiation emitted by the flame.

Underwriter's Laboratories (UL)   An independent testing organization. Any equipment, device, or system manufacturer who wishes to display the UL label indicating the product is "UL listed" submits samples to the UL for extensive testing. For acceptance by the local fire official, fire protection equipment that is specified by the designer must be either UL or FM approved. (See *Factory Mutual.*)

Water flow alarm device (flow switch)   A pressure switch that is activated when water flows in the system, and water is applied to the switch to initiate an electrical contact.

REFERENCES

151. FACTORY MUTUAL ENGINEERING CORPORATION. 1967. *Handbook of Industrial Loss Prevention.* 2nd ed. New York: McGraw-Hill Book Co.

152. _____. 1994. *Uniform Fire Code.* Austin, TX: International Fire Code Institute, Vol. I, pp. 1-349–1-361.

153. STORM, D.R. 1990. Winery fire protection systems, Part II. Winery Water and Waste, *Pract. Winery Vineyard* 10(6).

154. STORM, D.R. 1990. Winery fire protection systems, Part I. Winery Water and Waste. *Pract. Winery Vineyard* 10(5).

155. _____. 1993. *Fire in the U.S. 1983–90.* Washington, DC: National Fire Incidence Reporting System, Federal Emergency Management Agency.

156. _____. 1995. Justice Department cracks down on ADA compliance. *Consulting-Specifying Eng.* April 1995. P.R.

157. _____. 1980. *Life Safety Code Handbook,* 4th ed. Quincy, MA: National Fire Protection Association.

158. _____. 1994. *Fire Resistance Design Manual.* Washington, DC: Gypsum Association, Publ. GA-600-94.

159. _____. 1995. *Fire Protection Equipment Directory.* Northbrook, IL: Underwriters Lab.

160. _____. 1995. *Factory Mutual Approved Fire Protection Equipment Directory.* Norwood, MA: Factory Mutual Engineering Corp.

161. JENSEN, R. 1975. *Fire Protection for the Design Professional.* Boston, MA: Cahners Books.

162. _____. 1988. *Fire Protection Requirements for Occupancy Permit, Napa County, CA.* Resol. of Board of Supervisors No. 88-98, August 8.

163. FOOTE, E. 1995. *Personal communication.* California Department of Forestry and Fire Protection, St. Helena, CA.

164. _____. 1995. Personal communication. Stevens Water Monitoring Systems, A Division of Leupold & Stevens, Inc., Beaverton, OR.

165. COON, J. 1991. *Fire Protection Design Criteria, Options, Selection.* Kingston, MA: R.S. Means Co.

166. _____. 1991. Fairbanks-Morse Pump Division. 1991. *Fire Pump Products.* Kansas City, KA: pp. 280–281.

167. DESROCHERS, M. 1993. Design parameters for fail-safe fire pump systems. *Consulting-Specifying Eng.* 13(6):42–46.

168. _____. 1990. *National Fire Protection Association Publication No. 20.* National Fire Protection Association. Sec. 11-4.5, pp. 20–37.

169. INGRASSIA, E. 1992. Fire alarm systems: Seven factors influencing selection. *Consulting-Specifying Eng.* 12(1):60–62.

170. DUNN, J.E. and FRY, J.E. 1966. *Fires Fought with Five or More Jets.* Fire Research Technical Paper No. 16, London: Her Majesty's Stationery Office.

171. RUTSTEIN, R. 1979. Elimination of the fire hazard in different occupancies. *The Fire Surveyor.*

172. THOR, J. and SEDIN, G. 1979. *Fire Risk Evaluation and Cost Benefit of Fire Protective Measures in Industrial Buildings.* Stockholm: Swedish Institute of Steel Construction.

173. SCHWED, R.L. 1992. The marriage of fire alarm and building management systems. *Engineered Syst.*

174. CORSLEY, W.A. 1987. Protecting hazardous locations from fire disasters. *Consulting-Specifying Eng.*

175. McGREAL, M. 1995. Changes to the 1994 edition of NFPA 13. *Plumb. and Mech. Eng.*

176. McGREAL, M. 1995. Piping materials for automatic sprinkler systems. *Plumb. Mech. Eng.*

177. BYCHOWSKI, J.A. 1994. Automatic sprinklers: Simple, successful fire suppression devices. *Consulting-Specifying Eng.* 16(5):71–80.

178. WILCOX, W.E. 1990. Early suppression, fast-response sprinklers reduce losses. *Consulting-Specifying Eng.* 48–49.

179. MOORE, A.P. 1995. Choosing an alternative to Halon 1301. *Plumb. Mech. Eng.* 23–31.

180. BECK, P.E. 1990 Fire protection: Halons. *Consulting-Specifying Eng.* 43–45.

CHAPTER 11

SOLID WASTE SYSTEMS

OVERVIEW

In the United States, no winery site utility subsystem has changed more dramatically in the last decade than the management of solid waste. Other wine producing regions of the world are also reexamining their solid waste handling and disposal practices. The principal forces which have brought about the changes were as follows:

- Landfill sites were being filled at an alarming rate.
- Untapped opportunities for waste-to-energy projects emerged.
- Concern was expressed at all levels of government on the poor management of natural resources.
- The strong and sensible voice of the environmental movement was finally being heard around the world.

Before the radical changes in waste management policies and legislation took hold, the United States had for many years been rather complacent about solid waste generation and disposal methods, whether it occurred on

349

land, in the oceans, or by conventional incineration within large urban centers. Europe and Japan on the other hand, not being blessed with an abundance of land area, became the leaders in the philosophy and practice of waste minimization at the source. The scarcity of land in those two densely populated areas, coupled with the need for ever-increasing blocks of electrical energy to satisfy domestic and industrial demands, stimulated the development of waste-as-fuel technology that produced either steam or electrical power as an essential by-product. This technology is also synonymous with power co-generation. Power plants fueled by nuclear fission, during this period, were in a down cycle, and the two-pronged waste-to-energy systems had just too many benefits to be ignored.

Waste minimization and segregation, together with salvage and recycling of waste materials, are the cornerstones of contemporary solid waste management systems. They will be discussed in this chapter, as they relate to the solid wastes generated by a typical winery. When significant, the differences in the solid waste stream of a table winery brandy distillery, a bonded cellar, and a sparkling winery will be noted. Vineyard operations also produce solid waste, with the major component almost entirely in the biomass category of vineyard prunings.

The disposal of residual solids from both sanitary and process wastewater systems has been previously discussed in Chapter 8. Reference is made to the appropriate sections of Chapter 8 for the optional methods of disposal for septage (septic tank sludge) and settled sludge from process wastewater settling cells or clarifiers.

LAWS AND REGULATIONS GOVERNING THE HANDLING OF SOLID WASTE

The U.S. Environmental Protection Agency (U.S. EPA) was created circa 1970 by Presidential Order. Prior to 1970, all solid waste policy issues were the responsibility of the U.S. Public Health Service (U.S. PHS) through its National Office of Solid Waste Management (181). A report on the "Recovery and Utilization of Solid Waste" was completed by the U.S. PHS in 1971, just in time to transfer the task of implementation to the newly formed Environmental Protection Agency (182). The 1971 law began a long series of strong national policy movements away from the discard and bury philosophy, to a new era of waste material segregation, recycling, and reuse. Five years later, the Resource Conservation and Recovery Act (RCRA) further defined the methods and controls for the acceptable storage, transport, handling, and disposal of solid waste. The RCRA has undergone extensive modification through almost annual reviews and amendments up to the present. The RCRA is before Congress in 1996 for reauthorization. During the spring of 1995, the longest federal budget battle in the history of the United States almost stopped all government operations. With cost cutting and national debt down-sizing on the minds of all congressmen and congresswomen coupled with shrinking dollars for federal programs across the board, the likelihood of radical changes to the RCRA are hard to imagine. One of the 1993 amendments to the RCRA focused on waste lubricating oil. The United States generates, at current rates, 1.4 billion gallons of waste oil annually (183). The so-called Used Oil Management Standards (UOMS) establish guidelines for the proper handling and disposal of waste oil. Waste oil generators, transporters, reprocessors, refiners, and marketeers of used oil all fall under the purview of this statute. It also includes entities that use waste oil for fuel purposes in steam and power generation. The two exemptions from the provisions of the UOMS are private individuals, who change oil in their personal vehicles, and waste oils which contain PCB (polychlorinated biphenols), such as pre-1985 transformer cooling oils. The latter oils fall under the toxic waste category and must be dealt with under the provisions of the federal Toxic Substances Control Act. As of January 1, 1995, 26 states had adopted the federal waste oil model law or prepared similar regulations that were the equal to or more inclusive than the EPA version.

The other piece of significant solid waste law, which hopefully no winery will ever encounter, is the Comprehensive Environmental Response, Compensation and Liability Act of 1980, better known as CERCLA or Superfund Law. The possibility of a winery discovering an abandoned, buried fuel tank on a parcel of rural land once intensively farmed has been previously described in Chapter 6 in the subsection entitled "Prepurchase Winery Parcel Assessment." The groundwater or soil contamination from a chronic fuel tank leak might become a CERCLA issue, if environmental damage and cleanup is significant.

California's Integrated Waste Management Act of 1992 provided for, among other things, the source reduction of solid wastes on a time schedule to ease the volume of materials reaching landfills statewide by 25% over 1992 levels. Such reductions were to have been achieved by 1995. A 50% reduction or diversion to recycling from the solid waste stream earmarked for landfills is mandated for the year 2000 (184). The Office of the State of California Integrated Waste Management Board stated that the analysis of the waste diversion programs for achieving the 1995 goal of 25% reduction in waste to landfills was probably achieved (185). The next 25% increment of reduction mandated in the statute for year 2000 will not be as easy to achieve, according to the waste reduction program administrators. One way of diverting unrecyclable items away form landfills is to have them reengineered by the manufacturer. A case in point is the very common VCR film housing. First-generation VCR cases contained no fewer than six parts, each made from a different plastic material. Reengineering added two additional parts, but all components of the case were fabricated from the same type of plastic resin and thus became recyclable through the adaptive process (185).

The European Union (EU), formerly known as the European Economic Community (EEC), has drafted its own set of solid waste handling and recycling guidelines, with which EU members must comply. There are built-in escape clauses, however, that allow a member country to circumvent a solid waste management edict that is obviously an environmental or natural resource danger (188). Recycling and waste minimization are now part of the lexicon in most all of the wine producing regions of the world. The sometimes harsh economic realities of recycling have arrived, along with the technology, and for some recycling programs to survive, they must operate at a loss for a period of time. Government subsidies or tax credit support must be provided to the operators of recycling programs to both "prime-the-pump" and to make a more environmentally manageable problem of discarded materials than has previously been required in all of recorded history.

Chapter 10 explored many facets of the Uniform Fire Code (UFC) as related to wineries. One additional section of the UFC is devoted to solid waste and must therefore be reckoned with by wineries in planning for solid waste management. (See Section 1103 of the UFC, entitled "Combustible Materials.") Because of the danger of fires occurring in solid waste containers, the UFC is very specific. For example, Section 1103.2.1.3, "Oily Rags," states that oily rags must be stored in metal containers with tight-fitting covers. Section 1103.2.1.5, "Removal," specifies that combustible rubbish stored in containers outside of non-combustible vaults or rooms shall be removed from the building structure at least once each working day.

WINERY SOLID WASTE CHARACTERIZATION

Wineries have three basic sources of solid waste: the vineyard, the winery production facility, and the water and wastewater treatment facilities. The later two sources produce residual solids that require special handling and disposal methods. Residual solids disposal from sanitary and process wastewater treatment systems have been discussed previously in Chapter 8. If conventional water treatment is required to obtain potable water for winery process use, the solids resulting from the flocculation, coagulation, and sedimentation process (sludge) must receive special disposal as a potentially biohazardous waste material. Thus, under most public health and safety codes, incorporation of water treatment process sludge into vineyard or other soils would not be permitted. Also, the backwash water from pressure filters often used in water treatment, which contains entrapped organic and inorganic solids and associated microflora, cannot be discharged in an uncontrolled manner to natural streams or drainage courses. Buried concrete or fiberglass tanks can be used as storage vaults for both filter backwash discharges and for the settling tank sludge. Periodic servicing of the vault by a commercial septic tank pumper will ensure its proper disposal and statutory compliance for the winery with existing public health codes and regulations.

Winery Solid Waste

The solid waste stream from winery production spaces is most easily seen graphically by analyzing the major raw material inputs and waste discard outputs for a typical operation. Figure 11.1 lists 23 separate input items and 23 descriptors of the waste residual for the output item (if any) that must be dealt with as solid waste for the following:

- Unconditional disposal
- Segregation for recycling or destruction as hazardous waste
- Reusable containers returned to company of origin
- Segregation for recycling

The additional output item occurs because of the different solids content of machine-picked or hand-picked grapes (see output item 6).

The national goals for reducing the volume of solid waste reaching landfills or requiring incineration is mirrored in state and local statutes which set reduction rates of 25% for year 1995 and 50% by year 2000, as described earlier in this chapter. Lawmakers and their technical support staffs have generated what has to be a universal framework for reducing solid waste volumes:

1. Reduce waste at the source or waste minimization: A typical example might be in the packaging of breakfast cereals. Some companies have eliminated the "box" and use only a durable plastic bag for the product. The production cost savings in cardboard used and the elimination of a like amount from the wastestream is a significant waste reduction at the source.

Fig. 11.1. Solid waste balance for table winery operations.

IN — WINERY — OUT

IN:

1 Drums of juice concentrate or wine spirits
2 Consumable food and beverages
3 General cleaning products
4 Office supplies
5 Bathroom supplies
6 Grapes and juice
7 Laboratory supplies
8 Cooperage
9 Pallets
10 Wine bottles and cartons
11 Filter pads, cartridges
12 Corks
13 Yeast and ML cultures
14 Capsules
15 Labels and label glue
16 Soda ash, tank cleaner, citric of tartaric acid, D.E., bentonite
17 Styroshippers, retail bags and gift wrapping
18 LPG, Oils, Lubricants, Batteries, Tires
19 Incandescents, and fluorescent bulbs and ballasts
20 Recharge portable fire extinguishers
21 Stretch wrap and pallet straps
22 Other (lees and potassium bitartrates)
23 Winery landscape residuals

OUT:

1 Recycle steel or polyethylene drums
2 Recycle aluminum cans
3 Recycle HDPE containers
4 Recycle waste paper
5 Paper towels to trash
6 Machine harvest-Incorporate pomace into vineyard Hand pick-Incorporate stems and pomace into vineyard soils or deliver to compost facility
7 Recycle glass and polyethylene bottles
8 Cooperage for resale to other wineries or to compost or recycling as planters
9 Broken pallets to waste to fuel system
10 Broken glass and tasting room bottles for recycling; fiberboard cartons for recycling after baling
11 Filter pads and cartridges to trash
12 Tasting room corks and bottling line broken corks to trash
13 Yeast and ML bacteria metabolites in lees sold to ethanol recovery firm
14 Tasting room capsules and bottling line broken capsules segregated and recycled for tin, antimony or aluminum
15 Bottling line label rejections recycled with office paper after baling. Unusable label glue to drain.
16 Fiberboard, steel or polyethylene drums for recycling. DE and bentonite tees to trash after dewatering.
17 Styroshippers, retail bags and gift wrapping do not enter the winery solid waste stream
18 Spent oils, lubricants, batteries, and tires recycled as hazardous waste
19 Incandescents, and fluorescent bulbs to trash (ballasts are toxic waste)
20 Disposal of spent fire chemicals by service company
21 Stretch wrap and pallet straps to recycle
22 High moisture lees and potassium bitartrates for recycling for ethanol and tartrates
23 Lawn cuttings and chipped brush to compost

2. Diversion: If a winery can incorporate all of its pomace into vineyard soils, that major solid waste item becomes a "zero" in solid waste accounting practice, because although it is generated on site, it is also safely disposed of on site.

3. Recycling and reuse: In the last 10 years, an entirely new industry has been created that collects, processes, and packages waste aluminum, plastic of all kinds, paper of all kinds, glass, and waste oil and lubricants for resale. Wineries, which segregate their waste products for recycling, only represent a portion of the entire recycling equation. Deliberate purchase of products and supplies for the winery which contain all or part of recycled material completes the endless circle from consumer to postconsumer and back to consumer again.

When the winery concept plan is being developed, winery owners should, in concert with the professional design team, create a model for the handling of winery and vineyard solid waste. The model should respect the aesthetic values that will be created for the winery by the building and landscape architects. For example, careful siting and orientation of the solid waste container and storage area can facilitate movement of waste to the storage site and make it easier for the commercial refuse company to service the containers. Segregation of waste materials that are generated will begin in the winery, so that employees will find it easy to separate paper, soft drink bottles and cans, and food waste near the points where the generation and discarding occur.

In the recent past, where at least one small waste container per desk would be required in administrative spaces, two or more might be appropriate in this era of recycling to prevent cross-contamination of one segregated waste with another. Decisions should be made early in the winery planning process to standardize the type of drinking containers that will be used throughout the facility. This action not only simplifies purchasing, but it is the first step in helping to make the solid waste model both convenient, economical, and user friendly. For example, if "styrofoam" (aka: polystyrene) cups are chosen, the recycling potential is low for used cups, but high for polystyrene packing materials such as styrofoam peanuts. For guidance in planning for plastic material segregation and recycling, a tabulation of plastic resin types and the market potential for recycling is shown in Table 11.1. The U.S. plastic container industry requires that the identifying code for each type be stamped or embossed on the bottom of the container. These code numbers are shown in column 1 of Table 11.1. Also, the universal recycling symbol is often shown by manufacturers to indicate that the material is recyclable. If that same symbol is shown in solid print, it indicates that the product is made from recycled materials.

- Is a symbol that is commonly used by manufacturers. It does not necessarily mean that a product is made from recycled materials; it only means that the product is *technically recyclable.* A plastic product is nonrecyclable if no recycling center accepts it.

- It means that a product is made from recycled materials.

A number of hazardous waste materials will be generated by wineries. These discarded items must be segregated from the solid waste stream and disposed of as prescribed by the federal Toxic Substances Control Act, the Resource Conservation and Recovery Act (RCRA), and/or the state or local version of those laws:

- Tires
- Batteries (both wet cell and dry cell)
- Lubricants and oils
- Paint and paint thinner residuals (cans, brushes, and rags)
- Aerosol cans
- Fluorescent light ballasts

Toxic substance recognition and identification has been simplified, and since about May of 1989, all U.S. manufacturers are required by the Occupational Safety and Health Administration, OSHA, to provide so-called MSDS (Material Safety Data Sheets) data sheets for their products. The winery's purchasing agent can request and receive an MSDS data sheet for any product destined for use on the winery premises. Curiously, even felt-tip pens and the typists' friend, "white-out," have MSDS sheets, which describe the potentially toxic fumes emitted by those substances. The MSDS forms also contain a section on safe disposal practices. The winery must have these data sheets readily accessible to all employees, who can determine what, if any, precautions must be taken in using a certain product. A prudent winery general manager will have several looseleaf binders prepared containing MSDS pages for all products in current or expected future use on the premises. A typical MSDS form for wine tank cleaner is shown as Table 11.2. Surprisingly, not one computer software company has attempted to compile all the product data sheets into a CD-ROM format.

For wineries which use large quantities of aerosol cans for touch-up painting, specialty lubrication, belt-dressings, and landscape pesticides and herbicides, an aerosol can puncture-and-residual-material-capture-device might be economically justified. This allows the spent aerosol can to be segregated as scrap metal, not requiring the special handling of an unpunctured can still containing toxic or combustible substances. Figure 11.2 shows a commercially available system for recycling aerosol cans and for

Table 11.1. Guidelines for Recycling the Most and the Least Desirable Plastics Under California Integrated Waste Management Law

Symbol	Plastic type and typical use	Recycling potential and limitations
1	**Polyethylene terephthalate (PET)** Common uses: 2-L soda bottles, cooking oil bottles, peanut butter jars	This is the most widely recycled plastic and the only one with a redemption value under the California "Bottle Bill." Many recycling programs and centers require that you remove caps and flatten the bottles.
2	**High-density polyethylene (HDPE)** Common uses: detergent bottles, milk jugs, "crinkly" plastic bags	Many grocery and drug stores have programs to recycle their plastic bags. Call stores for details. Some curbside programs accept rigid containers. Check with your local service provider about whether they take #2 plastics and if containers need to be sorted by color.
3	**Polyvinyl chloride (PVC)** Common uses: plastic pipes, outdoor furniture, shrink-wrap, water bottles, salad dressing and liquid detergent containers	Recycling centers almost never take #3 plastic. Develop alternatives whenever possible.
4	**Low-density polyethylenes (LDPE)** Common uses: dry cleaning bags, produce bags, trash can liners, food storage containers	Some stores that accept HDPE (#2) bags also accept LDPE bags.
5	**Polypropylene (PP)** Common uses: aerosol caps, drinking straws	Recycling centers almost never take #5 plastic. Look for alternatives, whenever possible.
6	**Polystyrene ("styrofoam")** Common uses: packaging pellets or "peanuts," cups, plastic tableware, meat trays	Many packaging stores will accept polystyrene peanuts and other packaging materials for reuse. Cups, meat trays, and other containers that have come in contact with food are more difficult to recycle. If you are a business and have large quantities of these types of containers, call the Alameda County Recycling Hotline.
7	**Other** Common uses: certain kinds of food containers and Tupperware	This plastic usually contains a combination of several different types of plastics/polymers. Recycling centers do not take #7 plastic. Look for alternatives.

Source: Adapted from Ref. 189.

Table 11.2. Typical MSDS Format

MATERIAL SAFETY DATA SHEET:
WINE TANK CLEANER

Sierra Chemical Company
788 Northport Drive
West Sacramento, California 95691

Phone:
   Inquiry: (916) 371-5943
   Emergency: (404) 874-3326
IDENTITY: (As used on label and list)

HAZARD RATING		
Health	2	0 = Lease
		1 = Slight
Flammability	0	2 = Moderate
		3 = High
Reactivity	2	4 = Extreme

WINE TANK CLEANER

SECTION I

Manufacturer's Name:	Sierra Chemical Company
Address:	788 Northport Drive
	West Sacramento, California 95691
Emergency Phone Number:	(404) 874-3326
Phone Number for Information:	(916) 371-5943
Date Prepared:	August 30, 1990
Name of Preparer:	Clifford M. Cantrell

SECTION II—HAZARDOUS INGREDIENTS/INFORMATION

Hazardous Components	OSHA PEL	ACGIH TLV	Other Ceil	% or Range	SK
Sodium Hydroxide CAS# 1310-73-2	2mg/m3	2mg/m3	2mg/m3	< 95	

NOTE: An '*' under 'SK' (skin) indicates that OSHA requires skin protection to prevent or reduce exposure to this ingredient. Sodium hydroxide is a CERCLA reportable substance and also subject to SARA Title III, Section 313 and may require reporting by manufacturers.

EPA SARA TITLE III HAZARD CATEGORIES
ACUTE HEALTH—CHRONIC HEALTH—REACTIVE

SECTION III—PHYSICAL/CHEMICAL CHARACTERISTICS

Boiling Point _____ N/A
Vapor Pressure (mm Hg) _____ NIL
Vapor Density (Air = 1) _____ NIL

Bulk Density (Approx) _____ 1.2
Melting Point _____ N/A
Evaporation Rate
(Butyl Acetate = 1) _____ NIL

Solubility in Water: 20%
Appearance and odor: Tan granules with bland odor.

SECTION IV—FIRE AND EXPLOSION HAZARD DATA

Flash Point (Method Used): _____ None Flammable Limits: _____ LEL: N/A UEL: N/A
Extinguishing Media: N/A
Special Fire Fighting Procedures: Solutions and mists are corrosive.
   Firefighters should wear self-contained breathing apparatus and full protective clothing
   when fighting fires that involve this product.
Unusual Fire and Explosion Hazards: Reacts with aluminum and other soft metals to release
   explosive hydrogen

SECTION V—REACTIVITY DATA

Stability: Stable
Incompatibility (Materials to Avoid): Do not mix with acids or oxidizing agents.
Hazardous Decomposition or Byproducts: Heat and carbon dioxide.
Hazardous Polymerization: Will not occur.

SECTION VI—HEALTH HAZARD DATA

Route(s) of entry:
Inhalation?	Mists are corrosive.
Skin?	Solutions and mists are corrosive.
Ingestion?	Corrosive to gastrointestinal tract.
Eyes?	Corrosive and irritating.

Health Hazards (Acute and Chronic): Can cause chemical burns to skin and eyes.
Carcinogenicity:
NTP?	Ingredients not listed.
IARC Monographs?	Ingredients not listed.
OSHA Regulated?	Ingredients not listed.

Signs and Symptoms of Exposure:
Inhalation:	Corrosive to respiratory systems with possible difficulty in breathing.
Ingestion:	Small amounts may cause gastric distress, diarrhea and vomiting.
Skin Contact:	Chemical burns and severe irritation.
Eye Contact:	Chemical burns and severe irritation.

Medical Conditions Generally Aggravated by Exposure: Unknown
Emergency and First Aid Procedures:
Ingestion:	DO NOT INDUCE VOMITING. Administer citrus juice or vinegar; if these are not available, give large quantities of water. Do not administer fluids to an unconscious person. GET MEDICAL ATTENTION IMMEDIATELY.
Inhalation:	Remove victim to fresh air. If breathing does not return to normal, GET MEDICAL ATTENTION.
Eye Contact:	Immediately flush eyes in clear running water for at least 15 minutes lifting upper and lower lids periodically. GET MEDICAL ATTENTION IMMEDIATELY.
Skin Contact:	Flush skin with plenty of water for 15 minutes. If burns or irritation result, GET MEDICAL ATTENTION.

SECTION VII—PRECAUTIONS FOR SAFE HANDLING AND USE

Steps to be Taken in Case Material is Released or Spilled:
Small Spills:	Flush to sanitary sewer.
Large Spills:	This product contains materials which are CERCLA hazardous wastes. Dike area and transfer spill to closed containers for disposal in an approved hazardous waste site. Rinse spill area.
Waste Disposal Method:	Dispose of waste in a hazardous waste facility approved by federal, state and local regulations.
Precautions to be Taken in Handling and Storing:	Keep container closed.

Other Precautions: KEEP THIS AND ALL CHEMICALS OUT OF REACH OF CHILDREN.

(*continued*)

SECTION VIII—CONTROL MEASURES

Respiratory Protection (Specify Type):	Dust mask recommended for minor mists. Local
Ventilation:	Exhaust: To maintain exposure below control levels (see Section II).
Protective Gloves:	Rubber or neoprene.
Eye Protection:	Chemical goggles.
Other Protective Clothing or Equipment:	As required to prevent skin and eye contact. Wash product from clothes and skin. Use good
Work/Hygienic Practices:	housekeeping practices.

SECTION IX—ADDITIONAL INFORMATION

ENVIRONMENTAL

Biodegradability:	Biodegradable.
Waste Disposal Methods:	Dispose of in an authorized waste facility in accordance with local, state and federal regulations.

ADDITIONAL INFORMATION

Empty Container Handling:	WARNING! EMPTIED CONTAINER RETAINS PRODUCT RESIDUE. Observe all precautions even after container is emptied. Keep empty container closed tightly.

SARA TITLE III REPORTING REQUIREMENTS

Section 302—Extremely Hazardous Substances	Reporting not required.
Section 304—Hazardous Releases	Reporting required for releases of this product in excess of 10, 000 pounds.
Section 311—Community Right to Know (R-T-K)	Reporting required for inventory above TPQ.
Section 312—R-T-K Inventory Data	Reporting required for inventory above TPQ.
Section 313—Emissions and Release	Reporting may be required for users in manufacturing sector.
CERCLA	Same as Section 304.

REFERENCES

Toxic Substance Control Act List (TSCA)—Ingredients listed.
Permissible Exposure References:
>   Registry of Toxic Effects of Chemical Substances
>   Title 29 Code of Federal Regulations
>   National Toxicology Program (NTP) Report on Carcinogens
>   International Agency for Research on Cancer (IARC) Monographs

Regulatory Standards:
>   DOT Title 49 Code of Federal Regulations 172.101
>   SARA Title III
>   Nuclear Regulatory Agency

THE INFORMATION CONTAINED HEREIN is believed to be accurate but is not warranted to be so. Users are advised to confirm in advance of need that information is current, applicable, and suited to the circumstances of use. Vendor assumes no responsibility for injury to vendee or third persons proximately caused by the material if reasonable safety precautions are not adhered to as stipulated in the data sheet. Furthermore, vendor assumes no responsibility for injury caused by abnormal use of this material even if reasonable safety procedures are followed.

Fig. 11.2.   Waste aerosol can propellant and residual contents capturing system
(Photograph: Courtesy Krylon/Sprayon Ind. Products Group, Solon, OH.)

storing the residual substances while entrapping the vented propellant
vapors and hydrocarbon residuals in a dual-purpose filter mounted on the
storage drum.

Fluorescent light ballasts contain nearly 1 ounce (28 g.) of pure polychlorinated byphenyls (PCBs). This toxic element has been classified by
the U.S. EPA as a probable human carcinogen. All fluorescent light ballasts

manufactured before 1979 will contain PCBs. Ballasts made after 1979 will have an identifying "No PCB" on the label. Older wineries would be the most likely candidates for having PCB-containing light fixture ballasts. The two federal laws which deal with the safe disposal of ballasts conflict in the definition of a small waste generator (188). Because PCBs are of a particular danger to public health, prudence would have the winery transfer custody of the ballasts to a licensed toxic waste company, who would see to the disposal of the ballasts in a special chemical landfill or destruction by incineration in a licensed toxic chemical incinerator. The formality of a paper-custody transfer of the toxic material is essential to protect the winery from future legal action if the toxic waste contractor defaults on the proper disposal of the ballasts by contaminating soil or waters of the United States. The U.S. EPA Form 8700-22 (Rev. 9-88) entitled "Uniform Hazardous Waste Manifest" contains the necessary information to permit a cradle-to-the-grave tracking of the material through final disposal and destruction. The licensed hazardous waste contractor must have Form 8700. Once the ballasts are consigned to the licensed toxic waste handler, the winery's liability for future problems ceases to exist.

Non-PCB ballasts manufactured after 1979 still pose a problem for disposal. All ballasts produced after 1979 until 1991 contain diethylhexyl phthalate (DEHP); although not as potentially toxic to humans as PCBs, the compound is nonetheless classified as a carcinogen. DEHP is regulated under the Superfund Law (CERCLA) and is classified a hazardous substance and it must receive handling and disposal in the same manner as the PCB-containing ballasts previously described (190).

Waste oil is governed by the so-called "Used Oil Management Standards," which were adopted under the March 8, 1993 amendments to the Resource Conservation and Recovery Act (Title 40 CFR, Part 279), previously cited. If routine vineyard and winery equipment maintenance is performed on site, then the problem of storage and handling of used oils and lubricants must be addressed by the winery. Not all states list used oil as a hazardous waste—California does. For example, a California winery can elect to dispose of its waste oil, but then faces the economic consequences of the legal restrictions placed on hazardous waste contractors to guarantee end-point disposal in a hazardous waste landfill. The special licensing of transport vehicles and the scarcity of sanctioned hazardous waste dump sites makes the waste oil disposal alternative considerably more costly than the recycling option. Even though the waste oil has possibly received the hazardous designation (depending on the statutory view of the jurisdiction for the particular winery location), under the U.S. EPA Used Oil Management Standards, the used oil can be recycled. Just as with the disposal of fluorescent light fixture ballasts previously described, the winery must ensure that the service company to which the used oil will be formally consigned has a U.S. EPA identification number and permit to purchase, transport, and/or process waste oil. Spent oil filters can be disposed of as scrap meal if the free-draining oil is removed and stored in the designated waste oil container on the winery property. If waste hydraulic fluids are also to be recycled, the question of separate storage or mixing of waste fluids in the same container should be discussed with the used oil recycling agent. If the oil is destined for re-refined lubrication stock, it is unlikely that mixing of the oils would be acceptable. If it is to be used as fuel oil, the mixing may be acceptable. Under current oil recycling industry practice, approximately 85% of used oil is burned as fuel and 15% is re-refined for lubrication stock oil (183).

Used tires have created a solid waste problem of enormous proportions worldwide. The disposition of used tires can be placed into three distinct categories (not among them is burial in a sanitary landfill, in the United States at least):

- Retreading or remanufacture
- Shredding and reuse for boiler fuel
- Rubber-modified asphalt

Retreading of tires received negative publicity from U.S. consumers in the span of years 1960–1975 when quality control and workmanship of tire retreading companies was at a very low ebb. Currently, only truck and off-road construction vehicle tires are manufactured with retreading as a distinct objective. Data referenced to the year 1990 suggests that approximately 39% of all replacement truck tires are retreaded products, with some 60% of off-road construction equipment tire replacements as retreads (191). Tire-derived fuel uses require some 33 million tires annually to fire boilers in thermoelectric power plants located at some six geographical locations (181). Rubber-augmented asphalt is used in two ways to provide a road-surfacing material that is apparently of inferior quality to conventional hot-mix road asphalt. Economic incentives to paving contractors and persistent governmental pressure to find new uses for the never-ending supply of tires may eventually change user attitudes about rubber-modified asphalt paving. U.S. tire shops generally receive the spent casings when new tires are purchased, so the disposal problem for the winery, which deals directly with a retail or wholesale tire outlet, is non-existent. For larger wineries with on-site vehicle maintenance shops, the tires must be transferred to a licensed tire recycler, who will then vector them to the best used tire marketplace, which gives the greatest economic return. Tire disposal costs can range from $1 (U.S.) to $3 (U.S.), depending on the proximity of the nearest market buyer and the haul distance to service the winery.

Batteries of both the wet cell and dry cell type contain toxic heavy metals of one kind or another. Automotive or farm equipment batteries are listed as hazardous waste because of the large mass of lead contained in the plates. As with new tire purchases, battery purchases usually involve an exchange, with the old battery going to the retailer for disposition. Battery manufacturers in the United States are required by statute to use a certain percentage of recycled lead in their new products. If all virgin lead is used, the manufacturer pays a penalty fee for noncompliance. Dry cell batteries, although smaller than their lead-plate, wet cell counterparts, contain small quantities of some very toxic metals (i.e., mercury, silver, and cadmium). Even the harmless-appearing button batteries used in hearing aids have cores of mercuric oxide. There is currently legislation in several states which will require household-type battery manufacturers to recover old batteries through a redemption program or other means just to keep them under control and out of sanitary landfills. A winery can generate a significant number of waste batteries with cellular phones, pagers, flashlights, answering machines, emergency lights, and portable gas detectors. It is a simple enough matter to provide for their safe disposal when the winery's master solid waste model is developed. A properly labeled plastic bucket or two, strategically placed, and an ongoing employee training and education program in managing discarded material in the winery are all that is needed.

Non-toxic waste products generated by the winery also require the segregation process step previously described. As shown in Table 11.1, the most desirable plastic container for recycling and reuse is the clear plastic (PET, aka: polyethylene terephthalate) bottle favored by juice and bottled water manufacturers. In California and other states throughout the United States with container return incentive programs, a redemption deposit makes the recycling value of that particular container very attractive.

Glass from broken wine bottles and empty bottles from the tasting room are recyclable as makeup for glass manufacture or, if the bottle is still intact, it can be cleaned, delabeled, sterilized, and re-cartoned by wine bottle recycling companies for use as wine vessels again. The only glass from a winery that is not recyclable is Pyrex glass from the winery's quality control laboratory. Mixing post-consumer glass from wine bottle discards and broken laboratory glassware can render the mix unacceptable for recycling.

Oak barrels are not a disposal problem. There always seems to be a ready market for used barrels that can be shaved, toasted, and reused for wine aging. Depending on the wall thickness of the barrel, shaving can be performed at least two times, if only ⅛ in. of material is removed. For shaving operations that go deeper, say ¼ in., the barrel can only be processed one time. For wineries which add wood shavings to their wines, in lieu of oak barrel aging, the spent shavings can be incorporated into the winery's composting operation. Direct incorporation of the cellulose material into vineyard soils is not advisable because of the nitrogen demand placed on the soil by the decomposition process. Wine barrel and wine tank oak lattice inserts, which have reached the end of their useful lives, can be cut into appropriate lengths for stove or fireplace kindling. Barrels which have reached the shaving cycle limit still have recycle value as half-barrel planters.

Handling of the winery's paper and fiberboard waste becomes a problem of mere bulk and storage space. Wineries in the large to very large size are likely to produce enough waste fiberboard to justify the installation of a cardboard baler. The balers reduce the bulk volume 25–30% and can be accessorized with either a manual or automatic strapping feature to produce a neat and easily handled bale. A good question to ask baling machinery vendors is how well does the equipment comply with American National Standards Institute (ANSI) Standard Z245.5. Capital costs for balers can range from $10,000 to $15,000 for used baling units to $35,000 to $40,000 for new automatic strapping balers (1996 price levels) (192).

Crusher–compactor machines are also available that consolidate mixed industrial waste with published compaction ratios of up to 50%. This device for facilitating the in-house handling of waste seems to defeat the principles of waste segregation and recycling at the source. If all the compacted and crushed waste could be kept in the combustible category and earmarked for a waste to energy unit, the reduction in volume to reduce the buyer's handling and transport costs could make the unit economically attractive (193).

Wineries can be considered to be a very clean group in the industrial class, with solid waste disposal presenting problems only with sheer volume, as for example, with grape pomace. For medium to small wineries, pomace becomes an internal waste handling task if vineyard soils are available for pomace incorporation. Large to very large wineries are faced with pomace composting options, usually off site. California's Gallo Winery has a pomace composting operation in Fresno, which encompasses some 40 acres (186). A 1991 amendment to California's Integrated Waste Management Act added a degree of bureaucratic precision that was not entirely favored by the wine industry. In summary, the statute states that if a winery composts pomace and utilizes all of the finished product on site, then no solid waste facility permit is required. If the volume of pomace exceeds 2500 cubic yards or any of the pomace is sold, a solid waste facility operator's permit is required (186).

Pomace production can range from about 5 to nearly 17 lbs (2.3 to 7.7 kg) per case of finished wine production (54,187). If stems are included, the total grape waste solids component unit generation values range from 7 to about 19 lbs (3.2 to 8.6 kg) per case of finished wine production, depending on the source of grape material composition data used. Machine harvesting of grapes leaves the stems on the vine with only the excised grapes and a minor amount of leaves, which the air blower unit on the harvester sometimes fails to remove. Therefore, machine harvesting only shifts the stem disposal problem but does not necessarily eliminate it from the solid waste stream.

Napa Valley, California is blessed with a commercial solid waste service company that hauls pomace from many wineries in both Napa and neighboring Sonoma County. The pomace is mixed with limestone and gypsum, composted in wind rows, which are turned mechanically to provide aerobicity, and after 9–2 months, the compost is harvested, stockpiled, and sold to vineyards, landscape contractors, and commercial organic gardeners. Nutrient content is in the 3-3-6 range (nitrogen, phosphorus, potassium) with a near-neutral pH. The optimum moisture content of the pomace for composting is 50–60% (181). A chronic problem with grape pomace composting is the high moisture content, particularly from the white grape species. Wente Brothers winery in Livermore, CA, compost a high percentage of their pomace, but go a step further and incorporate so-called "green waste" into their mix. Lawn cuttings and chipped landscape cuttings not only add to the bulk but fortify the nutrient content as well. The art and science of composting is well described in the literature. A detailed description of the "dos" and "don'ts" in pomace composting are beyond the scope of this book. The following references are particularly good sources for the beginner, as well as the composting practitioner wishing to enrich his/her knowledge:

(a) *On-Farm Composting Handbook.* 1993. G. Rynk, Midwest Plan Service, Iowa State University, Ames, IA.
(b) *Integrated Solid Waste Management.* 1993, G. Tchobanoglous, H. Theisen, and S. Vigil, McGraw-Hill, Inc., New York.
(c) *Compost Engineering: Principles and Practices.* 1980. R.T. Haug, Ann Arbor Science Publ., Ann Arbor, MI.

Vineyard Solid Waste

Vineyard managers are well aware of the prescribed disposal methods for empty pesticide and herbicide bags and containers. Under California practice, the local county Farm Bureau receives a copy of the sale of any

EPA-registered agricultural chemical from the chemical supply whole-saler/retailer. The Farm Bureau, in turn, tracks the containers until they receive the correct disposal at a licensed hazardous waste facility. Organically farmed vineyards are spared this tightly controlled solid waste disposal practice. The non-organic vineyard is also intimately tied, in U.S. practice at least, to the Pesticide Control Adviser (PCA) who recommends and monitors agriculture chemicals and their designed safe application rates. PCAs also ensure that the vineyard workers have access to and wear the recommended protective clothing and breathing apparatus.

Specialty fertilizers and soil amendments such as soil sulfur and gypsum are generally bagged products. The packaging residuals are empty kraft paper sacks, often with an inner plastic-film liner to prevent moisture accretions during storage. If the bags are split open and the plastic liner is removed, the kraft paper portion can be recycled with other brown paper waste from the winery. The plastic liner would be contaminated with product and should be disposed of as unrecyclable waste material. As the bagged material is usually palletized, the pallets are either returned to the agricultural material supplier or stacked and stored until a sufficient number are accumulated to attract a pallet recycler.

Contaminated diesel fuel, oily rags, waste oil, lubricants, and batteries must be disposed of through a licensed hazardous waste services company, as described previously under the subsection entitled "Winery Solid Waste."

Protective face masks, respirators with replaceable vapor/dust cartridges that are worn by vineyard personnel, should be placed in plastic bag for disposal as hazardous waste. If time is taken to remove the filter media from the plastic or metal cartridge casing, only the media would require special handling and the cartridge casing could be recycled as scrap metal or plastic.

For machine-harvested grapes, there is no residual vegetative waste to cope with, as the stems remain on the vine until incised during the vine pruning cycle. They become part of the pruned cane mass that can be:

- Chipped or hogged for biomass waste to energy fuel systems or earmarked for green landscape waste composting systems
- Burned in the field:
  - in loose piles at the end of vine row
  - in commercially available, tractor-drawn, wheeled incinerators (propane gas fired)
  - for small vineyard operations, recycled 55-gal (208 L) drums can be adapted for transport and used as burn containers for the collected canes and other vineyard cuttings (see Fig. 11.3)

Fig. 11.3.   Portable vineyard prunings transport and field incinerator in Burgundy, France
(Photo reproduced courtesy of *Practical Winery and Vineyard*. Photo by Joseph Drouhin,
Beanne, France.)

The current practice in the phylloxera-ravaged California vineyards is to
remove the vines and consolidate them in piles for field burning. This
practice allows for the destruction of large masses of cellulose along with
other vine disease vectors, including the destructive root louse, phylloxera.
Metal vineyard stakes are replaced at the time of replanting to accommo-

date the closer vine spacing favored by most contemporary viticulturists. Pulling the stakes with chains and backhoe usually render them unusable, but they still represent a considerable tonnage of scrap metal for recycling.

New vineyard plantings have their share of waste. Bench grafts, which are shipped from the nurseries in special growing containers, are planted with their biodegradable container intact. Rootstock bud stick bundles are usually packed in 4-ft × 4-ft × 2-ft fruit bins or strapped to pallets in twine-wrapped packets of 500 or more. The insignificant amount of waste twine is easily disposed of and the fruit bins and pallets, with a probable cash deposit surcharge, will be returned to the vineyard nursery.

Miscellaneous small quantities of plastic tie-tape and other cane control devices are usually left in the vineyard to decompose. The tie-tape is designed to resist weathering and ultraviolet ray damage and usually remains near the soil surface for an extended period of time. Small sections of drip hose and broken emitters are often left in the vineyard by the field crews, unless the owner or vineyard manager is meticulous about the appearance of the vineyard.

Bird netting is too expensive to receive only a single season use; therefore, it is rolled into convenient bundles for storage and several seasons of reuse until the mere handling and placement of the nets eventually produces tears and cuts.

Portable toilets and wash stations for field crews can be served by a portable toilet leasing company, which sees to the maintenance and paper supply needs of the unit. The disposal of used paper towels is left to the winery. Because of possible blood contamination of the wash station and bathroom paper waste, the materials should not enter the recycling waste stream, but should be bagged in plastic garbage bags for disposal in a landfill, along with any food waste that is generated in the vineyard by the field crew.

Figure 11.4 is a solid waste balance for a typical vineyard, showing vineyard inputs and the nature and disposition of outputs.

WINERY WASTE GENERATION RATE ESTIMATES

For planning purposes, it is essential for winery developers to have some idea of the volumes of waste material that have recycling potential. Storage bins and space set-asides can be more precisely dimensioned and located if waste generation values are used. The data reported herein were based on the operation of three large wineries located in an upper Napa Valley, CA municipal jurisdiction that conveniently contained within its boundaries the only industries known to exist. The raw industrial waste generation data

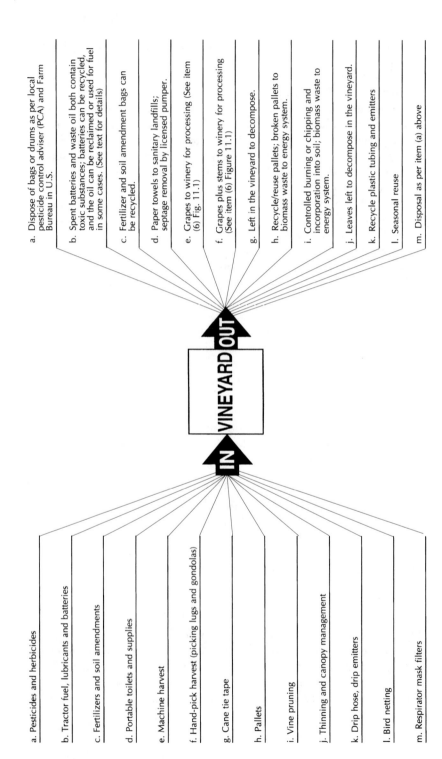

a. Pesticides and herbicides

b. Tractor fuel, lubricants and batteries

c. Fertilizers and soil amendments

d. Portable toilets and supplies

e. Machine harvest

f. Hand-pick harvest (picking lugs and gondolas)

g. Cane tie tape

h. Pallets

i. Vine pruning

j. Thinning and canopy management

k. Drip hose, drip emitters

l. Bird netting

m. Respirator mask filters

a. Dispose of bags or drums as per local pesticide control adviser (PCA) and Farm Bureau in U.S.

b. Spent batteries and waste oil both contain toxic substances; batteries can be recycled, and the oil can be reclaimed or used for fuel in some cases. (See text for details)

c. Fertilizer and soil amendment bags can be recycled.

d. Paper towels to sanitary landfills; septage removal by licensed pumper.

e. Grapes to winery for processing (See item (6) Fig. 11.1)

f. Grapes plus stems to winery for processing (See item (6) Figure 11.1)

g. Left in the vineyard to decompose.

h. Recycle/reuse pallets; broken pallets to biomass waste to energy system.

i. Controlled burning or chipping and incorporation into soil; biomass waste to energy system.

j. Leaves left to decompose in the vineyard.

k. Recycle plastic tubing and emitters

l. Seasonal reuse

m. Disposal as per item (a) above

Fig. 11.4. Solid waste balance for vineyard operations.

reported in Ref. 187 was adjusted to an annual unit production per case of finished wine. Table 11.3 reports the estimated unit generation values for seven categories of winery waste material.

Not included in Table 11.3 is a category for so-called "inert" wastes. Inerts are generally construction debris such as rock, soil, concrete, or asphalt pavement. Inert waste production would, for a table winery, be an infrequent event that would produce inert waste in proportion to the scale and scope of the winery project [i.e., addition of building space, expansion of winery site utilities, or maintenance and repair of winery infrastructure (roads, storm drains, etc.)].

Thus, for a 50, 000-case annual production winery, approximately 390, 000 lbs (177 Tm) of recyclable solid waste could be generated annually. Pomace could add an additional annual mass of approximately 800, 000 lbs. (364 Tm) Fortunately, the appropriate frequency of pickup for segregated recyclable solid waste will allow the storage space for materials to remain relatively modest. Part of the preparation of the solid waste management model must include searching in the immediate area of the winery for a company that is willing to work with the winery on a long-term contractual basis to pick up and haul the recyclables as well as the waste, which must be disposed of in a landfill or by other approved methods. The segregated hazardous waste from the winery may or may not be handled by the same disposal contractor. Exploring the issue with the waste contractor at the time of contract negotiations will reveal whether the proper hazardous waste licenses for hauling and disposal are held.

Planning for segregation of hazardous solid waste has been previously

Table 11.3. Estimates of Unit Solid Waste Generation for Three California Wineries

Waste category	Unit waste generation values (pounds per case of finished wine production per year)
Corrugated cardboard (aka: fiberboard)	1.6
Mixed and high-grade paper	2.96
Plastic (film plastic and other)	0.63
Glass	0.23
Ferrous Metals	0.50
Wood (pallets, etc.)	1.73
Textiles (rags)	0.15
TOTAL	7.80

Notes: (1) Rates of pomace production were reported in a previous subsection of this chapter.

(2) Pomace from all three wineries included in the study database is composted by a licensed solid waste disposal company.

discussed, with suggestions for receptacles to be strategically located within the winery and at vehicle and vineyard equipment maintenance stations. The centralized storage for hazardous wastes needs to be more carefully designed and placed. Figure 11.5 is a concept plan for a centralized hazardous waste storage area. Modifying the dimensions and geometry of the storage pad is the province of the winery architect in concert with the winery general manager. Security fencing, landscape screening, and even a simple roof structure can simplify the spill containment features and storm water handling system.

SOLID WASTE CONTAINERS AND EQUIPMENT

The American National Standards Institute has among its many roles as a certifier of manufactured products, the task of publishing a standard for the safety of waste containers (ANSI Z245.30, ''Waste Container Safety Requirements'') (194). Wineries should not concern themselves with the functional appropriateness of the containers that are leased or purchased but only that they are compatible with the solid waste contractor's trucks that will perform the pickups. The winery is also responsible to see that the containers are not loaded to exceed the rated capacity. Containers larger than 1.0 cubic yard must have warning signs to caution children to neither play near the container nor to occupy the inside at any time. There have apparently been a large number of deaths and serious injuries to people who have either entered the waste container to salvage recyclables, children seeking lost balls, or homeless people looking for a place to sleep. Locking containers is a method of preventing accidents that the solid waste industry would just as soon avoid. Unlocking and locking a container would double the time it takes to service a container (195). White reflective tape must be placed on the ends and long sides of the containers to warn vehicles at night of the container and its location (194).

Figure 11.6 depicts the four common types of waste containers, which are described by their direction for pickup and loading into the refuse collection truck.

Containerization of waste does not provide any insurance that nuisance insect vectors will not be a problem. Flies and yellow jackets are the most troublesome insects, and if food wastes are packaged into garbage bags before loading into the container, both objectionable odors and nuisance insects can be kept under control. If allowed to propagate freely in and around a winery, ants can be a veritable thorn-in-the-side, even for wineries that have an exceptionally well-run and executed winery sanitation program. Ant control around the waste containers again depends on the care-

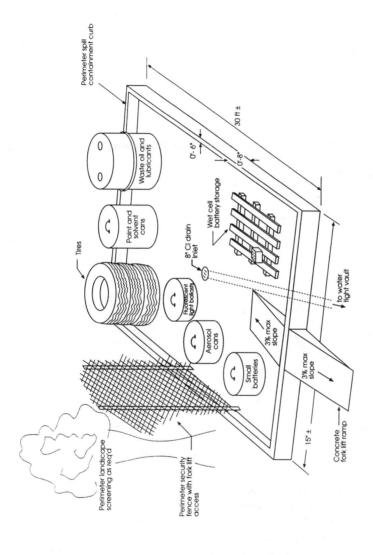

NOTE:

The toxic waste container pad and structure needs to be designed by the winery architect with guidance from the site utilities engineer, so that it can be integrated into the overall winery site plan and design objectives.

Fig. 11.5. Concept design for a typical waste segregation container arrangement for toxic materials generated by winery and vineyard operations.

373

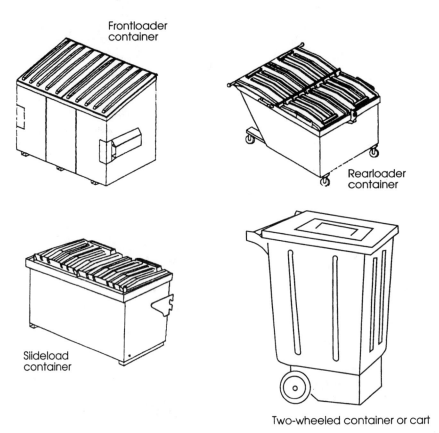

Frontloader container

Rearloader container

Slideload container

Two-wheeled container or cart

NOTE:
Not shown are the large roll-off containers that are used principally for construction debris.

Fig. 11.6.   Solid waste container configurations and nomenclature.

ful packaging of food wastes, which will deny the sought-after food source to the ubiquitous ants. Rodents must also be denied access to the containers, as the plastic bags will not deter rats and mice from food scraps. Tight-fitting covers for the containers are essential to combat rodents. It is the responsibility of the refuse collection company to maintain the containers in good mechanical condition.

Figure 11.7 is an hypothetical winery with points of waste generation indicated. To make segregation of waste by employees work effectively within the winery envelope, separate containers for paper, aluminum cans, plastic, and waste food must be conveniently located and easily serviced by the winery personnel responsible for pickup and delivery to the waste storage area. To optimize the labor and equipment used for waste pickup

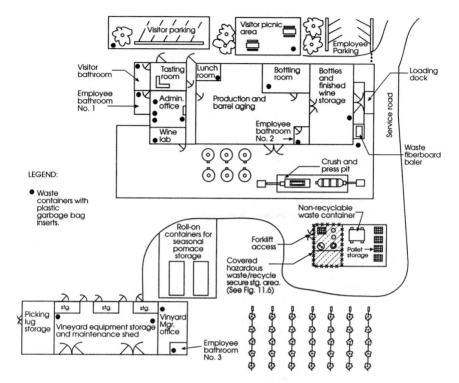

Fig. 11.7. Winery solid waste plan for hypothetical winery complex.

from the illustrative winery and separate vineyard equipment/vineyard manager structure, the waste should be as close to an external exit door as possible to shorten the haul distance to the waste pickup vehicle. If a building maintenance service company is utilized by the winery, the same logic applies, because of the reduced time of pickup for the service company employees and, hence, monthly charges to the winery.

Figure 11.8 is a solid waste origin and destination model based upon the winery configuration and spatial plan depicted in Fig. 11.7. The model as shown is qualitative in function but can be enhanced to provide estimates of waste produced by space categories using unit waste generation values previously reported in Table 11.3. The development of a set of waste production values from this analysis can provide the owners and the winery's design team with a database on which waste disposal and recycling contractors can more accurately estimate the costs to provide the pickup and disposal services. Also, with competition increasing yearly among the new breed of waste recycling companies, the waste generation data can provide an excellent framework for soliciting a formal proposal from a number of

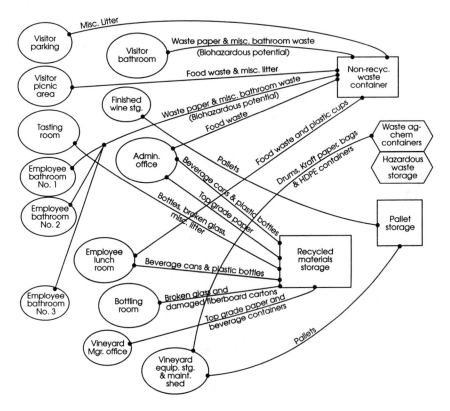

NOTES:
1. This model can produce a quantitative algorithm. If the unit solid waste generation values reported in Table 11.3 are proportioned among the various sources.
2. See Figure 11.7 for winery special arrangement that corresponds to this model.

Fig. 11.8.   Winery solid waste management model: Source and destination matrix.

solid waste disposal companies, which could bid competitively for the winery's solid waste products.

SOLID WASTE SYSTEM OPERATION AND MAINTENANCE

Solid waste systems for wineries have a much greater emphasis on "operations" than on maintenance than any of the utility systems described previously in any of the 10 chapters of this book.

Waste containers are virtually indestructible and with the exception of small containers within the winery, the maintenance and safety inspections of waste containers are the responsibility of the solid waste service contractor. Control of waste-storage-associated insects and rodents is a major maintenance task previously described. Solid waste storage and picnic areas must be continually policed to see that random litter does not escape and become dispersed on the winery grounds.

The single most important piece of machinery associated with the winery's solid waste management program is the fiberboard baler, if the winery size and scale of operations justify the purchase of such a machine. Balers are no different than any other piece of electrical–mechanical–hydraulic equipment. Winery maintenance personnel, skilled in the necessary trades, are usually found only in large to very large wineries. Thus, the medium to small winery is forced to use a service/maintenance contract, usually from the company that supplied the new or used baling machine.

Employee safety is of paramount importance in any job tasks related to the handling of solid waste, both benign and hazardous. Proper equipment including gloves, goggles, and disposable coveralls should be the minimum uniform for in-winery waste handlers. Any employee accidents or injuries resulting from solid waste management activities should receive proper first aid and/or medical treatment. Employee safety and health issues, as seen from the winery utility system perspective, are discussed in detail in Chapter 12.

WINERIES AND ADOPT-A-HIGHWAY PROGRAMS

Almost all states of the United States have created adopt-a-highway programs that allow a winery, fraternal organization, or church group to maintain a section of interstate highway or secondary road in return for a sign listing the name of the responsible group (196). For wineries, the program has the advantage of getting name recognition on a well-traveled thoroughfare, where federal/state highway beautification statutes often prohibit commercial signage of any type. The author's winery has adopted a 2-mile section of a nearby interstate highway from which discarded solid waste from vehicles fills up 50–60, 1-cubic-yard capacity bags 3–4 times yearly.

The safety and waste handling procedures for waste-pickup along a freeway are strictly enforced, with the state of California providing gloves, orange vests, hard hats, and waste storage bags. Along with saving the state large sums of money in litter removal, the event can be turned into a semisocial activity for the winery, with lunch and beverages provided for the environmentally conscious winery staff who donate 2–3 h of work on a day convenient to all. The orange safety vests are also emblazoned with the winery name, which the freeway drivers acknowledge with a friendly wave or honk of the horn. The additional recognition for a winery doing their share to make the roadways of the world less unsightly does not go unnoticed to the general public.

GLOSSARY

Collection (waste)   The act of picking of wastes at homes, businesses, commercial, and industrial plants and other locations, loading them into a collection vehicle (usually enclosed), and hauling them to a facility for further processing or transfer or to a disposal site.

Collection routes   The established routes followed in the collection of commingled and source-separated wastes from homes, businesses, commercial, and industrial plants and other locations.

Collection systems   Collectors and equipment used for the collection of commingled and source-separated waste. Waste collection systems may be classified from several points of view, such as the mode of operation, the equipment used, and the types of wastes collected. In this text, collection systems have been classified according to their mode of operation into two categories: hauled container systems and stationary container systems.

Combustible materials   Various materials in the waste stream that are combustible. In general, they are organic in nature (e.g., food waste, paper, cardboard, plastics, yard wastes).

Commingled recyclables   A mixture of several recyclable materials in one container.

Commingled waste   Mixture of all waste components in one container.

Compactor   Any power-driven mechanical equipment designed to compress and thereby reduce the volume of wastes.

Compactor collection vehicle   A large vehicle with an enclosed body having special hydraulically driven equipment for loading, compressing, and distributing the waste load.

Component separation   The separation or sorting of wastes into components or categories.

Compost   A mixture of organic wastes partially decomposed by aerobic and/or anaerobic bacteria to an intermediate state. Compost can be used as a soil conditioner.

Composting   The controlled biological decomposition of organic solid waste materials under aerobic conditions. Composting can be accomplished in windrows, static piles, and enclosed vessels (known as in-vessel composting).

Construction wastes   Wastes produced in the course of construction of homes, office buildings, dams, industrial plants, schools, and other structures. The materials usually include used lumber, miscellaneous metal parts, packaging materials, cans, boxes, wire, excess sheet metal, and other materials. Construction and demolition wastes are usually grouped together. (Sometimes called "inerts" or demolition waste.)

Container   A receptacle used for the storage of solid wastes until they are collected.

Disposal   The activities associated with the long-term handling of (1) solid wastes that are collected and of no further use and (2) the residual matter after solid wastes have been processed and the recovery of conversion products or energy has been accomplished. Normally, disposal is accomplished by means of sanitary landfilling.

Diversion rate   A measure of the amount of material now being diverted for reuse and recycling compared to the total amount of waste that is thrown away previously.

Drop-off center   A location where residents or businesses bring source-separate recyclable materials. Drop-off centers range from single-material collection points (e.g., easy access "igloo" containers) to staffed, multimaterial collection centers.

Energy recovery   The process of recovering energy from the conversion products derived from solid wastes, such as the heat produced from the burning of solid wastes.

Ferrous metals   Metals composed predominantly of iron. In the waste materials stream, these metals usually include tin cans, automobiles, refrigerators, stoves, and other appliances.

Fiberboard compactors   Hydraulic or mechanical devices that compact bale and strap waste fibreboard.

Green waste   Genetic term for landscape and agricultural waste that has a high potential for composting and reuse as a soil conditioner.

Hazardous wastes   Wastes that by their nature may pose a threat to human health or the environment, the handling and disposal of which is regulated by federal law. Hazardous wastes include radioactive substances, toxic chemicals, biological wastes, flammable wastes, and explosives.

Heavy metals   Metals such as cadmium, lead, and mercury which may be found in municipal solid waste MSW in discarded items such as batteries, lighting fixtures, colorants, and inks.

High-density polyethylene (HDPE)   One of the more desirable plastic postconsumer products that has a relatively stable market for recycling (milk bottles and lubricating oil containers).

Industrial wastes   Wastes generally discarded from industrial operations or derived from manufacturing processes. A distinction should be made between scrap (those materials that can be recycled at a profit) and solid wastes (those that are beyond the reach of economical reclamation).

Integrated solid waste management   The management of solid waste based on a consideration of source reduction, recycling, waste transformation, and disposal arranged in a hierarchical order. The purposeful, systematic control of the functional elements of generation; waste handling, separation, and processing at the source; collection; separation and processing and transformation of solid waste; transfer and transport; and disposal associated with the management of solid wastes from the point of generation to final disposal.

PET   Clear polyethylene containers (juice and soft drinks) that have a high potential for recycling.

Reclamation   The restoration to a better or useful state, such as land reclamation by sanitary landfilling or the extraction of useful materials from solid wastes.

Recoverable resources   Materials that still have useful physical or chemical properties after serving a specific purpose and can, therefore, be reused or recycled for the same or other purposes.

Recycling   Separating a given waste material (e.g., glass) from the waste stream and processing it so that it may be used again as a useful material for products which may or may not be similar to the original.

Refuse   A term often used interchangeably with the term *solid waste.*

Resource recovery   General term used to describe the extraction of economically usable materials or energy from wastes. The concept may involve recycling or conversion into different and sometimes unrelated uses.

Reuse   The use of a waste material or product more than once.

Rubbish   A general term for solid wastes—excluding food wastes and ashes—taken from residences, commercial/industrial establishments, and institutions.

Sanitary landfill   An engineered method of disposing of solid wastes on land in a manner that protects human health and the environment. Waste is spread in thin layers, compacted to the smallest practical volume, and covered with soil or other suitable material at the end of each working day.

Separation   To divide wastes into groups of similar materials, such as paper products, glass, food wastes, and metals. Also used to describe the further sorting of materials into more specific categories, such as clear glass and dark glass. Separation may be done manually or mechanically with specialized equipment.

Shredding   Mechanical operation used to reduce the size of solid wastes.

Solid wastes   Any of a wide variety of solid materials, as well as some liquids in containers, which are discarded or rejected as being spent, useless, worthless, or in excess. Does not usually include waste solids from wastewater treatment facilities.

Source reduction   The design, manufacture, acquisition, and reuse of materials so as to minimize the quantity or toxicity of the waste generated.

Source-separated materials   Waste materials that have been separated at the point of generation. Source-separated materials are normally collected separately.

Source separation   The separation of waste materials from other commingled wastes at the point of generation.

Special wastes   Special wastes include bulky items, consumer electronics, white goods, yard wastes that are collected separately, batteries, oil, and tires. Special wastes are usually handled separately from other residential and commercial wastes.

White goods   Large unusable household, commercial, and industrial appliances, such as stoves, refrigerators, dishwashers, and clothes washers and dryers.

REFERENCES

181. TCHOBANOGLOUS, G., THEISEN, H., and VIGIL, S. 1993. *Integrated Solid Waste Management.* New York: McGraw-Hill.

182. DROBNY, N.L., HULL, H.E., and TESTIN, R.F., 1971. *Recovery and Utilization of Municipal Solid Waste.* Washington, DC: U.S. EPA. Publ. No. SW-10C.

183. RAY, G. 1995. *Managing Used Oil.* Cahnevs Publ. Oakpark, IL. Plant Services, pp. 94–96.

184. STORM, D.R. 1992. Landfills as an endangered species. Winery Water and Waste. *Pract. Winery Vineyard.* 12(5).

185. SCHMIDLE, C. 1996. Personal communication. Sacramento, CA: California Integrated Waste Management Board.

186. NAKITA, W. 1996. Personal communication. Delicato Vineyards.

187. _____. 1995. *Updated Waste Generation Study and SRRE Diversion Projections for Upper Valley Waste Management Agency, City of American Canyon, and Remaining Unincorporated Napa County Areas.*

188. BRADSHAW, A.D., SOUTHWOOD, R., and WARNER, F. 1992. *The Treatment and Handling of Wastes.* London: Chapman & Hall.

189. _____. 1994. *Guidelines for Recycling Waste Materials in California.* Sacramento, CA: California Integrated Waste Management Board, p. 5.

190. DONG, M. and McCAGG, B. 1995. *A Practical Guide to Ballast Disposal.* Bronx, NY: Full Circle Ballast Recyclers, pp. 1–33.

191. SIKORA, M. 1990. A little retreading goes a lot of miles. *Resource Recycling.* 9(1): 50–58.

192. _____. 1995. Personal communication, Case Compactors, Inc.

193. _____. 1995. *Waste Reduction Technology Services,* Cahners Publ., Oakpark, IL. Plant Services, p. 56.

194. CURTIS, R.B. 1994. New safety standards for waste containers. *Waste Age.* 73–78.

195. MALLOY, M.G. 1994. Locks or checks: Worth the trouble. *Waste Age.* 67–70.

196. STORM, D.R. 1995. One on one with highway trash. Winery Water and Waste. *Pract. Winery Vineyard.* 80–81.

CHAPTER 12

WINERY UTILITIES AND HEALTH AND SAFETY PROGRAMS

INTRODUCTION

A winery's health and safety program interacts with all levels of wine production and viticultural activities, including the maintenance and operation of its associated utility systems. In this chapter, the principal focus will be on the relationships of the utility systems and subsystems that have been previously detailed in this text and how they influence the safety and health program and overall winery risk. New technology generally brings with it the promise of easier methods, faster production times to finished, end products and hopefully lower production costs. Often with the new technology are new risks to be reckoned with for both managers and employees alike. It is therefore profoundly important that wineries have a program framework into which the old as well as the new systems and their associated risks can be integrated and in-winery rules and regulations developed to protect the winery's most valuable resource—its employees.

As with most specialty disciplines, professional safety engineers can now be licensed by state boards of registration for engineers. For small to medium wineries, the cost to have a safety engineer and/or an industrial hygienist prepare a plan is probably not justified. If the winery is blessed with an assistant winemaker or production manager with good writing and organizational skills, a reasonably good plan can probably be created that would serve both statutory requirements and act as a working document for in-winery health and safety program familiarization and employee training programs. Negotiating a contract with a safety engineer and/or industrial hygienist to merely review the winery's draft plan would, for a very little expenditure of operating funds, produce a more streamlined document which precisely fits the winery's peculiar set of geographical circumstances and potential risk factors.

Because top management decisions for wineries must be equated to profit and loss, there is little incentive to create programs to protect employees, unless there are tangible benefits (excluding pure humanitarianism). If, for example, a reduction in wine production and vineyard-related accidents can be estimated and eventually proven because of the success of the health and safety program, by means of statistical data analysis, then the Workers' Compensation insurance and/or Health Maintenance Organization premiums can be reduced, the plan becomes an "easy sell." Accountants may carry the injury accident and illness costs as a direct cost, so the data are easily factored out of a winery's books. In the last 10 years, Workers' Compensation premiums for wineries amount to about 2–3% of the total winery payroll. Although the direct costs of reduced injuries and illnesses can be calculated, it is another matter to place a "hard number" on the indirect costs of accident/illness events. The National Safety Council's "Accident Prevention Manual for Business and Industrial Operations," 1992 edition, lists a number of indirect costs elements that would be inversely related to a reduction in the frequency and number of winery illness and accident occurrences. In other words, the indirect cost categories suggested in the following summary would be winery expenses foregone. Other indirect cost elements could be added to the list to fit special circumstances of winery scale and position in the marketplace.

- Wages lost by workers who were not injured, who may have stopped working to assist the injured worker, or whose job(s) were negatively affected because of equipment damage associated with an accident (overturned tractor, failure of grape gondola tailer hitch, etc.)
- Repair of damaged equipment (replacement cost less accrued depreciation, if no salvage value; must agree with value carried by the accountant in the winery's books)
- Wages loss by sick or injured employee that is paid by the winery in addition to Workers' Compensation benefits
- Cost of overtime wages to compensate for disruption of normal wine production or vineyard task

- Extra cost to winery supervisors to train and monitor replacement employee
- Uninsured medical costs paid by the winery
- Loss in work output for period of time after worker returns

HUMAN BEHAVIOR AND HIGH-RISK JOB ACTIVITIES

Wineries provide a large number of opportunities for unsafe, work-associated acts. Worldwide, often the people most admired are those in very risky and life-threatening jobs. Whether it is for recognition, status, or peer esteem, winery cellar and vineyard workers will often place themselves in serious jeopardy performing jobs that require the assistance of a fellow worker or a simple risk assessment to judge quickly what the consequences of a certain act might be. It may be expedient to accomplish the task in the most rapid manner with disregard for danger or safety. It is the lack of knowledge about specific job hazards that becomes the cornerstone of the winery's safety and health program. The knowledge can come from training programs that the winery health and safety manager designs and executes.

An industry-by-industry analysis of accidents and job-associated sickness is not normally the way in which the statistical data are arranged. The labor statistic base that is closest to the industry's category would be "agriculture." Using the State of California's Compensation Insurance Fund published data, the following tabulation was prepared, which gives an indication of the nature of agricultural accidents during the period 1991–1993 (197):

Category	Percent
Motor vehicle	3.9
Machine or tools	8.4
Lifting and strains	32.5
Slipping and falling	20.8
Cut or struck by an object	15.2
Other*	19.2
Total	100.0

*"Other" includes cumulative electrical and eye injuries, occupational disease, as well as bites, stings, and plant poisons.

As simple a procedure as lifting a field-crew drinking water container [at approximately 42 lbs (19kg)] into the back of a pickup truck has enormous potential for a lower back strain. Overcoming the male ego is a great challenge for the safety training officer. In the real-life situation, the field hand lifting the water container, would probably not consider asking a fellow worker for assistance, fearing a loss of stature or demonstration of weakness in front of his peers.

Next to creating an award-winning Merlot or Chardonnay, the most demanding job in the winery may be the creation and implementation of a successful health and safety plan, that produces tangible results.

WORKERS' COMPENSATION LAW

This chapter would not be complete without a focused discussion of the 50-state requirements for accident insurance coverage for all winery workers, regardless of winery size.

Almost all U.S. model laws administered by the states that are designed to protect injured workers have a limited liability provision for employers, unless gross employer negligence can be shown. This provision protects the winery from enormous accident claims, but it does not find great favor with organized labor groups and employee bargaining associations. Because the winery does not actually disburse the injury compensation payments [the Workers' Compensation Law (WCL) insurance underwriter makes the disbursements from its claims funds], the insurance company is keenly interested in the scope and quality of the winery's health and safety program.

"Self-insuring" was often an option voiced by a wine industry that pays a substantial overhead cost for general business liability coverage. In a lower-accident-risk industry, group self-insurance might succeed. The need for self-insurance coverage administration is responsible for spawning a new professional called "a loss control representative" (198) whose job it is to filter out the fraudulent from the legitimate claims and to determine if the accident occurred at the winery or off the job. The loss control representative must also fill the insurance carrier void by assisting the winery in its preparation of the health and safety document and to ensure that employee accident prevention training is vigorously pursued.

An important aspect of winery accident and sickness incidents is the necessity of accurate record keeping. In early 1980, the federal Occupational Safety and Health Administration promulgated record-keeping requirements. If the winery is not self-insured, the Workers' Compensation entity in the state in which the winery is situated will ensure that the necessary claims and accident reports are provided and that the reports are factual and contain all the necessary statistical information on the nature of and circumstances surrounding the injury. The winery's health and safety officer should become familiar with OSHA forms, No. 200, "Log of Occupational Injuries and Illnesses," and No. 101, "Supplementary Record of Occupational Injuries and Illnesses." Form 101 is a record of a more detailed account of the injury or illness suffered by the employee.

WINERY OPERATIONS AND AIR RESOURCES

The problem of air quality and winery operations has three components. First and foremost is the indoor air quality (IAQ) issue, which was discussed briefly in Chapters 1 and 5. Second and more precisely, a subcategory of IAQ is confined space air quality [tanks, vats, boilers, ducts, sewers (including manholes), vaults and pits]. Third are the fugitive emissions which enter the earth's atmosphere as the products of fermentation, boiler exhaust, and halogenated compounds (chlorine, bromine and fluorocarbon refrigerants). The latter group includes ethyl alcohol in the vapor phase, SO_2, CO_2, CO, and other minor constituents that escape during the fermentation process. Production gas emissions will be discussed under the subsection entitled "Indoor Air Quality." Formaldehyde is also a component of the grape fermentation process (199–201), but is apparently emitted at such low concentrations as to present no hazard to winery workers or atmospheric resources.

Ethanol emissions from wineries in critical air basin portions of California came under serious scrutiny in the early 1980s, as there was a rather strong belief that ethyl alcohol vapors escaping along with CO_2 and other fermentation products was providing a significant source of photochemical reactive hydrocarbons, which would contribute to the creation of additional levels of smog and a concurrent seasonal degradation in air quality when fermentation activity in wineries was at its peak (202). The Wine Institute and its member wineries expended a considerable amount of research dollars to disprove the San Joaquin Valley Regional Air Pollution Control district's contention. To date, no fugitive ethanol vapor recovery systems have been mandated nor is it probable to see the issue being raised again seriously in the near future.

Utility Systems as Air Pollution Sources

Utility systems have the potential to produce emissions that can contribute to the pollutant loading in the air basin that is tributary to the winery. Table 12.1 lists the specific utility system and the potential air pollutants and their possible environmental or health consequences.

Indoor Air Quality (IAQ)

Control of building air quality in wineries is no different in principle, at least, with maintaining a healthy work space in any industrial structure. In Chapter 5, reference was made to the current U.S. standard for ventilation

Table 12.1. Winery Utility Systems and Potential Air Pollutants

Chapter	System	Accidental or incidental emissions and consequences
4	Sanitation, steam and hot water	Steam boiler stack emissions of particulate matter and oxides of nitrogen (also CO and CO_2); NO_x emissions.
5	Refrigeration, ventilation and air conditioning	Fluorocarbon refrigerant emissions; damage to earth's stratospheric ozone layer. Ammonia as refrigerant is hazardous to humans. Forms oxides of nitrogen and becomes a significant air pollutant.
6	Potable water supply systems	Chlorine gas even in low concentrations is hazardous to health; additional chlorine loading to atmosphere has potential for ozone layer destruction.
9	Liquified petroleum gas systems	Hydrocarbon release provides additional pollutant loading to create photochemical smog.
10	Solid waste systems	Improper handling of spent propellant spray cans or other wastes containing VOCs (volatile organic compounds), creates additional pollutant loading for creation of photochemical smog.

of about 20 cubic feet per minute (cfm) of outside air per person. The rather "open" design of most winery spaces (caves and earth-bermed structures excepted) allows some outside air introduction without the use of mechanical ventilation. Doors, windows, and loading dock closures are common for most conventionally designed winery buildings. What the ventilation system designer must consider is the so-called "worst-case" condition when fermentation tanks are located inside the building and are evolving CO_2 at a rate of 9400 ft^3 (266 m^3) for every 1000 gal (3786L) of 22° Brix fermentable must or juice. Exhaust fans will be able to maintain a safe level of oxygen in the wine production spaces with supply-side air diluting the combined effects of ethanol vapor, SO_2, and CO_2. Outside (fresh) air has a CO_2 concentration in the range of 350–450 ppm. Even with CO_2 levels at 500 ppm to as high as 3000 ppm, no health threat results, but in some individuals noticeable discomfort—feelings of stuffiness, lethargy, inattention, and the amplification of unpleasant odors—are reported (203). Gas detectors are now available that can be located at critical locations in the winery production spaces and barrel-aging rooms to provide a signal to a master programmed linear controller (PLC) or ventilation system microprocessor when oxygen levels reach a critically low stage or CO_2 levels reach a critically high stage. Such a system can optimize power consumption and save valuable overhead dollars by preventing over-ventilation. Ventilation systems of this type are called Demand-Controlled

Ventilation Units (DCVU). Combining the sensor-controlled principles of the DCVU with the night-air-cooling concept, described previously in Chapter 5, would create a dual-use system that could show a rapid investment write-off. Personal, belt-attached, battery-powered gas detectors are also an inexpensive method of preventing accidents in tanks, where cellar workers must enter and perform work tasks such as pomace removal and tank cleaning. (Such devices are an indispensable monitoring device when a winery is establishing a program of compliance for Confined Space Regulations, which will be discussed in detail in later.) One of the belt-clip units that is currently available commercially weighs about 5 oz (142 g) and can be equipped to sense and provide an audiovisual alarm for the following:

- Oxygen (activates at 19.5% O_2)
- Carbon monoxide (activates at 35 ppm)
- Hydrogen sulfide (activates at 10 ppm)
- Chlorine (activates at 0.5 ppm)
- Ammonia (activates at 25 ppm)
- Sulfur dioxide (activates at 2 ppm)
- Four other gases (factory calibration and alarm limits set for each gas)

The lithium battery has a 30-month operational life in the no-alarm mode (204). Unit price (1996 cost levels) is about $450 (U.S.).

Dust is an IAQ problem for most wineries in Mediterranean climates, with grape harvest coinciding with a dry fall season. The winery's ventilation engineer can propose filters on the ventilation supply to trap a high percentage of dust and plant pollens. Good housekeeping and overall winery sanitation are made simpler if fugitive dust can be controlled to some acceptable level.

Compressed gas storage safety is generally a matter of common sense, tempered by Uniform Fire Code (UFC) regulations that provide a more precise definition of safe storage of gas cylinders. Nitrogen, carbon dioxide, and sulfur dioxide are the gases of choice in most wineries for use as biocides/wine-preserving agents, sparging agents, or for creating inert gas blankets in wine storage tanks with potentially oxidizing headspace. Oxyacetylene welding tanks are also often part of the maintenance equipment inventory for the vineyard and winery. These tanks of compressed gases fall under the same UFC safety regulations as the inert gas cylinders used in wine production.

Some of the pertinent UFC-mandated compressed gas cylinder handling and safety precautions that are often overlooked by wineries are as follows:

Section of UFC, 1994 edition	Summary statement of compressed gas safety provisions
7401.5.3	Permanently installed piping shall state contents of pipe, direction of flow at each valve, wall, floor, and ceiling penetration, and at each change in piping direction not to be less than 20-ft spacing for the entire pipe run.
7401.6.3	Gas cylinders and associated piping that are vulnerable to damage from winery equipment operations (pallet jacks, forklifts, etc.) shall be protected by pipe bollards or other means.
7401.6.4	Gas cylinders shall be secured from falling from vibration, seismic event, or other occurrence in a manner acceptable to the local fire jurisdiction.
7401.7.2	Protective caps shall be installed and in place except when the gas cylinders are being serviced or filled.
7401.8.3	No combustible material shall be placed within 10 ft of compressed gas cylinders.
7401.8.5	No gas cylinder shall be exposed to temperatures exceeding 125°F.
7402.1.2	All compressed gas cylinders shall be stored in the upright or valve-end-up position.

Other liquified gases, such as liquified petroleum gas (LPG), that are part of the winery equipment inventory and contained customarily in 10-gal capacity fuel tanks for forklifts are stored under lower pressures and present a different kind of fire/explosion hazard than the inert gases. LPG safety was discussed previously in Chapter 9 and is dealt with in the UFC in Article 82, "Liquified Petroleum Gases." The Material Safety Data Sheets (MSDS) that were described previously in Chapter 11, and attached as an illustrative example as Table 11.2, are an important part of the winery's safety program as related to industrial compressed gases. Just posting them on the employee information and labor law bulletin board is not sufficient. Every employee should be familiar with the compressed gas hazards and be ready to act quickly and appropriately in the event of an accident.

Respiratory Protection

In wineries or other industrial activities, it is impossible to totally eliminate air contaminants. Employees must be able to discriminate between an acceptable and a hazardous situation, and take appropriate precautions. The extremes of those situations would be a chlorine gas or ammonia gas leak, both of which require only very low levels of concentration to present a health hazard. [The OSHA standard short-term exposure limit (STEL) is the method by which allowable health hazard risks are established.] Chlorine has an STEL of 10 ppm and ammonia 35 ppm. Respiratory protection must be donned immediately in the case of chlorine, and as soon as ammonia is detected by the human nose or about 1 ppm. Thus, for the mask protection to be effective, the travel distance to a properly maintained mask or masks must be less than the time interval between detection and the time it takes to locate and put on the mask. Winery employees with past military experience will excel in donning protective masks, as their collective training, regardless of branch of service, will have included learning to remove, put on, and adjust a mask for life-threatening chemical or biological agents, usually in the presence of tear gas, just to make the experience memorable.

There are a number of issues which arise that make the use of mask-type respirators by winery employees more of a "custom fit" than just standardizing purchases to fit the everyman/everywoman. Eyeglasses or have facial hair create a mask-sealing problem that needs to be addressed by the winery's health and safety officer. Regular, noncontact-type eyeglasses and full facepiece respirators are incompatible. The temple bars on the glasses prevent the sealing surface of the mask from creating a bottle-tight fit. The same holds true for beards. The eyeglass problem is solvable; the winery need only request that the mask be equipped with an accessory frame that holds the eyeglasses in place without the temple pieces. The beard problem can be overcome without forcing the employee to transform his facial appearance to clean shaven. Masks can be obtained that are either full-face (covers mouth, nose and eyes), one-half-face (covers nose and mouth), or a mouthpiece respirator that covers the mouth only (this type requires the use of a nose clip to ensure that breathing is exclusively through the mouth). Depending on the extent of the facial hair, one of the alternative mask types can be found to make an adequate skin-to-mask seal.

Contact lenses were once a problem because of the likelihood of a lens becoming dislodged. The wearer might remove the mask in a hazardous atmosphere or in the anxiety to replace the lens, contaminate it, so that the eye would sustain injury. A federal OSHA (Occupational Safety and Health Administration) regulation was originally issued in 1970 which detailed the several responsibilities of employers and employees regarding airborne hazardous substances. The rule was entitled "Respiratory Protection" and is contained in the OSHA statutes as Part 1910.134. The standard has been revised several times and is currently under extensive revision (205). The current U.S. trend in the administration of health and safety programs is away from the strictly compliance-oriented plans with a movement toward what has been called "Voluntary Protection Program" or VPP (206). Adopting the VPP does not exempt a winery from OSHA compliance, and OSHA and their state equivalents should be in a position to provide guidelines for the establishment of VPP. Terrell (206) makes a strong statement for such programs, with statistical data to support reductions in work loss accidents and illnesses. OSHA encourages the VPP system but does not mandate it.

One final note about the purchase of respiratory equipment, that is tailored for your winery and as it pertains more specifically to the "Confined Space Regulations" will be covered in a subsequent section. Rescue plans that would require someone from the winery staff to enter a space (wine tank, vault, etc.) where an employee has been overcome generally specify that the rescuer be equipped with an oxygen breathing apparatus (OBA), not just an air-purifying respirator. The OBA apparatus is bulky and is strapped to the body in either a backpack or frontpack fashion. If the OBA-equipped rescuer cannot enter the manhole of a large fermenter easily, the OBA chosen is obviously the incorrect piece of equipment. Most suppliers of industrial safety equipment would be alert to that requirement, but it is a good test for all emergency rescue gear that it fits the winery's special conditions of access geometry and dimensions.

All of the respiratory protection training and drills are not going to save one cellar worker from harm in the event of airborne hazardous substance accident if the masks are not properly cleaned, repaired, and stored, and ready 100% of the time. Respirator service companies can maintain the winery's mask inventory; however, the mask is a personal piece of equipment. It is much like the lore associated with parachutes—if you pack your own parachute, you are going to be a lot more confident about it opening when you pull the "D"-ring. The local fire department on their periodic inspections of the winery's fire protection equipment (see Chapter 10) can also provide guidance on mask maintenance, as they are personally equipped to enter spaces that are filled with all descriptions of noxious gases and vapors.

NOISE

Wineries do not have sources that generally produce sound levels that can be damaging to the human ear. Ear damage is the result of both sound intensity and duration, with high frequencies being more damaging than low frequencies. Continuous noise is also more damaging than intermittent noise. There are different tolerances to sound and hearing damage among individuals, so the safe sound level standards are set for the most sensitive individual in a sample population. Table 12.2 lists the permitted industrial noise exposures by sound level and duration.

Of all the health hazards that exist in wineries, noise is one of the easiest to mitigate:

- Treat the noise source
- Obstruct the noise path
- Protect the noise target

Table 12.2.   OSHA-Permissible Noise Exposures

Duration of exposure (h)	Sound level (dBA slow response[a])
8	90
6	92
4	95
3	97
2	100
1.5	102
1	105
0.5	110
0.25 or less	1154
Impulse or impact noise	140 peak level

[a]dBA slow response = decibel average = sound pressure intensity.

Treating the noise source is most easily accomplished by wineries in the planning and design phase by having the equipment manufacturer certify that noise levels will be below a certain level. (Probably below the values reported in Table 12.2, because nuisance noise, although not damaging to the hearing system, is a work distraction that can be avoided.)

Obstruction of the noise path can be an architectural treatment consisting of sound baffles, curtain walls, or acoustical ceiling and wall finishes that attenuate sound. (Acoustical absorbing finishes are less likely to be practical in wine production spaces where sanitation requires that walls be impervious and washable.) Exterior noise sources can be attenuated by the use of landscape barriers and/or physical sound walls (114).

Disposable ear plugs or earmuff-type ear protectors are efficient and inexpensive methods of providing employee protection. The only problem that ear protection introduces is the potential for not being able to hear a warning of a work-associated problem. Baggage handlers at busy airports seem to have overcome this problem by creating an alphabet of hand signals to communicate while working in close proximity to the high-dBA-generating gas turbine engines.

A sound level meter is a "must" piece of equipment for the winery's safety officer to verify sound levels for comparison to OSHA-permitted levels and durations and to have a basis to act or not to act upon employee complaints about excessive noise in the workplace. The bottling room would be a good place to calibrate the instrument and to measure the composite effects of noise from several sources. A good approximation of excessive noise can be determined; if a "test employee" talks in a normal voice with a separation of an arm's length from the tester and the words cannot be comprehended, then the noise level is probably too high (or the tester's hearing is impaired).

CONFINED SPACE REGULATIONS

The entry into confined winery spaces can be hazardous, in particular, wine tanks at the end of fermentation, when cellar workers must enter the tanks to remove pomace or to facilitate cleaning. Under a relatively new California statute, an in-winery permit procedure has been instituted to prevent worker accidents in a most responsible way (207). The regulations encompass winery employees and service personnel or contractors, whose work tasks might require their entry into confined spaces. Not all confined spaces will be classified as permit-required confined spaces (PRCS). Using the wine tank as an illustrative example, the following step process must be followed to comply with the OSHA statute:

	YES	NO
Does the winery contain PRCS as defined by the confined space statute?	☐	☐
Have cellar workers going into tank been infomed of potential hazards?	☐	☐
Will space be entered by employees?	☐	☐
Can task be accomplished from outside the space?	☐	☐
Will an outside contractor enter space?	☐	☐
Does space have known or potential hazards?	☐	☐
Can the hazards be eliminated? *Note:* Forced mechanical ventilation to introduce fresh air and vent CO_2 and ethanol vapors.	☐	☐
Can space be maintained in a safe condition?	☐	☐

If the space could not be rendered safe, then the "permit" process would have to be initiated:

- Monitor and record hazardous conditions in tank
- Issue permit listing conditions in tank, work to be performed, equipment to be used, (signed by three responsible winery supervisors)
- Permit issued and signed by supervisor and space tester (kept at tank site until work is completed)

Confined space entry requirements for sewer systems will be described in a subsequent section entitled "Site Utilities and Health and Safety Planning."

MACHINERY GUARDS, WALKING AND WORKING SURFACES

The second most frequent cause of accidents in the agricultural sector, as reported previously in the section entitled "Human Behavior and High-

Risk Job Activities," is "slipping and falling," representing some 21% of all accidents reported for 1991–1993. Whether this statistic is higher or lower when measured for the wine industry alone is unknown. As wine production is a "wet" agricultural product process, where floors and working surfaces are almost always wet, the potential exists for more slipping and falling accidents than in a flour mill or a walnut drying facility.

Surface irregularities in winery production spaces, both inside and outside the winery building envelope, can be a source of tripping, where the motion of the foot is interrupted during a step. Combine those concrete floor chips with wine transfer hoses (particularly in small wineries where fixed, stainless-steel transfer piping networks are less likely to be found), power cords, potable water hoses for washdown, and a wet surface, and even the most nimble and athletic of cellar workers will occasionally stumble and fall. The footwear of preference for most work is the durable, rubber boot. The lug-type soles on almost all rubber boots do not give nonskid traction on wet concrete surfaces. Certainly, the type of floor finish can assist in providing the needed friction (see Chapter 4, section entitled "Winery Design Concepts, Winery Age, and Sanitation"). Both rubber boots and so-called deck shoes are sometimes the only thing between a recreational sailor staying on a boat or going overboard. The magic, "razor-cut" soles have an extraordinary ability to adhere to wet surfaces. Why such adaptive footwear is not universally used in wineries remains a mystery.

Remarkably, there are instruments to measure the slipperiness of floors, so that surface treatments or alternative footwear can be selected to make the working environment safer. All of the instruments are calibrated on a scale of 0 to 1. A slipperiness rating of 0.5 is safe, whereas a reading of greater than 0.5 would indicate that the floor–shoe combination is slip resistant. There are no fewer than seven separate, approved tests for slipperiness measurement. The prestigious and very conservative American Society for Testing Materials (ASTM) details the seven test methods (208):

ASTM C 1028	ASTM P 127
ASTM F 609	ASTM P 2047
ASTM P 125	ASTM P 126
ASTM P 128	

Guard rails, catwalks, and ladders are all part of the necessary accessory structures to access fermentation and storage tanks and to prevent an employee from stepping or falling into a recessed crush/press pit or pump vault. A 3 ft, 6 in. (1.1 m) guard rail will prevent 99% of the world's population from rotating over the top rail. A cellar worker who is 6 ft

(1.8 m) tall will have a center of gravity at about the 39-in. (1 m) height; thus, the 42-in. (1.1 m)-high railing would prevent an incidental over-the-top-rail rotation (209). Catwalks and ladders installed in wineries are constructed of stainless steel or fiberglass. If a winery elects to design and construct its own version of a catwalk-ladder system, the fabricator must be intimately familiar with the OSHA design guidelines for such structures, so that guard rails, handholds, tread dimensions, and toe board requirements are met or exceeded. Small wineries with tanks not exceeding 15 ft (4.6 m) in height, often choose a wheeled, platform ladder that can be moved from tank to tank. The platform ladders meet OSHA and ANSI safety standards and present a minimum investment while providing safe ascent and descent for tank inspection, filling, and cleaning.

Figure 12.1 is a typical configuration for a platform ladder. Models can be equipped to fit any required height with the limit being about 12 ft (3.7 m) to the top platform, which requires about 15 steps.

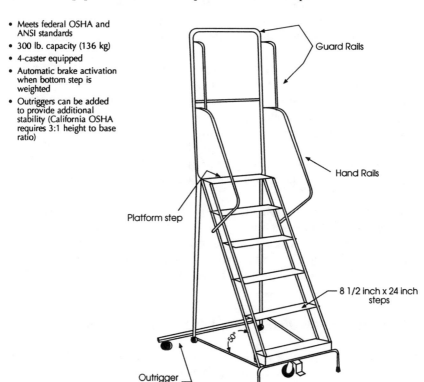

- Meets federal OSHA and ANSI standards
- 300 lb. capacity (136 kg)
- 4-caster equipped
- Automatic brake activation when bottom step is weighted
- Outriggers can be added to provide additional stability (California OSHA requires 3:1 height to base ratio)

Guard Rails

Hand Rails

Platform step

8 1/2 inch x 24 inch steps

Outrigger accessory

Fig. 12.1.   Typical platform ladder which meets or exceeds OSHA and ANSI standards for safety.

Getting an arm, leg, or finger caught in a piece of machinery or having a loose piece of clothing become entangled in an unguarded belt drive is a serious form of industrial accident. New winery equipment is developed almost constantly—if not a new machine to make wine production more efficient, it will be a modification of the existing device, possibly with added microelectronic control features, or more efficient electric drive motors. With all the modernization that has taken place since the beginning of the industrial revolution, the moving parts guarding principles have changed very little. Guards on winery equipment are designed to prevent workers and their wearing apparel from incidentally or accidentally coming in contact with a moving belt, shaft, gear train, or chain. The design philosophy of machinery guard systems is as follows:

- Prevent access to the machinery danger zone.
- The guards must not create a hazard by having sharp or protruding structural components.
- Guards should be designed to remain in place and permit routine maintenance, cleaning, lubricating, and repair operations.

If new machinery and equipment is installed in the winery, the winery safety officer should inspect the guard safety features for adequacy and appropriateness for the particular space orientation in the winery. Screw feed augers are used universally in the wine industry to transfer grapes and to convey pomace to trucks or containers for ultimate disposition. Accessible sections of the auger and trough often need to be cleared of debris. Safety screens in some wineries are intentionally removed, so that frequent visual checks can be made by the press operator of the full length of the device.

Finally, emergency stop controls must be easily seen and located on an accessible part of the machine. For example, each member of a bottling line crew should know the location and function of the stop controls, so that line inactivation can occur within milliseconds if a cellar worker is accidentally caught on the conveyor or conveyor drive train.

SITE UTILITIES AND HEALTH AND SAFETY PLANNING

In late October 1989, the State of California Senate signed into law a program for establishing an injury and illness prevention program. This statute is contained in the California Labor Code Section 6400 and establishes the methods by which the health and accident safety plan shall be structured. The law crosses all boundaries of commercial and industrial activities regardless of size. In California, the law is administered by Cal-OSHA, a division of the Department of Labor. The California statute requirements will be used as an illustrative example in this section; however, similar laws are in place in all states, which have modeled their regulations after a federal OSHA workplace safety program that was mandated for adoption nationwide.

The discussion which follows focuses on the potential injury and sickness hazards from the nine utility systems described herein and the methods for mitigating or eliminating such hazards.

Electrical Systems

Electrical shock from an improperly grounded piece of winery equipment is one method by which employees can sustain injury or death in the case of electrocution. Equipment which requires frequent adjustments, such as variable-speed pumps, remote control pumps for barrel topping, or must pumps, are generally found in wet production spaces and present the greatest hazard. Ground fault protection devices [sometimes called Ground Fault Circuit Interrupters (GFCI)], installed on the power supply cable to the motor, inactivates the source of power in the event of an accidental grounding of the motor on the pump frame or other current load malfunction. The GFCI is a very-fast-acting circuit breaker that can detect current differences of as low as 2 mA between the power supply conductor and the neutral groundwire. The results of electrical shock can vary greatly from individual to individual from (a) mild shock, (b) painful shock, (c) no-release threshold, (d) muscular paralysis, (e) asphyxiation, (f) heart fibrillation, and (g) heart paralysis and tissue destruction (208). In general, alternating current (ac) is more dangerous than direct current (dc), with 60 Hz (cycles per second) being more dangerous than higher frequencies.

Heating of conductors and fires are caused by current overloading or wire sizes that are too small for imposed carrier current.

All winery production machinery of both foreign and domestic manufacturer must meet the rigid requirements of the National Electric Code.

Extension cords must have waterproof, twist-lock-type connections. Cords should be frequently checked for breaks or cuts in the insulation. Junction boxes and switches subject to environmental moisture must be of at least a NEMA 5 (waterproof) rating (National Electrical Manufacturers Association).

Portable power tools that are used from time to time on winery equipment maintenance and repair tasks need to be of the highest commercial/industrial quality, with carefully insulated interior electrical components that cannot under any circumstance energize the exterior metal housing. Forklifts can be equipped with nonconductive isolation systems that will prevent current passing from the forks to the chassis, frame, and control levers.

Only qualified electricians should be permitted to perform maintenance tasks on the interior or exterior power supply distribution system. So-called lock-out devices must be installed on circuit breaker boxes and disconnect switches, so that when a system requires de-energizing, no one can have access to those circuits until the maintenance electrician has completed the repair.

One piece of electrical equipment that is common to wineries with battery-powered pallet jacks or forklifts is the battery charger. Lead–acid batteries can explode by two different modes. During the charge cycle, lead–acid batteries produce hydrogen gas from the sulfuric acid electrolyte. If concentrated in an enclosed space, the spark from connecting the charger leads to the battery terminals can be enough to ignite the very explosive hydrogen gas. Connecting batteries of different voltages can also cause an explosion. In connecting and disconnecting batteries, the negative (ground) should always be attached last (when connecting for service) and the ground should be the first to be disconnected when taking the battery out of service.

A deluge shower and eye flush fountain should be installed near the battery-charging station, and safety goggles, rubber gloves, and a rubber apron to prevent acid burns and damage to clothing should be the mandated uniform for battery maintenance tasks.

Refrigeration, Ventilation, and Air Conditioning

This important winery subsystem essentially consists of electric motors, compressors, and fan motors, and electrical master control panels and disconnects. Reference is made to Chapter 5, section entitled "Refrigerants," for a detailed treatment of refrigerants ad their potentials for stratospheric ozone depletion. Among the substitute refrigerants being opted for by wineries is ammonia, which is a superior refrigerant and thermodynamically efficient working substance but which is also very toxic to humans, and in concentrations of 15–30% in air, it is a serious explosion risk. Contemporary ammonia gas detectors are set to alarm at gas concentrations of 25 ppm (204). Detection of ammonia by the human nose will occur at levels of around 50 ppm; thus, the gas detection alarm will activate even before the gas is detected by the employee. Ammonia is lighter than air, so it will tend to rise in an enclosed space making detection somewhat easier.

As wineries begin their conversions from chlorofluorocarbon (CFC) refrigerants to non-ozone-depleting substitutes, accidental escape of CFCs is likely to occur. Only licensed heating, ventilation, and air conditioning contractors with permits and CFC salvage equipment will be allowed to perform the transfers. Ethylene glycol is the heat-transfer substance of choice for most winery secondary-cooling-loop tank chillers. Ethylene glycol is not extremely toxic, but OSHA reports a "ceiling" limit in air standard of 50 ppm (208). Leakage in secondary cooling loops is less frequent because of considerably lower glycol pressures than are contained in the primary loop of refrigerant.

The same electrical hazards prevail as described in Chapter 2 "Electrical Systems." Reference is made to that portion of the text for electrical hazards and precautions to prevent injury.

Telecommunications

Little or no risks exist to winery employees from operation of cellular telephones, PCs, linear programmable controllers, VHF radios, or pagers. Each one of the systems is part of the safety program accessories, which permit the rapid notification of personnel injury or accident and thus the quickest medical response that is possible.

Sanitation, Steam, and Hot Water

The safe handling of cleaning equipment and chemical cleaning agents requires a carefully programmed training schedule to achieve the optimum level of personnel safety for cleaning system operators and fellow cellar workers as well.

Portable steam or high-pressure hot- or cold-water cleaning units present the highest risk of injury from scalding burns to physical body damage from the high-pressure streams. Steam hoses, fittings, and nozzles require frequent inspections to ensure that steam leakage will not occur. Mixing valves and their functions have been described previously in Chapter 4. Visual checks of the gauges will verify that the water supply blend is of the proper temperature and prevent, to the greatest extent possible, accidental burns to the cleaning hose operator and fellow cellar workers. The caustic cleaning agents are quite hazardous to eyes and soft tissue and the illustrative MSDS information regarding safety precautions and protective clothing should be strictly observed. (See Table 11.2.)

Potable Water Systems

Potable water systems are generally accessed only by the winery's utilities maintenance technician, so the entire winery employee population is not exposed to machinery, electrical, or hazardous gas risks. Chlorine in its gaseous form is still the disinfectant of choice of most conventional water treatment systems. The chlorine is often housed in a separate room with a gas cylinder scale to permit tracking gas use volumes, along with a gas regulator and mixing device for achieving the proper concentration of an aqueous chlorine solution. Chlorine gas at very low concentrations is lethal; thus, if a leak occurs in the piping system or injector, the operator must first don the proper respiratory protection (mask) before entering the chlorine room. If the emergency equipment is located inside the room with the chlorine cylinder, it is useless and inaccessible during an emergency as protective equipment for the operator. For other water treatment chemicals such as aluminum sulfate (alum), lime, activated carbon, and polyelectrolyte flocculants, if used in some way by the winery for water treatment, the MSDS sheets for each compound should be consulted and available for reference by the staff utilities operator.

Entering water storage tanks requires compliance with the California Confined Space Law or that of other jurisdictions. The details of that nearly universal statute were described previously.

Wastewater Systems

The two major subdivisions of winery produced wastestreams are sanitary wastewater and water that is a by-product of wine production activities or process wastewater.

Sanitary wastewater in its raw form contains billions of pathogens capable of producing disease and illness. As with water systems, the general winery employee population will not be involved in the maintenance and repair of the sewers and sewage treatment and disposal works; thus, it is the winery's maintenance staff or a commercial sewer service company that will perform periodic sewer cleaning or septage removal from septic tank sludge compartments. Therefore, the greatest worker risk for sanitary wastewater systems results from accidental or incidental contact with the raw sewage. Proper disposable, protective coveralls, gloves, face mask, and head covering are a minimum for sewer maintenance workers, who must remove a sewer blockage or enter a manhole for inspections or servicing a stage recorder. The winery's safety officer should ensure that the workers, and the families of workers, who will be exposed to sewage receive inoculations for polio, Type "B" tuberculosis, cholera, typhoid, and other transmittable infectious diseases that may be endemic to the winery's particular location.

Entering a manhole brings into action the confined space law, previously discussed. The maintenance crew supervisor must be sure that the winery complies with the special provisions of that law. (Sewer System entry protocol is contained in App. E of California Confined Space Regulations.) Sewer overflows should be given the same respect for illness risk, as described above, and winery areas that have been contaminated should be liberally dosed with high test hypochlorite (powder form) to inactivate any pathogens.

Process wastewater from the winery presents a negligible risk for illness or disease. During certain periods in the winery's production cycle, a large percentage of the process wastestream could contain caustic chemicals that could cause eye injury. These risks should be brought to the attention of the process wastewater treatment plant operator, so that risk assessments can be made and contact with the influent stream minimized during the period of intense tank cleaning activity. If nutrient addition and pH neutralization are part of the treatment works, special safety precautions must be taken by the operator in the handling of sodium hydroxide, ammonium phosphate, and/or anhydrous ammonia. As described earlier, the MSDS sheets become the training aid and guidelines for safe handling of the chemicals.

Rotary drum screens present a moderate hazard, but all models have machinery guards, as described in the section entitled "Machinery Guards, Walking and Working Surfaces." There are occasions when the slots in the drum must be physically cleaned. The electric lock-out device previously described must be used to prevent accidental energizing of the screen before maintenance is completed. Some of the author's designs have included flow-switch-activated drum screens to permit operation of the unit only when wastewater is flowing. Such a device would have to be overridden to prevent an accident if separate switches and circuits served the drum screen and the flow switch system.

All of the electrical hazard precautions that have been described for other utility systems apply even more emphatically for process wastewater systems. Higher current draws and line voltages prevail with the process wastewater treatment equipment, because of generally higher installed horsepower/kilovolt-ampere ratings for the necessarily larger pumps and aeration devices. The wastewater treatment ponds also present a formidable drowning threat to utility operators and outside service personnel, who may be called upon for repair assistance from time to time. Life jackets, better known in nautical circles as personal flotation devices (PFDs), should be worn by operators working in and around aeration ponds, sedimentation cells, or clarifiers. The drowning hazard also exists for trespassers (generally curious children) who are drawn to the water sources. A perimeter fence for waste treatment works is mandatory, although fencing with appropriate warning signage will not deter a determined trespasser. Intrusion alarm systems have been described previously in Chapter 2.

Liquified Petroleum Gas Systems

Safety requirements for the safe storage and handling of liquid petroleum gases (LPG) were detailed in Chapter 9. Reference is made to that chapter for safety details, including forklift fueling and worker safety. The Uniform Fire Code (UFC) also is specific on the correct methods for the storage and use of LPG. Reference is made to the 1994 edition of the UFC, Article 82 and Article 25 for fire safety requirements.

Liquid petroleum gas contains artificial odorants to make their presence easily known. LPG leaks and the explosion hazards associated with ignition of this highly combustible gas must be given high priority by the winery safety officer.

Fire Protection Systems

Winery personnel safety that could be directly related to fire protection systems becomes almost a non-issue. As it is the function of the fire protection system to provide winery management and staff employees protection from fires, the hazard from the system is negligible, having received design attention and scrutiny to avoid any risks from the equipment, fire suppression, or alarm systems and subsystems. The only risks to winery personnel from fire protection systems were previously identified in Chapter 10. Reference is made to that chapter for minor hazards and dangers associated with the use of 2½-in. fire hoses on fixed hose reels by untrained winery personnel.

Solid Waste Systems

All personnel come in contact with refuse and discarded materials. The sensible precautions are the same as prescribed in most family situations (i.e., be careful of cuts from broken glass, jagged can tops and edges, and hazardous biomedical wastes). Truly hazardous wastes of the types found in wineries have been described previously in Chapter 11. Reference is made to that chapter for the safe handling and storage of hazardous wastes.

All winery personnel should receive training on the solid waste management program that is in effect, and their roles identified in keeping disease and injury risks to an absolute minimum.

Accident Prevention Plan

Table 12.3 is a model Accident Prevention Program developed for a small table winery. Its content and format can be modified to fit a winery's unique situation or equipment and end product mix.

The following are forms/notices comprising the plan:
> Management Policy Statement
> Identification of Plan Administrators
> Responsibilities
> Safety Rules
> Disciplinary Procedures
> Safety Training
> Inspections
> Safety Committee & Safety Meetings
> Accident Investigations and Reports
> Hazard Communication
> Right-to-Know Training Program
> Emergency Action Plan
> Exhibits

Table 12.3   Model Accident Prevention Plan

WINTERS WINERY ACCIDENT PREVENTION PLAN

WINTERS WINERY, INC.
SAFETY POLICY

May 7, 1991

Winters Winery, Inc. shall establish an accident and illness prevention program to cover both seasonal part-time workers and private contractors (mobile bottling service companies, barrel service companies, etc.).

The winemaker shall be designated as the winery safety officer who shall be responsible for structuring the program and monitoring results. The assistant winemaker shall be the safety training officer and allow sufficient time during working hours for field and winery demonstrations on the specific subjects of:

1. Safe lifting techniques
2. Forklift operations
3. Electrical hazards
4. First-aid and treatment for minor cuts and abrasions
5. Platform ladder fundamentals
6. Protective clothing and footwear for specific job tasks
7. Steam cleaner operations
8. Compressed gas cylinder handling
9. Pump fundamentals and electrical energy (220v 3ϕ, 220v 1ϕ, and 115v 1ϕ)
10. Portable toilets and field sanitation
11. Alcohol, drugs and tobacco policy
12. Fire hazards

13. Hazardous chemicals and their use
14. Confined space laws
15. MSDS familiarity

This policy shall be implemented with a winery specific accident and illness prevention plan with training and education of employees as the principal precept.

D. R. Storm
President
WINTERS WINERY

IDENTIFICATION OF PLAN ADMINISTRATORS

The following persons are responsible for implementing the accident prevention plan for Winters Winery:

NAME	TITLE
David R. Storm	President/Winemaker
J. P. Mack	Vice President
Michael Petersen	Assistant Winemaker and Safety Officer
L. H. Storm	Treasurer

RESPONSIBILITIES

MANAGERS:

In effectively executing their safety responsibilities, managers will:

1. Familiarize themselves with the safety program and ensure its effective implementation.
2. Be aware of all safety considerations when introducing a new process, procedure, machine or material to the workplace.
3. Give maximum support to all programs and committees whose function is to promote safety and health.
4. Actively participate in safety committees as required.
5. Review serious accidents to ensure that proper reports are completed and appropriate action is taken to prevent repetition.

SUPERVISORS:

Our supervisors are the foundation of the safety program. Their responsibilities are to:

1. Familiarize themselves with company safety policies, programs and procedures.
2. Provide complete safety training to employees prior to the assignment of duties.
3. Consistently and fairly enforce all company safety rules.
4. Investigate injuries to determine cause, then take action to prevent repetition.
5. See that all injuries, no matter how minor, are treated immediately and referred to the personnel office to ensure prompt reporting to the insurance carrier.
6. Inspect work areas often to detect unsafe conditions and work practices. Utilize company self-inspection checklists as required.

EMPLOYEES:

Employee responsibilities for safety include the following:

1. Adhere to all safety rules and regulations.
2. Wear appropriate safety equipment as required.
3. Maintain equipment in good condition, with all safety guards in place when in operation.
4. Report all injuries, no matter how minor, immediately to a supervisor.
5. Encourage co-workers to work safely.
6. Report unsafe acts and conditions to the supervisor.

SAFETY RULES

For the protection and safety of all employees, Winters Winery has established the following rules designed to prevent accidents and injuries. Compliance with these rules is mandatory. Documentation will be made when the rules are distributed to new employees.

1. Proper footwear and clothing will be worn at all times.
2. Do not wear loose clothing, jewelry or keep long hair in a down position where there is a danger of catching such articles in moving machinery.
3. Horseplay, running, fighting or any activity that may result in injury or waste will not be tolerated.
4. Eye protection is required when performing any task that could produce flying particles.

5. Operate machinery with all guards in place. Tampering with safety devices is cause for immediate disciplinary action.

6. Do not operate any machine you are not familiar with.

7. Machines must never be cleaned, adjusted or repaired until after the machine is turned off, the circuit is broken at the power source (including lock-out) and a warning tag is placed at the controls. Each person involved in maintenance must have his/her own personal padlock to ensure total lack of power until all work has been completed.

8. Any defects in materials, machinery, tools and equipment must be reported immediately to a supervisor.

9. Do not leave tools, materials or other objects on the floor which might cause others to trip and fall.

10. Do not block exits, fire doors, aisles, fire extinguishers, gas meters, electrical panels or traffic lanes.

11. Avoid risk of rupture, internal injury or back injury in attempting to lift or push excessive loads. If an object is too heavy to move without strain, ASK FOR HELP.   Observe the correct position for lifting. Stand with your feet slightly apart, assume a squatting position with knees bent and tuck your chin. Tilt head forwards, grasp the load with both hands and gradually push up with your legs, keeping your back straight and avoiding any abrupt movement.

12. Do not distract others while working. When approaching a machine operator for any purpose, do so from the front or the side in a way that he or she will see you coming and will not be shocked or surprised. If conversation is necessary, first make sure the machine is turned off.

13. Do not allow oil, wax, water, or any other material to remain in the floor where you or others may slip. Report any spills to your supervisor.

14. When handling hazardous materials, ensure you follow prescribed safety procedures and use required safety equipment. When using secondary containers filled by others, ensure that they are labeled as to their contents and hazards.

15. Use appropriate gloves when handling materials with sharp or jagged edges which may result in lacerations.

16. Do not attempt to operate machinery for which you are not trained.

17. Unnecessary and excessive haste is the cause of many accidents. Exercise caution at all times. WALK, DO NOT RUN!   The use of hot production equipment or materials for the purpose of cooking or heating food is strictly prohibited.

18. All work related injuries and accidents, no matter how minor, must be reported immediately to your supervisor.

It is imperative that all employees become thoroughly familiar with the above safety rules. Failure to comply with safety rules or procedures or failure to wear the appropriate safety equipment will result in disciplinary action up to and including termination.

DISCIPLINARY PROCEDURES

Employees who fail to comply with safety rules will be subject to disciplinary action up to and including termination. Supervisors will follow the normal disciplinary procedures as follows:

1. Verbal counseling—the fist step. Must be documented in the employee's personnel file.
2. Written warning—outlining nature of offense and necessary corrective action.
3. Suspension without pay—the third step or a separate disciplinary action resulting from a serious violation.
4. Termination—if an employee is to be terminated, specific and documented communication between the supervisor and the employee, as outlined, must have occurred.

Supervisors will be subject to disciplinary action for the following reasons:

1. Repeated safety rule violation by their department employees.
2. Failure to provide adequate training prior to job assignment.
3. Failure to report accidents and provide medical attention to employees injured at work.
4. Failure to control unsafe conditions or work practices.
5. Failure to maintain good housekeeping standards and cleanliness in their departments.

Supervisor who fail to maintain high standards of safety within their departments will be demoted or terminated after three documented warnings have been levied during any calendar year.

SAFETY TRAINING

The goal of our safety training program is to develop safe work habits and attitudes. It is critical that new workers understand work rules and procedures prior to being assigned a job. Supervisors are responsible for providing safety training to their department employees utilizing the job instruction training (JIT) method described below.

HOW TO GET READY TO INSTRUCT	JOB INSTRUCTION TRAINING (JIT) HOW TO INSTRUCT
Have a Timetable— how much skill you expect them to have, by what date.	**1. Prepare** Put the worker at ease. Define the job and find out what is already know about it. Get the person interested in learning job. Place in correct position.
Break Down the Job— list important steps, pick out the key points. (Safety is always a key point.) **Have Everything Ready—** the right equipment, materials and supplies.	**2. Present** Tell, show, and illustrate one IMPORTANT STEP at a time.
Have the Workplace Properly Arranged— just as the worker will be expected to keep it.	**3. Try Out Performance** Have person do the job—correct errors.
Remember, when teaching adults, the following points are important: 1. Adults learn best in a warm, friendly atmosphere. 2. Adults don't like to waste time. 3. Adults respond quickly to praise ad attention.	**4. Follow-Up** Put them on their own. Designate to whom to go for help. Check frequently. Encourage questions. Taper off extra coaching and close follow-up. Safety is always a key point. **SAFETY TRAINING INSTITUTE NATIONAL SAFETY COUNCIL**

The fundamentals of safety practices will be reviewed prior to a new employee's first job assignment. Our safety orientation checklist (see Exhibit C) will be utilized to document this training.

INSPECTIONS

Inspection works because it is an essential part of hazard control. It is an important management tool, not a gimmick. We will view inspections as a fact-finding process, not fault-finding. We will emphasize locating potential hazards that can adversely affect safety and health.

All personnel will be responsible for continuous, outgoing inspection of the workplace. When uncovered, potentially hazardous conditions will be corrected immediately or a report will be filed (see Exhibit A) to initiate corrective action.

Periodic, planned inspections will be made by members of the safety committee (or other designated individuals) utilizing the company self-inspection form (see Exhibit B). The report will be reviewed by the safety com mittee and action will be taken to eliminate uncovered potential hazards. Assignments, target dates for completion and actual completion dates will be documented in the minutes of the safety committee.

SAFETY COMMITTEE & SAFETY MEETINGS

Our company safety committee will be comprised of members (supervisors and/or employees) of the various departments and management. They will meet on a monthly basis and review the following:

1. Minutes of the previous meeting.
2. Unfinished business of the previous meeting.
3. Self-inspection reports.
4. Discussion of accidents and corrective action taken.
5. Accident trends
6. New and outstanding recommendations submitted by outside agencies (insurance carrier, fire department, Cal-OSHA, etc.).
7. New business.

All meetings will be documented.

Group safety meetings—supervisors will be responsible for holding department safety meetings on a regular basis. Employee attendance and discussion topics will be documented.

ACCIDENT INVESTIGATIONS AND REPORTS

It is the policy of Winters Winery to carry out a thorough program of accident investigation. Supervisory personnel will be primarily responsible for making an investigation of all accidents in their areas of responsibility. Accidents involving fire, death, serious injury, or extensive property damage will be investigated jointly by the Supervisor, the Plant Manager, and the Personnel Manager.

The primary goal of the accident investigation program is the prevention of future similar accidents through the use of knowledge derived form the investigations. Additionally, the investigation will be used to prepare reports required by Federal and State law as well as the Worker's Compensation Insurance Carrier. These reports are critical in establishing the Company's and the Supervisor's liability under the law.

When an employee is injured at work, the supervisor is responsible for taking emergency action to have first aid administered, to obtain professional medical attention as soon as possible and protect other employees and equipment. The supervisor must then begin to investigate the circumstances of the accident. The following procedures have been found to be effective when investigating accidents:

1. **GO** to the scene of the accident at once.
2. **TALK** with the injured person, if possible. Talk to witnesses. Stress getting the facts, not placing blame or responsibility. Ask open-ended questions.
3. **LISTEN** for clues in the conversations around you. Unsolicited comments often have merit.
4. **ENCOURAGE** people to give their ideas for preventing a similar accident.
5. **STUDY** possible causes—unsafe conditions, unsafe practices.
6. **CONFER** with interested persons about possible solutions.
7. **WRITE** your accident report giving a complete, accurate account of the accident.
8. **FOLLOW UP** to make sure conditions are corrected. If they cannot be corrected immediately, report this to your supervisor.
9. **PUBLICIZE** corrective action taken so that all may benefit from the experience.

In order for the Supervisor's Report to be effective, it should contain as a minimum, a detailed answer to the following questions:

1. **What Was the Employee Doing?**—Explain in detail the activity of the employee at the time of the accident.
2. **What Happened?**—Indicate in detail what took place, describe the accident, the type of injury, the part or parts of the body affected, and whether the employee was wearing appropriate safety equipment.
3. **What Caused the Accident?**—Explain in detail the condition, act, malfunction, etc., that caused the accident. Remember that it is possible to have more than one reason or cause for an accident.
4. **What Can Be Done to Prevent a Similar Accident?**—Indicate corrective action to prevent recurrence.

The supervisor's Report, along with the Employee Report, must be submitted to the Personnel Office not later than 24 hours after the accident. Each supervisor must maintain an adequate supply of the Supervisor's Report and the Employee's Report forms which may be obtained from the Personnel Office.

HAZARD COMMUNICATION

HAZARD EVALUATION

Chemical manufacturers and importers are required to review the available scientific evidence concerning the hazards of the chemicals they produce, then report that information to employers who purchase their product. In most cases, Winters Winery will choose to rely on the evaluations performed by our suppliers. If, for some reason, we do not trust the evaluation of the manufacturer, we will arrange for additional testing.

We will consider any chemicals listed in one of the following sources to be hazardous:

- 29 CFR 1910, Subpart Z, Toxic and Hazardous Substances, Occupational Safety and Health Administration (OSHA),
- Threshold Limit Values for Chemical Substances and Physical Agents in the Work Environment, American Conference of Governmental Industrial Hygienists (ACGIH),
- Those hazardous substances prepared pursuant to Labor Code Section 6382, or
- California Proposition 68 List of Toxic Substances.

LABELS & OTHER FORMS OF WARNING

We will make certain that containers are adequately labeled to identify the hazardous chemicals contained therein, and will show hazard warnings appropriate for employee protection. The warnings will utilize a combination of words, pictures and symbols which will convey the hazards of the chemical(s) in the container. The labels will be legible and prominently displayed.

Exceptions to this rule are as follows:

- We are permitted to post signs which convey the hazard information if there are a number of sanitary containers in a given area which have similar contents and hazards.
- Operating procedures, process sheets, batch tickets, blend tickets and similar written materials can be substituted for container labels on stationary process equipment if they contain the same information and are readily available to employees.
- We are not required to label portable containers, as long as the trans-

ferred chemical is for immediate use by the employee who made the transfer.
- We are not required to label pipes or piping systems.

Our employee training program will include instruction on how to read and interpret information.

MATERIAL SAFETY DATA SHEETS (MSDS)

The management of Winters Winery is responsible for obtaining or developing a MSDS for each chemical used in the workplace. Each MSDS will include the specific chemical identity of the chemical involved and the common names.

Each data sheet will provide information on the physical and chemical characteristics of the chemical; known acute and chronic health effects and related health information; exposure limits; whether the chemical is considered to be a carcinogen; precautionary measures; emergency and first aid procedures; and the identification of the organization responsible for preparing the sheet.

Each department supervisor will be responsible for maintaining the MSDSs describing chemicals used in his/her department and for keeping them readily available to employees. The Program Coordinator will maintain a master file for all departments.

Our employee training program will include instruction on how to read and interpret information on a MSDS, and how employees can obtain and use the available hazard information.

EMPLOYEE TRAINING

It is the goal of Winters Winery to provide hazard communication training during the first 30 days of employment and whenever a new chemical is introduced to a given work area. Training will be done in a classroom setting and will be conducted by the Program Coordinator or another who has been properly trained.

The training program will consist of:

- How the hazard communication program is implemented, how to read and interpret information on labels and MSDS, and how employees can obtain and use the available hazard information.

- The hazards of the chemicals in the work area.
- Measures employees can take to protect themselves from the hazards.
- Specific procedures put into effect by the company to provide protection, such as personal protective equipment.
- Methods and observations, such as visual appearance or smell, workers can use to detect presence of a hazardous chemical they may be exposed to.

RIGHT-TO-KNOW TRAINING PROGRAM

1. Introduce the Right-to-Know coordinator and explain his role.
2. Review the company's written hazcom program and explain how to obtain and use the document.
3. Explain applicable safety and health requirements mandated by OSHA and state standards.
4. Identify locations where hazardous chemicals are stored, handled, dispensed or transported, and the location of each process and operation that uses them.
5. Explain how to recognize potential health and safety hazards and review monitoring techniques used to detect potential health hazards.
6. Explain how to read MSDSs and related information (labels).
7. Explain safety precautions to be taken by the individual worker.
8. Explain in detail the labeling system used by the company.
9. Use audiovisuals to teach basic hazcom information to the general plant population.
10. Warn about specific work activities that increase the likelihood of a loss.

EMERGENCY ACTION PLAN

Major disasters must be anticipated and procedures must be developed and mastered if the well-being of our personnel is to be protected and if we are ready to serve our community.

The following pages detail the organizational structure of our plan and outlines emergency measures to be taken in the event of fire or other emergency.

Remember, your conduct and actions during the first few minutes of any emergency may not only save your life, but the lives of your fellow workers and other members of the community as well.

GENERAL INFORMATION

Two important telephone calls need to be made if the facility is to be evacuated for any of the following reasons:

1. A fire or disaster within the facility.
2. An external hazardous condition threatening the facility.

If either of these two situations occur, notify these agencies:

1. Name and phone number of local fire department.
2. Name and phone number of local Office of Emergency Services.

Upon order of management or other person(s) in charge of totally evaluate the facility, the following action will be taken:

1. Initiate evacuation center receiving plan. (It may be necessary to transport company personnel to a local evacuation center.)
2. Priority of evacuation may be a necessity if there are handicapped employees.
3. Materials and supplies to be evaluated:
   a) First aid kits
   b) Personnel roster

RESPONSIBILITIES

The Safety Committee will:

1. Coordinate the Emergency Evacuation Plan throughout the facility.
2. Make certain the Program is familiar to all personnel and that all new employees are promptly oriented.
3. Schedule fire classes as necessary.
4. Arrange and execute fire drills within the facility.
5. Maintain a log of fire drills conducted. The log shall include the date and time of each drill, the time required to evacuate the building, and the initials of the person making the recording.
6. Report any deficiencies noted during the fire drill.
7. Correct any deficiencies noted during the fire drill.
8. Maintain a file of committee meetings and activities, including committee minutes. All documents are to be signed by the committee chairman.

The Safety Committee will be aided by Supervisors who will:

1. Facilitate the Emergency Evacuation Plan.
2. Keep constant check on all personnel to be sure that they are completely familiar with all phases of the Plan which they are required to know.

3. See that all personnel participate in **ALL** fire drills, fire classes, and other practice sessions.
4. Be certain that all personnel are familiar with, and make thorough fire prevention inspections when they are assigned to do so.
5. Take the necessary steps required to correct any fire hazards discovered.

It is the duty of every employee to:

1. Be completely familiar with the Emergency Evacuation Plan and his or her duties and responsibilities in the program.
2. Participate in all fire drills and practice sessions.
3. Attend all fire training classes when assigned.
4. Learn the location of and how to operate fire alarm systems and all fire extinguishing equipment.
5. Report any fire and/or safety hazard located any place on company property.

FIRE PROCEDURE

"Keep Calm . . . Report all fires and smoke."

Personnel have been assigned to:

1. Sound internal fire alarm.
2. Notify office staff.
3. Remove personnel from the building.
4. Close all doors and windows in the fire area, ONLY if this can be done safely.
5. Notify the fire department.

The person reporting the fire to the fire department will provide them with the following information:

1. Company name.
2. Address.
3. What is burning (machines, paper, etc.)
4. Location of fire (roof, plant, office, etc.)
5. Type of fire (electrical, liquid, etc.)

Additional assignments have been made to:

1. Attempt to extinguish the fire with the use of on-premises equipment (extinguishers, hoses, etc.). A minimum of two persons is required to fight a fire. To ensure employee safety, this is to be done only during the early stages of the fire.

Working away from the involved area, personnel will be assigned to:

1. Clear the aisles, hallways and other areas of personnel and visitors.
2. Close all doors and windows.
3. Check driveways to see that they are clear for entry of firefighting equipment. See that gates are unlocked and open.
4. Wait at the front entrance for arrival of fire fighting equipment. Direct the firemen to the fire, if necessary.

Re-entry onto the property will not be permitted until it is declared safe to do so by someone with executive authority or by the local fire/law enforcement officials.

EARTHQUAKE

In the event of an earthquake, the following procedures shall be followed:

1. Assess damage and injuries.
2. Give first aid as needed. Remember, after an earthquake, utilities, police and fire agencies may not be readily available. DO NOT ATTEMPT TO TELEPHONE UNLESS ESSENTIAL.
3. Notify executive management if any are away from the premises.
4. Call the Fire Department only in the case of fire.
5. The nearest hospital for treatment is:

   Sutter Davis Hospital
   1100 Covell Boulevard
   Davis, California
   (916) 756-6440

6. Have damaged or potentially damaged utilities shut off at the main controls.
7. Personnel are to be instructed during orientation that they are to take shelter under a sturdy table or equipment during an earthquake and remain there until all shaking has ceased.
8. Evacuate as necessary. Supervisors shall be responsible for seeing that employees are evacuated to a safe area outside the building and clear of overhead electrical lines, utility posts, block walls, etc., which might fall during aftershocks. Supervisors are cautioned to be alert for fallen high tension lines which may be touching metal objects on the ground.
9. Have all areas of the building inspected for damage before allowing personnel to return to the building(s).
10. Have gas, electrical, water and fuel systems checked for damage before allowing personnel to return to the building(s).
11. Drinking water should be checked to determine that it is not contaminated. Water contained in toilet tanks can be boiled and used if absolutely necessary for drinking or treating injuries.

EXHIBITS

EXHIBIT A
MAINTENANCE REQUEST FORM

Plant _____ Dept._____ Date _____

Submitted by _____

Location of Unsafe Condition (Describe in Detail) _____

_____

Explanation of Unsafe Condition (Explain in Detail) _____

_____

_____

_____

_____

_____

If Necessary, Draw a Diagram in the Following Box

```
┌─────────────────────────────────────────────────┐
│                                                   │
│                                                   │
│                                                   │
│                                                   │
└─────────────────────────────────────────────────┘
```

Recommendations to Correct This Condition _____

_____

_____

_____

_____

_____

_____

_____

_____   _____
Employee Signature   Supervisor Signature

===

(Maintenance Department Use Only)

Recommended Corrective Action: _____

_____

_____

_____

Estimated Cost: _____   Approved By: _____

Estimated Completion Date: _____

EXHIBIT B
SAFETY INSPECTION REPORT

Inspection Conducted By _____

Date _____ Dept. _____ Plant _____

SAFETY PRACTICES	Yes	No	Explain
Are employees wearing the required safety equipment?			
Are employees using adequate footwear and clothing?			
Are employees following safety rules and procedures?			
Are food or drinks present in the work areas?			
Other comments:			
HOUSEKEEPING	Yes	No	Explain
Are floors kept clean?			
Are floors slippery?			
Is equipment and material neatly and safely kept and stored?			
Are working tables kept neatly and clean?			
Are hazardous materials present?			
Are hazardous materials being properly stored and labeled?			
Is hazardous waste being properly disposed and labeled?			
Are there adequate trash cans?			
Other comments:			
FIRE SAFETY	Yes	No	Explain
Are fire extinguishers accessible, serviced and tagged?			
Are fire alarms available and in working order?			
Are exit doors accessible and properly marked?			
Are flammable materials properly stored and labeled?			
Is flammable waste and rubbish being properly disposed?			
Are overhead fans clean?			
Are electrical wiring, connections, boxes and controls in good condition?			
Are fire doors free of obstructions?			
Other comments:			

MACHINERY AND EQUIPMENT	Yes	No	Explain
Are moving parts of machines and equipment properly guarded?			
Are points of operation properly guarded?			
Are safety controls and devices operating properly?			
Are dust collection and vacuum devices installed and in good working order?			
Are cylinders secured and properly stored?			
Are forklifts in good working order?			
Other comments:			
GENERAL CONDITION	**Yes**	**No**	**Explain**
Is there adequate ventilation?			
Is dust control adequate?			
Are hand tools properly maintained and in good condition?			
Are floors in good condition?			
Are storage racks in good condition and earthquake safe?			
Are employees aware of safety rules and procedures?			
Is the non-smoking policy being enforced?			
Are bathrooms clean and in good working order?			
Are required safety signs properly displayed?			
Is first aid cabinet properly stocked?			
Is emergency lighting available and in good working order?			
Does supervisor have a working flashlight?			
Are aisles properly marked and free of obstructions?			
Other comments:			
General comments and recommendations			

Signature:

EXHIBIT C
NEW EMPLOYEE SAFETY CHECKLIST

Employee: _____   Department: _____

Date Hired: _____   Supervisor: _____

Supervisor: Check off each item as you discuss it with the new employee prior to having that employee start work.

1) Employee provided company safety policy statement and safety rules. _____

2) Explained functions of company safety committee. _____

3) Reviewed injury reporting procedures. _____

4) Issued safety equipment—glasses, ear plugs, respirator, etc., and explained use and care. _____

5) Reviewed lock-out and tag procedure. _____

6) Reviewed safe lifting procedures. _____

7) Will forklift training be required? If yes, when? _____ _____

8) Reviewed housekeeping and clean-up procedures. _____

9) Located first aid kits and/or company hospital. _____

10) Reviewed hazard communication program, location of safety data sheets and how to read a MSDS. _____

11) Reviewed evacuation procedures and any specific duties. _____

12) Does the employee understand the above? _____

I acknowledge that information on the above subjects was furnished to me during my orientation.

Employee's Signature _____   Dept. _____

I have instructed the above named employee in the fundamentals of safety practices.

Supervisor's Signature _____   Date _____

Sign and return the original copy immediately to the Personnel Office following the employee's date of hire or transfer into your department. Retain a copy in the employee's departmental file.

GLOSSARY

Accident   An unplanned and sometimes injurious or damaging event which interrupts the normal progress of an activity and is invariably preceded by an unsafe act or unsafe condition or some combination thereof. An accident may be seen as resulting from a failure to identify a hazard or from some inadequacy in an existing system of hazard controls. Accidents are commonly classified according to the context in which they occur (e.g., work, motor vehicle, home, farm, school, recreation, public, etc.) and in each context there are numerous subclasses. Motor vehicle accidents are also known as collisions or crashes.

Accidental death and dismemberment benefits   Payments to dependents of workers who suffer accidental death, or to workers who are accidentally dismembered, such payments often being provided under the terms of a combined insurance policy financed in whole or in part by the employer. Dismemberment is the loss of one or more hands or feet through or above the wrist or ankle joints, or the entire and irrevocable loss of sight.

Accident analysis   A critical examination by one or more qualified persons of the information developed in an accident investigation for the purpose of identifying its causal factors and prescribing measures designed to prevent its recurrence.

Accident costs   Monetary losses associated with an accident.

- Direct costs—Monetary losses directly ensuing from an accident occurrence (e.g., costs of workers' compensation payments and medical expenses)
- Indirect costs—Costs not directly associated with an accident occurrence but which are real and measurable and would not have been incurred had the accident not occurred (e.g., wages paid above compensation costs, cost of supervisor's accident investigation, lost time of other workers stopping to assist the injured or to watch, etc.)
- Insured costs—Costs covered by compensation insurance and other insurance programs (e.g., medical, property damage, etc.)
- Uninsured costs—Definite, measurable costs not covered by existing insurance programs (e.g., property damage, wages paid for nonproductive hours by the injured workers, extra cost of overtime necessitated by the accident, cost of wages paid by supervisors for time required for activities necessitated by the accident, wage cost caused by decreased output of injured worker after return to work, cost of learning period of new worker, uninsured medical cost borne by the company, cost of time spent by higher supervision and by clerical workers, and other miscellaneous costs such as loss of profit on contracts canceled or orders lost, cost of excess spoilage by new employees, public liability claims, cost of renting replacement equipment, etc.).

Acute   Severe, usually critical, often dangerous exposure in which relatively rapid changes are occurring. An acute exposure normally runs a comparatively short course and its effects are easier to reverse in contrast to a chronic exposure.

Air-quality criteria   The varying levels of air pollution and lengths of exposure to them at which specific adverse effects to health and well-being occur.

ANSI   American National Standards Institute. Establishes design, safety and performance standards for manufactured goods.

Atmospheric sound absorption   Diminution of intensity of a sound wave in passing through the air, apart from the normal inverse square relation, and arising from true absorption.

Avoidable accident   An accident which could have or can be prevented by proper behavior, or by environmental or equipment modifications or controls. All accidents are theoretically avoidable within the limits of our understanding of scientific and behavioral phenomena. However, not all accidents can be avoided by all individuals involved (e.g., a sudden swerving of an oncoming automobile into the wrong lane too late for possible evasive action by a second driver).

Barrier guard   A device designed to protect operators and other individuals from hazard points on machinery and equipment. Major types are as follows:

- Adjustable—An enclosure attached to the frame of the machinery or equipment, with adjustable front and side sections
- Fixed—A point-of-operation enclosure attached to the machine or equipment by fasteners.
- Gate or movable—A device designed to enclose the point of operation completely before the clutch can be engaged.
- Interlocking—An enclosure attached to the frame of the machinery or equipment and interlocked with the power switch so that the operating cycle cannot normally be started unless the guard, including its hinged or movable sections, is in its proper position. In some applications, movement of the guard will interrupt the machine cycle.

Contingency plan   A document that sets forth an organized, planned, and coordinated course of action to be followed in case of a fire, explosion, or release of hazardous waste or constituents thereof which could threaten human health or the environment.

Density   The amount of matter contained in a unit of volume of a substance; that is, mass per unit volume. Usually expressed in pounds per cubic foot or grams per cubic centimeter.

Electrical grounding   An electrical connection between a conductive body and the earth which eliminates any difference in potential between the object and ground. An adequate ground will always discharges a charged conductive body.

Ergonomics   The study of human characteristics for the appropriate design of the living and work environments. Using primarily the methodologies of anthropometry, physiology, psychology, engineering, and their interrelationships, ergonomics attempts to "fit the job to the person."

Eye protection   A device which safeguards the eye in an eye-hazard environment. Refer to Practice for Occupational and Educational Eye and Face Protection—ANSI Z87.1.

Face shield A protective device designed to prevent hazardous substances, dust particles, sharp objects, and other materials from contacting the face. Refer to Practice for Occupational and Educational Eye and Face Protection—ANSI Z87.1.

Guardrail A device consisting of posts and rail members, or wall sections, erected to mark points of major hazard, and to prevent individuals from coming in contact with the hazard. Refer to Floor and Wall Openings, Railings, and Toeboards—ANSI A12.1.

Hard hat A metal or plastic helmet worn by a worker to provide head protection when the worker is subject to the hazard of falling or moving objects. Refer to Industrial Head Protection—ANSI Z89.1. Special hard hats for electrical workers protect also against electrical shock. Refer to Industrial Protective Helmets for Electrical Workers—ANSI Z89.2.

Human factors engineering The engineering application of information from the biological and behavioral sciences to the design or redesign of components and systems as a means of better fitting jobs to people in order to improve safety and performance. Operators who use machines and equipment that are so designed are not likely to be stressed beyond the limits of their capabilities.

Intensity level (noise) In relation to sound, the sound-energy flux density level, measured in decibels. Equals 10 times the logarithm to the base of 10 of the ratio of the intensity of a given sound to the reference intensity.

Machine guarding The installation of equipment or devices on machines to eliminate hazards created by operation of the machines.

Respirator A protective device for the human respiratory system designed to protect the wearer from inhaling contaminated air. The two types of respiratory protective devices are (a) air purifiers, which remove the contaminants form the air by filtering or chemical absorption before inhalation, and (b) air suppliers, which provide clean air from an outside source or oxygen from a tank.

Respiratory disease Any disease condition resulting from a toxic agent in the respiratory tract; e.g., pneumonitis, bronchitis, pharyngitis, rhinitis, or acute congestion due to chemicals, dusts, gases, or fumes.

Respiratory hazard Any toxic gas, vapor, organic or inorganic mist, dust, or fumes which can produce harmful effects if breathed by humans.

Respiratory irritant Any irritant that affects the respiratory tract; e.g., dusts, vapors, or gases.

Safety audit A period, methodical, in-depth examination of an organization, performed by one or more trained persons using a predetermined checklist of items that reflect good safety practice to provide the basis for management decisions affecting the organization's safety program. The audit, for example, could (a) review the record of accidents, injuries, and illnesses sustained by employees since the previous audit; (b) analyze the time and money devoted to identifying and controlling hazards, to training, and to safety motivation; (c) ascertain the extent to which various levels of management are involved in accident prevention; and (d) include the results of a physical inspection of the premises and observations of personnel performing operations which accident records show were hazardous in the past.

Safety officer   Member of winery management team that is responsible to the general manager or CEO for the design, implementation, and training functions for the winery's health and safety program.

Workers' compensation   A system of insurance required by state law and financed by employers which provides payments to employees and their families for occupational illnesses, injuries, or fatalities incurred while at work and resulting in loss of wage income, usually regardless of the employer's or employee's negligence.

REFERENCES

197. KEAN, R. 1996. New vineyard safety video available. *Grape Grower.*

198. ASTAHL, R.C. 1984. *Industrial Safety and Health Management.* Englewood Cliffs, NJ: Prentice-Hall.

199. RODOPUB, A.K. and EGOROV, I.A. 1965. *Karbonid'nye Soedineniia Khenesa.* Vinodeliei Vinogradarstvo, SSSR.

200. HENNIG, K. and VILLFORTH, F. 1943. *Les Sustances de L'arôme des Vins.* Bulletin O.I.V. No. 16, pp. 8–16.

201. DRAWERT, F. and RAPP, A. 1966. *Ueber In halts stoffe von Mosten und Weinen VI.* Vitis, Vol. 5, pp. 351–376.

202. _____. 1983. *Wine Institute Comments to Fresno Air Pollution Control District Proposal to Control Winery Ethanol Emissions.* 165 Post Street, San Francisco: Wine Institute, pp. 1–10.

203. SCHELL, M. 1995. *Solving the IAQ–Energy Dilemma.* Engineered Systems, pp. 40–52.

204. _____. 1995. *AIM Model 400 Gas Hazard Alarm.* Austin, TX: AIM Safety USA.

205. CONFORTI, J.V. 1995. *Respirator Pocket Guide.* Schenectady, NY: Genium Publishing Corp.

206. TERRELL, M.J. 1995. *Safety and Health Management in the Nineties.* New York: Van Nostrand Reinhold.

207. _____. 1994. *Confined Space Regulations.* State of California, Department of Ind. Relations, Div. of Occupational Safety and Health, Title 8, CCR GISO 5756, 5757 and 5758.

208. BRAUER, R.L. 1990. *Safety and Health for Engineers.* New York: Van Nostrand Reinhold.

209. _____. 1970. "Guard Rail Design Standards and Safety." U.S. OSHA, 29 CFR 1910.23(e).

210. _____. 1995. "Applicable Federal Standards Code of Federal Regulations," Subtitle 29, Part 1910.5

METRIC UNIT CONVERSION CHART—SI, U.S. AND COMMON

(Read across)

Temperature

$°C = {}^5/_9 (°F\text{-}32)$	$°F = {}^9/_5 (°C\text{+}32)$

Length

Inches	Feet	Millimeters	Meters
1	8.333×10^{-2}	25.4	2.54×10^{-2}
12.00	1	304.8	0.3048
3.937×10^{-2}	3.281×10^{-3}	1	1.000×10^{-3}
39.37	3.281	1000	1

Area

Square feet	Square meters
1	0.0929
10.76	1

Volume

Cubic feet	Gallon (U.S.)	1000 Gallons	Liters	Cubic meter
1	7.481	7.481×10^{-3}	28.32	2.832×10^{-2}
0.1337	1	1.000×10^{-3}	3.786	3.786×10^{-3}
133.7	1000	1	3786	3.786
3.531×10^{-2}	0.2642	2.642×10^{-4}	1	1.000×10^{-3}
35.31	264.2	0.2642	1000	1

Mass

Pounds	Tons (U.S.)	Grams	Kilograms	Tons (Metric) (T_m)
1	4.46×10^{-4}	453.6	0.4536	4.536×10^{-4}
2240	1	$1.016 \times 10^{+6}$	1016	1.016
2.205×10^{-3}	9.843×10^{-7}	1	1.000×10^{-3}	1.000×10^{-6}
2.205	9.843×10^{-4}	1000	1	1.000×10^{-3}
2205	0.9843	$1.000 \times 10^{+6}$	1000	1

Volumetric flow rate

ft^3/s	1000 gal./hr	gal./min	L/s	m^3/s
1	29.10	485.0	28.32	2.832×10^{-2}
3.447×10^{-2}	1	15.48	0.9737	9.737×10^{-4}
2.205×10^{-3}	6.460×10^{-2}	1	6.309×10^{-2}	6.309×10^{-5}
3.531×10^{-2}	1.027	15.85	1	1.000×10^{-3}
35.31	1027	1.585×10^4	1000	1

Density

lb/ft^3	g/cm^3	kg/m^3
1	1.602×10^{-2}	16.02
62.4	1	1000
6.240×10^{-2}	1.000×10^{-3}	1

Viscosity

lb/ft/hr	Centipoise	Pascal.second
1	0.4134	4.134×10^{-2}
2.419	1	1.000×10^{-3}
2419	1000	1

Thermal conductivity

Btu/hr/ft^2/(°F/ft)	W/m^2/(°C/m)
1	1.731
0.5778	1

Diffusivity

ft^2/hr	m^2/s
3.875×10^4	1
1	2.581×10^{-5}

Heat capacity

Btu/lb/°F	J/kg/°C
1	4.187
0.239	1

Heat transfer coefficient

Btu/hr/ft^2/°F	W/m^2/°C
1	5.68
0.1761	1

Mass transfer coefficient

lb-mole/hr/ft^2/mole fraction	kg-mole/s/m^2/mole fraction
1	1.356×10^{-3}
737.5	1

Force

lb$_f$	Newton
1	4.444
0.2250	1

Pressure

mm Hg	psi	Atmosphere	Bar	kPascal
1	1.934×10^{-2}	1.316×10^{-3}	1.333×10^{-3}	0.1333×10^{2}
51.71	1	6.805×10^{-2}	$.895 \times 10^{-2}$	6.895×10^{3}
760.0	14.70	1	1.013	101.3×10^{5}
750.1	14.50	0.9869	1	1000×10^{5}
75.01×10^{-5}	1.450×10^{-4}	9.870×10^{-6}	1.000×10^{-5}	1

Heat, energy, or work

BTU	kWhr	ftlb$_f$	Calorie	Joule
1	2.930×10^{-4}	778.2	252.0	1055
3413	1	$2.655 \times 10^{+6}$	$8.604 \times 10^{+5}$	$3.600 \times 10^{+6}$
1.285×10^{-5}	3.766×10^{-7}	1	0.3241	1.356
3.966×10^{-3}	1.162×10^{-6}	3.086	1	4.184
9.484×10^{-4}	2.772×10^{-7}	0.7376	0.2390	1

Power

Horsepower	Kilowatt	Ftlb$_f$/s	BTU/s	Watt/s
1	0.7457	550.0	0.7068	745.7
1.341	1	737.56	0.9478	1000
1.818×10^{-3}	1.356×10^{-3}	1	1.285×10^{-3}	1.356
1.415	1.055	778.16	1	1055
1.341×10^{-3}	1.00×10^{-3}	0.7376	9.478×10^{-4}	1

Volumetric Flux

gal./ft^2/hr	m^3/m^2/s	L/m^2/s
1	9.767×10^{-5}	9.767×10^{-2}
10,239	1	1000
10.24	1.000×10^{-3}	1

Energy flux

Btu/ft^2/hr	W/m^2
1	3.155
0.3170	1

Gas constant, R

8.314 kPa m^3/kg-mole K (J/gm-mole K)

Common unit abbreviations and prefixes

meter = m	Newton = N
Pascal = Pa	Watt = W
gram = g	liter = L
Joule = J	second = s

μ = micro 10^{-6} m = milli 10^{-3} k = kilo 10^{+3} M = mega 10^{+6}

INDEX